T0312151

Multivariate Analysis for the Behavioral Sciences

Second Edition

Chapman & Hall/CRC

Statistics in the Social and Behavioral Sciences Series

Aims and scope

Large and complex datasets are becoming prevalent in the social and behavioral sciences and statistical methods are crucial for the analysis and interpretation of such data. This series aims to capture new developments in statistical methodology with particular relevance to applications in the social and behavioral sciences. It seeks to promote appropriate use of statistical, econometric and psychometric methods in these applied sciences by publishing a broad range of reference works, textbooks and handbooks.

The scope of the series is wide, including applications of statistical methodology in sociology, psychology, economics, education, marketing research, political science, criminology, public policy, demography, survey methodology and official statistics. The titles included in the series are designed to appeal to applied statisticians, as well as students, researchers and practitioners from the above disciplines. The inclusion of real examples and case studies is therefore essential.

Recently Published Titles

Handbook of International Large-Scale Assessment: Background, Technical Issues, and Methods of Data Analysis
Leslie Rutkowski, Matthias von Davier, and David Rutkowski

Generalized Linear Models for Categorical and Continuous Limited Dependent Variables
Michael Smithson and Edgar C. Merkle

Incomplete Categorical Data Design: Non-Randomized Response Techniques for Sensitive Questions in Surveys
Guo-Liang Tian and Man-Lai Tang

Handbook of Item Response Theory, Volume One: Models
Wim J. van der Linden

Handbook of Item Response Theory, Volume Two: Statistical Tools
Wim J. van der Linden

Handbook of Item Response Theory, Volume Three: Applications
Wim J. van der Linden

Computerized Multistage Testing: Theory and Applications
Duanli Yan, Alina A. von Davier, and Charles Lewis

Multivariate Analysis for the Behavioral Sciences, Second Edition
Kimmo Vehkalahti and Brian S. Everitt

For more information about this series, please visit: https://www.crcpress.com/go/ssbs

Multivariate Analysis for the Behavioral Sciences

Second Edition

Kimmo Vehkalahti
Brian S. Everitt

CRC Press is an imprint of the
Taylor & Francis Group, an **informa** business
A CHAPMAN & HALL BOOK

First edition published as *Multivariable Modeling and Multivariate Analysis for the Behavioral Sciences*.

CRC Press
Taylor & Francis Group
6000 Broken Sound Parkway NW, Suite 300
Boca Raton, FL 33487-2742

First issued in paperback 2020

ISBN 13: 978-0-367-65675-1 (pbk)
ISBN 13: 978-0-8153-8515-8 (hbk)

Library of Congress Cataloging-in-Publication Data

Names: Everitt, Brian, author. | Vehkalahti, Kimmo, author.
Title: Multivariate analysis for the behavioral sciences / Kimmo Vehkalahti & Brian S. Everitt
Other titles: Multivariable modeling and multivariate analysis for the behavioral sciences
Description: Second edition. | Boca Raton, Florida : CRC Press [2019] | Earlier edition published as: Multivariable modeling and multivariate analysis for the behavioral sciences / [by] Brian S. Everitt. | Includes bibliographical references and index.
Identifiers: LCCN 2018041904| ISBN 9780815385158 (hardback : alk. paper) | ISBN 9781351202275 (e-book)
Subjects: LCSH: Social sciences—Statistical methods. | Multivariate analysis.
Classification: LCC HA31.35 .E94 2019 | DDC 519.5/35—dc23
LC record available at https://lccn.loc.gov/2018041904

Visit the Taylor & Francis Web site at
http://www.taylorandfrancis.com

and the CRC Press Web site at
http://www.crcpress.com

Dedication

Brian dedicates the book to the memory of Graham Dunn; a much admired, respected, and loved colleague and friend.

Kimmo dedicates the book to Sirpa, the love of his life.

Contents

Preface xiii

Preface to *Multivariable Modeling and Multivariate Analysis for the Behavioral Sciences* xv

Authors xix

Acknowledgments xxi

1 Data, Measurement, and Models 1

1.1 Introduction 1
1.2 Types of Study 2
1.2.1 Surveys 3
1.2.2 Experiments 4
1.2.3 Observational Studies 5
1.2.4 Quasi-Experiments 6
1.3 Types of Measurement 7
1.3.1 Nominal or Categorical Measurements 7
1.3.2 Ordinal Scale Measurements 8
1.3.3 Interval Scales 8
1.3.4 Ratio Scales 9
1.3.5 Response and Explanatory Variables 10
1.4 Missing Values 10
1.5 The Role of Models in the Analysis of Data 11
1.6 Determining Sample Size 14
1.7 Significance Tests, p-Values, and Confidence Intervals . . . 16
1.8 Summary 19
1.9 Exercises 20

2 Looking at Data 23

2.1 Introduction 23
2.2 Simple Graphics—Pie Charts, Bar Charts, Histograms, and Boxplots 24
2.2.1 Categorical Data 24
2.2.2 Interval/Quasi-Interval Data 32
2.3 The Scatterplot and beyond 37
2.3.1 The Bubbleplot 40
2.3.2 The Bivariate Boxplot 42
2.4 Scatterplot Matrices 45

2.5 Conditioning Plots and Trellis Graphics 48
2.6 Graphical Deception 55
2.7 Summary 59
2.8 Exercises 60

3 Simple Linear and Locally Weighted Regression **63**
3.1 Introduction 63
3.2 Simple Linear Regression 64
3.2.1 Fitting the Simple Linear Regression Model to the Pulse Rates and Heights Data 66
3.2.2 An Example from Kinesiology 67
3.3 Regression Diagnostics 69
3.4 Locally Weighted Regression 73
3.4.1 Scatterplot Smoothers 75
3.5 Summary 81
3.6 Exercises 82

4 Multiple Linear Regression **83**
4.1 Introduction 83
4.2 An Example of Multiple Linear Regression 85
4.3 Choosing the Most Parsimonious Model When Applying Multiple Linear Regression 90
4.3.1 Automatic Model Selection 95
4.3.2 Example of Application of the Backward Elimination 96
4.4 Regression Diagnostics 98
4.5 Multiple Linear Regression and Analysis of Variance 102
4.5.1 Analyzing the Fecundity of Fruit Flies by Regression 102
4.5.2 Multiple Linear Regression for Experimental Designs 104
4.5.3 Analyzing a Balanced Design 105
4.5.4 Analyzing an Unbalanced Design 106
4.6 Summary 109
4.7 Exercises 110

5 Generalized Linear Models **113**
5.1 Introduction 113
5.2 Binary Response Variables 115
5.3 Response Variables That Are Counts 117
5.3.1 Overdispersion and Quasi-Likelihood 119
5.4 Summary 120
5.5 Exercises 121

6 Applying Logistic Regression **123**
6.1 Introduction 123
6.2 Odds and Odds Ratios 123
6.3 Applying Logistic Regression to the GHQ Data 125
6.4 Selecting the Most Parsimonious Logistic Regression Model 130

6.5 Driving and Back Pain: A Matched Case–Control Study . . 134
6.6 Summary 136
6.7 Exercises 136

7 Survival Analysis 139
7.1 Introduction 139
7.2 The Survival Function 140
7.2.1 Age at First Sexual Intercourse for Women 142
7.3 The Hazard Function 144
7.4 Cox's Proportional Hazards Model 146
7.4.1 Retention of Heroin Addicts in Methadone Treatment 149
7.5 Summary 152
7.6 Exercises 153

8 Analysis of Longitudinal Data I: Graphical Displays and Summary Measure Approach 155
8.1 Introduction 155
8.2 Graphical Displays of Longitudinal Data 157
8.3 Summary Measure Analysis of Longitudinal Data 159
8.3.1 Choosing Summary Measures 160
8.3.2 Applying the Summary Measure Approach 162
8.3.3 Incorporating Pre-Treatment Outcome Values into the Summary Measure Approach 164
8.3.4 Dealing with Missing Values When Using the Summary Measure Approach 164
8.4 Summary 166
8.5 Exercises 167

9 Analysis of Longitudinal Data II: Linear Mixed Effects Models for Normal Response Variables 169
9.1 Introduction 169
9.2 Linear Mixed Effects Models for Repeated Measures Data . 170
9.3 How Do Rats Grow? 174
9.3.1 Fitting the Independence Model to the Rat Data . . 174
9.3.2 Fitting Linear Mixed Models to the Rat Data 176
9.4 Computerized Delivery of Cognitive Behavioral Therapy—Beat the Blues 181
9.5 Summary 186
9.6 Exercises 187

10 Analysis of Longitudinal Data III: Non-Normal Responses 189
10.1 Introduction 189
10.2 Marginal Models and Conditional Models 190
10.2.1 Marginal Models 190
10.2.2 Conditional Models 194

10.3 Using Generalized Estimating Equations to Fit Marginal Models . . . 196
10.3.1 Beat the Blues Revisited . . . 196
10.3.2 Respiratory Illness . . . 197
10.3.3 Epilepsy . . . 201
10.4 Using Generalized Linear Mixed Effects Models to Fit Conditional Models . . . 203
10.4.1 Respiratory Illness . . . 203
10.4.2 Epilepsy . . . 204
10.5 Summary . . . 206
10.6 Exercises . . . 207

11 Missing Values 209
11.1 Introduction . . . 209
11.2 Missing Data Mechanisms . . . 210
11.3 Dealing with Missing Values . . . 212
11.4 Imputing Missing Values . . . 213
11.5 Analyzing Multiply Imputed Data . . . 215
11.6 Example of the Application of Multiple Imputation . . . 216
11.6.1 Complete-Case Analysis . . . 217
11.6.2 Mean Imputation . . . 218
11.6.3 Multiple Imputation . . . 219
11.7 Beat the Blues Revisited (Again) . . . 219
11.8 Summary . . . 222
11.9 Exercises . . . 222

12 Multivariate Data and Multivariate Analysis 225
12.1 Introduction . . . 225
12.2 The Initial Analysis of Multivariate Data . . . 226
12.2.1 Summary Statistics for Multivariate Data . . . 226
12.2.2 Graphical Descriptions of the Body Measurement Data . . . 229
12.3 The Multivariate Normal Probability Density Function . . . 230
12.3.1 Assessing Multivariate Data for Normality . . . 233
12.4 Summary . . . 237
12.5 Exercises . . . 237

13 Principal Components Analysis 239
13.1 Introduction . . . 239
13.2 Principal Components Analysis (PCA) . . . 239
13.3 Finding the Sample Principal Components . . . 241
13.4 Should Principal Components be Extracted from the Covariance or the Correlation Matrix? . . . 244
13.5 Principal Components of Bivariate Data with Correlation Coefficient r . . . 246

13.6 Rescaling the Principal Components 248
13.7 How the Principal Components Predict
the Observed Covariance Matrix 248
13.8 Choosing the Number of Components 249
13.9 Calculating Principal Component Scores 251
13.10 Some Examples of the Application of PCA 252
13.10.1 Head Size of Brothers 252
13.10.2 Crime Rates in the United States 255
13.10.3 Drug Usage by American College Students 260
13.11 Using PCA to Select a Subset of the Variables 264
13.12 Summary 265
13.13 Exercises 266

14 Multidimensional Scaling and Correspondence Analysis 267
14.1 Introduction 267
14.2 Multidimensional Scaling 269
14.2.1 Classical Multidimensional Scaling 270
14.2.2 Connection to Principal Components 273
14.2.3 Road Distances in Finland 274
14.2.4 Mapping Composers of Classical Music 278
14.2.5 Nonmetric Multidimensional Scaling 280
14.2.6 Re-mapping Composers of Classical Music 281
14.3 Correspondence Analysis 284
14.3.1 Simple Example of the Application
of Correspondence Analysis 286
14.3.2 Connections of Work Activities and Job Advantages 288
14.4 Summary 291
14.5 Exercises 292

15 Exploratory Factor Analysis 295
15.1 Introduction 295
15.2 The Factor Analysis Model 296
15.3 Estimating the Parameters in the Factor Analysis Model . . 299
15.4 Determining the Number of Factors 301
15.5 Fitting the Factor Analysis Model: An Example 302
15.6 Rotation of Factors 304
15.6.1 A Simple Example of Graphical Rotation 306
15.6.2 Numerical Rotation Methods 309
15.6.3 A Simple Example of Numerical Rotation 311
15.7 Estimating Factor Scores 311
15.7.1 Analyzing the Crime Rates by Factor Analysis 312
15.8 Exploratory Factor Analysis and Principal Component
Analysis Compared 315
15.9 Summary 316
15.10 Exercises 317

16 Confirmatory Factor Analysis and Structural Equation Models 319
16.1 Introduction 319
16.2 Estimation, Identification, and Assessing the Fit for Confirmatory Factor Analysis and Structural Equation Models . . 320
16.2.1 Estimation 320
16.2.2 Identification 321
16.2.3 Assessing the Fit 322
16.3 Examples of Confirmatory Factor Analysis 324
16.3.1 Ability and Aspiration 325
16.3.2 Drug Usage among Students 327
16.4 Eight Factors of Systems Intelligence 331
16.4.1 Testing the Factorial Validity of the SI Inventory . . 333
16.5 Structural Equation Models 335
16.5.1 Example of a Structural Equation Model 335
16.6 Summary 337
16.7 Exercises 337

17 Cluster Analysis 341
17.1 Introduction 341
17.2 Cluster Analysis 343
17.3 Agglomerative Hierarchical Clustering 344
17.3.1 Clustering Individuals Based on Body Measurements 347
17.3.2 Clustering Countries on the Basis of Life Expectancy 348
17.4 k-Means Clustering 352
17.4.1 Clustering Crime Rates 355
17.5 Model-Based Clustering 356
17.5.1 Clustering European Countries 359
17.6 Summary 362
17.7 Exercises 363

18 Grouped Multivariate Data 365
18.1 Introduction 365
18.2 Two-Group Multivariate Data 366
18.2.1 Hotelling's T^2 Test 366
18.2.2 Fisher's Linear Discriminant Function 369
18.3 More Than Two Groups 374
18.3.1 Multivariate Analysis of Variance (MANOVA) 374
18.3.2 Classification Functions 378
18.4 Summary 382
18.5 Exercises 382

References 385

Index 401

Preface

In some respects this book is a second edition of *Multivariable Modeling and Multivariate Analysis for the Behavioral Sciences* but in others it is largely a new book with new chapters on missing values and the analysis of longitudinal data where the response variable can not be assumed to be normally distributed. The book also includes a wider account of generalized linear models as well as a new chapter on multidimensional scaling and correspondence analysis and separate chapters on exploratory and confirmatory factor analysis; in the latter there is also new coverage of structural equation models. A number of interesting, new examples and exercises have been added in several chapters.

The original lengthy title tried to explain that the book covered situations where the variables of interest consisted of a response variable and explanatory variables and where interest lies in finding suitable models relating the response to the explanatory variables, hence *multivariable modeling*, in addition to techniques that can be applied to what has historically been termed *multivariate data* where there is no division of the variables and the aim is to find a parsimonious description of the structure of the data. The current book contains extended coverage of both situations but has been given a somewhat shorter title, *Multivariate Analysis for the Behavioral Sciences*, because both types of data mentioned can be represented symbolically by an $n \times q$ matrix $\mathbf{X}$ containing the q variable values for n number of units (often subjects) in the data set. Chapters 3–11 will describe methods for dealing with data when one of the q variables is a response and the others explanatory, while the remaining chapters deal with methods for the analysis of data sets where there is no such division of the variables. We hope this makes it clear that this book covers accounts of a wider range of statistical methodology than covered in the conventional 'multivariate analysis' textbook.

The 'we' opening the last sentence above brings us to the most important change between this book and that mentioned in the first line of this Preface, namely the arrival of a co-author Dr. Kimmo Vehkalahti. It is Kimmo who is responsible for many of the changes and most of the new material in this book compared with the original on which it is based.

Most chapters include 'technical detail' sections which briefly describe the theory behind the methods of concern in the chapter. These sections often require a familiarity with some relatively advanced mathematics, for example, matrix algebra, for their understanding (see Puntanen et al., 2011, 2013). Readers without the necessary grounding in maths can largely ignore such sections and instead concentrate on the examples given to help them understand the practical implications of the methods.

There are exercises at the end of each chapter, some of which are 'starred' (*) to indicate that they are more challenging and could perhaps be used as the basis of student projects. Data sets for most exercises are not given in the text but are available on the associated web sites (see later) where they are identified by the relevant exercise number. For the starred exercises the web sites also contain pointers to the appropriate analysis.

The web site for the book is www.crcpress.com and in addition the book has a GitHub repository (https://github.com/KimmoVehkalahti/MABS) for distributing the complete data sets and the R code for reproducing the examples and for answering the exercises; consequently this allows us to abbreviate the listings of most data sets in the text.

We hope that this book will be found useful in a number of different ways, including:

- As the main part of a formal statistics course for advanced undergraduates and postgraduates in all areas of the behavioral sciences,
- As a supplement to an existing course,
- For self-study,
- For researchers in the behavioral sciences undertaking statistical analyses on their data,
- For statisticians teaching statistics to psychologists and others,
- For statisticians using R when teaching intermediate statistics courses both in the behavioral sciences and in other areas.

Brian S. Everitt
Dulwich, London

Kimmo Vehkalahti
Vuosaari, Helsinki

Preface to Multivariable Modeling and Multivariate Analysis for the Behavioral Sciences

The *Encyclopedia of Statistics in Behavioral Science* (Everitt and Howell, 2005) opens with the following paragraph:

> Forty years ago there was hardly a field called "behavioral science." In fact, psychology largely was the behavioral sciences, with some help from group theory in sociology and decision making in economics. Now, of course, psychology has expanded and developed in a myriad of ways, to the point where behavioral science is often the most useful term. Physiological psychology has become neuroscience, covering areas not previously part of psychology. Decision-making has become decision science, involving people from economics, marketing, and other disciplines. Learning theory has become cognitive science, again exploring problems that were not even considered 40 years ago. And developments in computing have brought forth a host of new techniques that were not possible in the days of manual and electronic calculators. With all these changes, there have been corresponding changes in the appropriate statistical methodologies.

Despite the changes mentioned in the last sentence of this quotation, many statistical books aimed at psychologists and others working in the behavioral sciences continue to cover primarily simple hypothesis testing, using a variety of parametric and nonparametric significance tests, simple linear regression, and analysis of variance. Such statistical methodology remains important in introductory courses, but represents only the first step in equipping behavioral science students with enough statistical tools to help them on their way to success in their later careers. The aim of this book is to encourage students and others to learn a little more about statistics and, equally important, how to apply statistical methods in a sensible fashion. It is hoped that the following features of the text will help it reach its target:

- The central theme is that statistics is about solving problems; data relevant to these problems are collected and analyzed to provide useful answers. To this end, the book contains a large number of real data sets arising from real problems. Numerical examples of the type that involve the skiing activities of belly dancers and politicians are avoided as far as possible.

- Mathematical details of methods are confined to numbered and separated Technical Sections. For the mathematically challenged, the most difficult of these displays can, at least as a last resort, be ignored. But the study of the relevant mathematical material (which on occasion will include the use of vectors and matrices) will undoubtedly help in the reader's appreciation of the corresponding technique.

- Although many statistical methods require considerable amounts of arithmetic for their application, the burden of actually performing the necessary calculations has been almost entirely removed by the development and wide availability of powerful and relatively cheap personal computers and associated statistical software packages. It is assumed, therefore, that all students will be using such tools when undertaking their own analyses. Consequently, arithmetic details are noticeable largely by their absence, although a little arithmetic is included where it is considered helpful in explaining a technique.

- There are many challenging data sets both in the text and in the exercises provided at the end of each chapter. All data sets, both in the body of the text and in the exercises, are given on the Web site associated with the book, as are the answers to all the exercises. (Because the majority of data sets used in the book are available on the book's Web site (http://www.crcpress.com/product/isbn/9781439807699), tables of data in the text only give a small subset of each data set.)

As mentioned in the penultimate bullet point above, the text assumes that readers will be using one or another of the many available statistical software packages for data analysis. This raises the thorny question for the author of what information should be provided in the text about software. Would, for example, screen dumps from SPSS be useful, or listings of STATA code? Perhaps, but neither are included here. Instead, all the computer code used to analyze the many examples to be found in the text is given on the book's Web site, and this code is in the R language, where R is a software system for statistical computing, data analysis, and graphics. This may appear a strange choice for a book aimed at behavioral scientists, but the rationale behind the choice is first that the author uses R in preference to other statistical software, second that R can be used to produce many interesting and informative graphics that are difficult if not impossible to produce with other software, third that R is free and can be easily downloaded by students, and fourth, R has a very active user community and recently developed statistical methods become available far more quickly than they do with other packages. The only downside with R is that it takes a little more time to learn than say using "point-and-click" SPSS. The initial extra effort, however, is rapidly rewarded. A useful book for learning more about R is Everitt and Hothorn (2009).

The material covered in the book assumes the reader is familiar with the topics covered in introductory statistics courses, for example, population, sample, variable, parameter, significance test, p-value, confidence interval, correlation, simple regression, and analysis of variance. The book is primarily about methods for analyzing data but some comments are made in Chapter 1 about the various types of study that behavioral researchers may use and their design. And it is in Chapter 1 that the distinction between multivariable and multivariate—both of which appear in the book's title—will be explained.

It is hoped that the text will be useful in a number of different ways, including:

- As the main part of a formal statistics course for advanced undergraduates and postgraduates in all areas of the behavioral sciences.
- As a supplement to an existing course.
- For self-study.
- For researchers in the behavioral sciences undertaking statistical analyses on their data.
- For statisticians teaching statistics to psychologists and others.
- For statisticians using R when teaching intermediate statistics courses both in the behavioral sciences and in other areas.

B. S. Everitt
Dulwich, U.K.

Authors

Kimmo Vehkalahti is a fellow of the Teachers' Academy, University of Helsinki, Finland. He has been a part of the faculty of Social Sciences for over 25 years, currently as senior lecturer of the Social Data Science in the Centre for Research Methods. He is author of a Finnish textbook on measurement and survey methods. The present book is his first international textbook on statistics. His research and teaching activities are related to open data science, multivariate analysis, and introductory statistics. His spare time is divided (unequally) between jogging and trail running, reading, watching ice hockey, holidays with his wife, and singing tenor in choir.

Brian S. Everitt is professor emeritus, King's College, London, UK. He worked at the Institute of Psychiatry, University of London for over 35 years, finally as head of the Biostatistics and Computing Department and professor of behavioural statistics. He is author or co-author of over 70 books on statistics and approximately 100 papers and other articles, and was a section editor for the *Encyclopedia of Biostatistics*, published by Wiley. In retirement, he divides his time between working as editor-in-chief of *Statistical Methods in Medical Research*, playing tennis, watching cricket, long walking holidays with his wife, and playing classical guitar in private.

Acknowledgments

Kimmo would like to thank the faculty of Social Sciences at the University of Helsinki for granting him a six-month sabbatical without which this book could not have been written. Three months of this sabbatical were spent at the Universitat Pompeu Fabra in Barcelona at the kind invitation and hospitality of Professor Michael Greenacre. The hard work needed to get to grips with R and LaTeX was greatly aided by the sharing of an office (and coffee breaks) with Michael at Campus de la Ciutadella. Gràcies!

Kimmo's thanks are also given to Maria Anna Donati, Ulla Palotie, Juha Törmänen, Raimo Hämäläinen, Esa Saarinen, and Seppo Mustonen, all of whom gave their permission to use their interesting data sets in this book; he has the warmest memories of staying in the beautiful home of the lovely Firenze family. Grazie Caterina Primi, Michele e Luisa!

Thanks are given to Dr. Deepayan Sarkar, the author of *Lattice: Multivariate Data Visualization with R*, and Springer, the publishers of the book, for permission to use Figures 2.8, 2.9, 4.5, 4.6, 4.7, and 5.16 in Chapter 2 of this book.

Brian and Kimmo thank each other for an extremely smooth collaboration that allowed this book to be completed on time whilst allowing them to become and remain good friends; they also agree that sincere thanks are given to Rob Calver of Taylor & Francis, who has been, as always, supportive and encouraging during the writing of the book.

1

Data, Measurement, and Models

1.1 Introduction

> Statistics is a general intellectual method that applies wherever data, variation, and chance appear. It is a fundamental method because data, variation and chance are omnipresent in modern life. It is an independent discipline with its own core ideas, rather than, for example, a branch of mathematics ... Statistics offers general, fundamental and independent ways of thinking.
>
> ***Journal of the American Statistical Association***

Quintessentially, statistics is about solving problems; data (measurements or observations) relevant to these problems are collected, and statistical analyses are used to provide useful answers. But the path from data collection to analysis and interpretation is often not straightforward. Most real-life applications of statistical methodology have one or more nonstandard features, meaning in practice that there are few routine statistical questions, although there are questionable statistical routines. Many statistical pitfalls lie in wait for the unwary. Indeed, statistics is perhaps more open to misuse than most other subjects, particularly by the nonstatistician with access to powerful statistical software. The misleading average, the graph with "fiddled axes," the inappropriate p-value, and the linear regression fitted to nonlinear data are just four examples of horror stories that are part of statistical folklore.

Statisticians often complain that many of those working in the behavioral sciences put undue faith in significance tests, use complex methods of analysis when the data merit only a relatively simple approach, and sometimes abuse the statistical techniques they are employing. Statisticians become upset (and perhaps feel a little insecure) when their advice to, say, "plot a few simple graphs," is ignored in favor of a multivariate analysis of covariance or similar statistical extravagance.

However, if statisticians are at times horrified by the way in which behavioral scientists apply statistical techniques, behavioral scientists may be no less horrified by many statisticians' apparent lack of awareness of what stresses behavioral research can place on an investigator. A statistician may, for example, demand a balanced design with 30 subjects in each cell so as to achieve some appropriate power for the analysis. But it is not the statistician who is faced with the frustration caused by a last-minute phone call from a subject who cannot take part in an experiment that has taken several hours to arrange. Again, the statistician advising on a longitudinal study may call for more effort in carrying out follow-up interviews so that the study avoids statistical problems produced by the presence of missing data. It is, however, the behavioral researcher who must continue to persuade people to talk about potentially distressing aspects of their lives, who must confront possibly dangerous respondents, or who arrives at a given (and often remote) address to conduct an interview, only to find that the person is not at home. Many statisticians often do not appear to appreciate the complex stories behind each data point in many behavioral studies. One way of improving the possible communication problems between behavioral scientist and statistician is for each to learn more about the language of the other. There is already available a plethora of, for example, "Statistics for Psychologists" books, but sadly, (as far as we know) no "Psychology for Statisticians" equivalent. Perhaps there should be?

Having outlined briefly a few caveats about the possible misuse of statistics and the equally possible conflict between statistician and behavioral scientist, it is time to move on to consider some of the basics of behavioral science studies and their implications for statistical analysis.

1.2 Types of Study

It is said that, when Gertrude Stein lay dying, she roused briefly and asked her assembled friends, "Well, what's the answer?" They remained uncomfortably quiet, at which she sighed, "In that case, what's the question?"

Research in the behavioral science, as in science in general, is about searching for the answers to particular questions of interest. Do politicians have higher IQs than university lecturers? Do men have faster reaction times than women? Should phobic patients be treated by psychotherapy or by a behavioral treatment such as flooding? Do children who are abused have more problems later in life than children who are not abused? Do children of divorced parents suffer more marital breakdowns themselves than children from more stable family backgrounds?

In more general terms, scientific research involves a sequence of asking and answering questions about the nature of relationships among variables (e.g., How does A affect B? Do A and B vary together? Is A significantly different from B? and so on). Scientific research is carried out at many levels that differ in the types of question asked and therefore in the procedures used to answer them. Thus, the choice of which methods to use in research is largely determined by the kinds of questions that are asked.

Of the many types of investigation used in behavioral research, the most common are perhaps the following:

- Surveys
- Experiments
- Observational studies
- Quasi-experiments

Some brief comments about each of these four types are given below; a more detailed account is available in the papers by Stretch, Raulin, and Graziano, and by Dane, all of which appear in the second volume of the excellent *Companion Encyclopedia of Psychology* (see Colman, 1994).

1.2.1 Surveys

Survey methods are based on the simple discovery that "asking questions is a remarkably efficient way to obtain information from and about people" (Schuman and Kalton, 1985, p. 635). Surveys involve an exchange of information between researcher and respondent; the researcher identifies topics of interest, and the respondent provides knowledge or opinion about these topics. Depending upon the length and content of the survey as well as the facilities available, this exchange can be accomplished via written questionnaires, in-person interviews, or telephone conversations; and, in the 21st century, surveys via the Internet are increasingly common.

Surveys conducted by behavioral scientists are usually designed to elicit information about the respondents' opinions, beliefs, attitudes, and values. Perhaps one of the most famous surveys of the 20th century was that conducted by Alfred Charles Kinsey, a student of human sexual behavior in the 1940s and 1950s. The first Kinsey report, *Sexual Behavior in the Human Male*, appeared in 1948 (see Kinsey et al., 1948), and the second, *Sexual Behavior in the Human Female*, in 1953 (see Kinsey et al., 1953). It is no exaggeration to say that both reports caused a sensation, and the first quickly became a bestseller.

Surveys are often a flexible and powerful approach to gathering information of interest, but careful consideration needs to be given to several aspects of the survey if the information is to be accurate, particularly when dealing with a sensitive topic. Having a representative sample, having a large-enough sample, minimizing nonresponse, and ensuring that the questions asked elicit accurate responses are just a few of the issues that the researcher thinking of carrying out a survey needs to consider. Readers are referred to Bradburn et al. (2004) and Tourangeau et al. (2000) for a broad coverage of practical advice for questionnaire construction and Laaksonen (2018), Groves et al. (2009), de Leeuw et al. (2008) and Lehtonen and Pahkinen (2004) for detailed accounts of survey sampling and survey methodology.

Examples of data collected in surveys and their analysis are given in several later chapters.

1.2.2 Experiments

According to Sir Ronald Fisher, perhaps the greatest statistician of the 20th century, "experiments are only experience carefully planned in advance and designed to form a secure basis of new knowledge." The essential feature of an experiment is the large degree of control in the hands of the experimenters, and in designed experiments the goal is to allow inferences to be drawn about the effects of an intervention of interest that are logically compelled by the data and hence allow assessment of a *causal relationship*. In many cases the "intervention" will be some form of therapy in which case the experiment is usually called a clinical trial.

In an experiment, the researcher controls the manner in which subjects are allocated to the different levels of the experimental factors. In a comparison of a new treatment with one used previously, for example, the researcher would have control over the scheme for allocating subjects to the two treatments. The manner in which this control is exercised is of vital importance if the results of the experiment are to lead to a largely unambiguous assessment of the effect of treatment. The objective in allocation is that the groups to be compared should be alike in all respects except the intervention (treatment) received. Comparable groups prior to the intervention ensure that differences in outcomes after the intervention reflect effects of the intervention in an unbiased fashion. Let us begin by considering two flawed allocation procedures that are unlikely to achieve the desired degree of similarity of the two groups.

- Perhaps the first subjects to volunteer to take part in the experiment should all be given the new treatment, for example, and the later ones the old treatment? The two groups formed in this way may differ in level of motivation and so subsequently in performance. Observed treatment differences would be confounded with differences produced by the allocation procedure. Alternatively, early volunteers might be more seriously ill, those desperate to find a new remedy that works, and again, this might lead to a bias in the measured difference between the two treatments.

- So what about putting alternate subjects into each group? The objection to this is that the experimenter will know who is receiving what treatment and may be tempted to "tinker" with the scheme to ensure that those patients who are most ill receive the new treatment.

So, how should we form the groups that will be used to assess an experimental intervention? The answer is deceptively simple—use randomization. The group to which a participant in the experiment is allocated is decided by chance. It could be arranged by flipping a coin each time a new eligible patient arrives, and allocating the patient to the new treatment if the result is a head, or to the old treatment if a tail appears. In practice, of course, a more sophisticated randomization procedure will be used. The essential feature, however, is randomization, rather than the mechanism used to achieve it. Randomization was introduced into scientific experiments far more recently, when in 1926 Fisher randomly assigned individual blocks or plots of land in agricultural experiments to receive particular types of "treatment"—different amounts of fertilizer. The primary benefit that randomization has is the chance (and therefore impartial) assignment of extraneous influences among the groups to be compared, and it offers this control over such influences whether or not they are known by the experimenter to exist. Note that randomization does not claim to render the two samples equal with regard to these influences; if, however, the same procedure was applied to repeated samples from the population, equality would be achieved in the long run. Thus, randomization ensures a lack of bias, whereas other methods of assignment may not. In a properly conducted, randomized, experiment the interpretation of an observed group difference is largely unambiguous; its cause is very likely to be the different treatments or conditions received by the groups.

Several of the data sets introduced and analyzed in later chapters arise from experimental studies, often clinical trials.

1.2.3 Observational Studies

Suppose a researcher is interested in investigating how smoking cigarettes affects a person's systolic blood pressure. Using the experimental approach described earlier, people would have to be allocated at random to two groups, the members of one group being asked to smoke some quantity of cigarettes per day, and the members of the other group required not to smoke at all. Clearly, no ethical committee would approve of such a study. So, what can be done? An approach that would get ethical approval is to measure the systolic blood pressure of naturally occurring groups of individuals who smoke, and those who do not, and then compare the results. This would then be what is known as an observational study, defined by Cochran (1965) as follows:

> An empiric comparison of "treated" and "control" groups in which the objective is to elucidate cause-and-effect relationships but where it is not possible to use controlled experimentation, in the sense of being able to impose the procedures or treatments whose effects it is desired to discover, or to assign patients at random to different procedures.

Many observational studies involve recording data on the members of naturally occurring groups, generally over a period of time, and comparing the rate at which a particular event of interest occurs in the different groups (such studies are often referred to as prospective). If, for example, an investigator was interested in the health effects of a natural disaster such as an earthquake, those who experienced the earthquake could be compared, on some outcome variable of interest, with a group of people who did not.

Another commonly used type of observational study is the case-control investigation. Here, a group of people (the cases) all having a particular characteristic (a certain disease perhaps) are compared with a group of people who do not have the characteristic (the controls), in terms of their past exposure to some event or risk factor. The cases and controls are usually matched one-to-one for possible confounding variables. An example of such a study is reported in Lehman et al. (1987). Here the researchers collected data following the sudden death of a spouse or a child in a car crash. They matched 80 bereaved spouses and parents to 80 controls drawn from 7582 individuals who came to renew their driver's license. Specifically, they matched for gender, age, family income before crash, education level, and number and ages of children.

The types of analyses suitable for observational studies are often the same as those used for experimental studies. Unlike experiments, however, the lack of control over the groups to be compared in an observational study makes the interpretation of any difference between the groups detected in the study open to a variety of interpretations. In the smoking and systolic blood pressure study, for example, any difference found between the blood pressures of the two groups would be open to three possible interpretations:

- Smoking causes a change in systolic blood pressure.
- Level of blood pressure has a tendency to encourage or discourage smoking.
- Some unidentified factors play a part in determining both the level of blood pressure and whether or not a person smokes.

In the design of an observational study, an attempt is made to reconstruct some of the structure and strengths of an experiment. But the possible ambiguity in interpretation of the results from an observational study, however well designed, means that the observational approach is not as powerful as a designed experiment. A detailed account of observational studies is given in Rosenbaum (2002).

1.2.4 Quasi-Experiments

Quasi-experimental designs resemble experiments proper but are weak on some of the characteristics. In particular (and as in the observational study), the ability to manipulate the groups to be compared is not under the investigator's control. But, unlike the observational study, the quasi-experiment involves the intervention of the investigator in the sense that he or she applies a variety of different "treatments" to naturally occurring groups. In investigating the effectiveness of three different methods of teaching mathematics to 15 year olds, for example, a method might be given to all the members of a particular class in a school. The three classes that receive the different teaching methods would be selected to be similar to each other on most relevant variables, and the methods would be assigned to classes on a chance basis.

For more details of quasi-experiments see Shadish et al. (2002).

1.3 Types of Measurement

The measurements and observations made on a set of subjects comprise the basic material that is the foundation of all behavioral science investigations. These measurements provide the data for statistical analysis from which the researcher will draw his or her conclusions. Clearly, not all measurements are the same. Measuring an individual's weight is qualitatively different from measuring that person's response to some treatment on a two-category scale: "improved" and "not improved," for example. Whatever measurements are made, they need to be objective, precise, and reproducible for reasons nicely summarized in the following quotation from Fleiss (1986):

> The most elegant design of a study will not overcome the damage caused by unreliable or imprecise measurement. The requirement that one's data be of high quality is at least as important a component of a proper study design as the requirement for randomization, double blinding, controlling where necessary for prognostic factors, and so on. Larger sample sizes than otherwise necessary, biased estimates, and even biased samples are some of the untoward consequences of unreliable measurements that can be demonstrated.

Measurements are often differentiated according to the degree of precision involved. If it is said that an individual has a high IQ, it is not as precise as the statement that the individual has an IQ of 151. The comment that a woman is tall is not as accurate as specifying that her height is 1.88 m. Certain characteristics of interest are more amenable to precise measurement than others. Given an accurate thermometer, a subject's temperature can be measured very precisely. Quantifying the level of anxiety or depression of a psychiatric patient or assessing the degree of pain of a migraine sufferer are, however, more difficult measurement tasks.

Four levels of measurement scales are generally distinguished.

1.3.1 Nominal or Categorical Measurements

Nominal measurements allow classification with respect to some characteristic. Examples of such measurements are marital status, sex, and blood group. The properties of a nominal measurement are

- The categories are mutually exclusive (an individual can belong to only one category).
- The categories have no logical order—numbers may be assigned to categories but merely as convenient labels.

1.3.2 Ordinal Scale Measurements

The next level of measurement is the ordinal scale. This scale has one additional property over those of a nominal measurement—a logical ordering of the categories. With such measurements, the numbers assigned to the categories indicate the amount of a characteristic possessed. A psychiatrist may, for example, grade patients on an anxiety scale as "not anxious," "mildly anxious," "moderately anxious," or "severely anxious," and use the numbers 0, 1, 2, and 3 to label the categories, with lower numbers indicating less anxiety. The psychiatrist cannot infer, however, that the difference in anxiety between patients with scores of, say, 0 and 1 is the same as the difference between patients assigned scores 2 and 3. The scores on an ordinal scale do, however, allow patients to be ranked with respect to the characteristic being assessed.

The following are the properties of an ordinal scale:

- The categories are mutually exclusive.
- The categories have some logical order.
- The categories are scaled according to the amount of a particular characteristic they indicate.

1.3.3 Interval Scales

The third level of measurement is the interval scale. Such scales possess all the properties of an ordinal scale plus the additional property that equal differences between category levels, on any part of the scale, reflect equal differences in the characteristic being measured. An example of such a scale is temperature on the Celsius (C) or Fahrenheit (F) scale; the difference between temperatures of 80°F and 90°F represents the same difference in heat as that between temperatures of 30° and 40° on the Fahrenheit scale. An important point to make about interval scales is that the zero point is simply another point on the scale; it does not represent the starting point of the scale or the total absence of the characteristic being measured. The properties of an interval scale are as follows:

- The categories are mutually exclusive.
- The categories have a logical order.
- The categories are scaled according to the amount of the characteristic they indicate.
- Equal differences in the characteristic are represented by equal differences in the numbers assigned to the categories.
- The zero point is completely arbitrary.

1.3.4 Ratio Scales

The final level of measurement is the ratio scale. This type of scale has one further property in addition to those listed for interval scales, namely, the possession of a true zero point that represents the absence of the characteristic being measured. Consequently, statements can be made about both the differences on the scale and the ratio of points on the scale. An example is weight, where not only is the difference between 100 and 50 kg the same as between 75 and 25 kg, but an object weighing 100 kg can be said to be twice as heavy as one weighing 50 kg. This is not true of, say, temperature on the Celsius or Fahrenheit scales, where a reading of 100° on either scale does not represent twice the warmth of a temperature of 50°. If, however, two temperatures are measured on the Kelvin scale, which does have a true zero point (absolute zero or -273°C), then statements about the ratio of the two temperatures can be made.

The properties of a ratio scale are

- The categories are mutually exclusive.
- The data categories have a logical order.
- The categories are scaled according to the amount of the characteristic they possess.
- Equal differences in the characteristic being measured are represented by equal differences in the numbers assigned to the categories.
- The zero point represents an absence of the characteristic being measured.

In many statistical textbooks, discussion of different types of measurements is often followed by recommendations as to which statistical techniques are suitable for each type. For example, analyses on nominal data should be limited to summary statistics such as the number of cases, the mode, etc., and for ordinal data, means and standard deviations are said to be not suitable. But Velleman and Wilkinson (1993) make the important point that restricting the choice of statistical methods in this way may be a dangerous practice for data analysis. In essence, the measurement taxonomy described is often too strict to apply to real-world data. This is not the place for a detailed discussion of measurement, but we take a fairly pragmatic approach to such problems. For example, we would not agonize too long over treating variables such as measures of depression, anxiety, or intelligence as if they were interval scaled, although strictly, they fit into the ordinal level described earlier.

1.3.5 Response and Explanatory Variables

This is a convenient point to mention a further classification of measurements that is used in many studies, and that is the division of measured variables into response or dependent variables (often also referred to as outcome variables), and independent (a misnomer; the variables are not independent of one another, and therefore a term to be avoided) or explanatory variables (also occasionally called predictor variables); in this book we shall stick to explanatory. Essentially, response variables are those that appear on the left-hand side of the equation defining the proposed model for the data, with the explanatory variables, thought to possibly affect the response variable, appearing on the right-hand side of the model equation. With such data sets only the response is considered to be a random variable and the analysis of the data aims to assess how the explanatory variables are related to the response. (In practice, of course, explanatory variables will also usually be random variables, so any analysis performed will be *conditional* on the observed values of these explanatory variables.)

But there are many data sets gathered in the behavioral sciences where the variables measured or observed are not divided into response and explanatory variables and in which all are random variables. In those types of research settings, analyses are carried out to discover how the variables relate to one another and to try to determine the underlying structure of the data.

Both these types of data are multivariate although to distinguish between them the term *multivariable* is often used for the response/explanatory variable type of data. In this book, methods for analysing both data types will be described.

1.4 Missing Values

Most researchers in the behavioral sciences undertake studies that involve human subjects and many of them will be able to immediately relate to much of the following quotation from Efron (1998) which was originally made in the context of clinical trials in medicine:

> There could be no worse experimental animals on earth than human beings; they complain, they go on vacations, they take things they are not supposed to take, they lead incredibly complicated lives, and sometimes, they do not take their medicine.

So situations will occur in almost all studies that observations or measurements that should have been recorded for a subject, but for some reason or another are not, for example, a subject may simply not turn up for a planned measurement session. The result is a missing value. When faced with missing values, many researchers simply resort to analyzing only complete cases since this is what most statistical software packages do automatically. If data are being collected on several variables, for example, the researcher might omit any case with a missing value on any of the variables. When the incomplete cases comprise only a small fraction of all cases (say, 5% or less), then case deletion may be a perfectly reasonable solution to the missing data problem. But when there are many cases with missing values, omitting them may cause large amounts of information, that is, the variable values on which a case *has* been measured, to be discarded, which would clearly be very inefficient. However, the main problem with complete-case analysis is that it can lead to serious biases in both estimation and inference unless the missing data are missing completely at random in the sense that the probabilities of response do not depend on any data values observed or missing (see Chapter 11 and Little and Rubin, 2002, for more details). In other words, complete-case analysis implicitly assumes that the discarded cases are like a random subsample. So, at the very least, complete-case analysis leads to a loss, and perhaps a substantial loss, in power (see Section 1.6), but worse, analyses based just on complete cases might in some cases be misleading.

Fortunately, there are now sophisticated techniques for dealing with missing values, and some of these will be the subject of Chapter 11. But researchers should remember that despite their sophistication, such methods will not rescue a study where a substantial proportion of subjects have missing values; in such a study it should perhaps be questioned whether any form of statistical analysis is worth undertaking. And it also still worth paying attention to that old axiom from the past, now largely forgotten in the growing enthusiasm for modern techniques for handling missing values, namely, the best solution to missing values is not to have any.

1.5 The Role of Models in the Analysis of Data

Models attempt to imitate the properties of "real" objects or situations in a simpler or more convenient form. A road map, for example, models part of the earth's surface, attempting to reproduce the relative positions of towns, roads, and other features. Chemists use models of molecules to mimic their theoretical properties, which, in turn, can be used to predict the behavior of real compounds. A good model follows as accurately as possible the relevant properties of the real object while being convenient to use.

Statistical models allow inferences to be made about an object, or activity, or a process by representing some associated observable data. Suppose, for example, a child has scored 20 points on a test of verbal ability, and after studying a dictionary for some time, scores 24 points on a similar test. If it is believed that studying the dictionary has caused an improvement, then a possible model of what is happening is

$$20 = \{\text{person's initial score}\}$$

$$24 = \{\text{person's initial score}\} + \{\text{improvement}\}$$

The improvement can now be found by simply subtracting the first score from the second. Such a model is, of course, very naive since it assumes that verbal ability can be measured exactly. A more realistic representation of the two scores, which allows for possible measurement error, is

$$x_1 = \gamma + \varepsilon_1$$

$$x_2 = \gamma + \delta + \varepsilon_1$$

where x_1 and x_2 represent the two verbal ability measurements, γ represents the "true" initial measure of verbal ability, and δ is the value of the improvement made in verbal ability. The terms ε_1 and ε_2 represent the measurement error for verbal ability made on the two occasions of testing. Here the improvement score can only be estimated as $\hat{\delta} = x_2 - x_1$. (The "hat" over a parameter indicates an estimate of that parameter.)

A model gives a precise description of what the investigator assumes is occurring in a particular situation; in the foregoing case it says that the improvement, δ, is considered to be independent of γ and is simply added to it. (An important point that needs to be noted here is that if you do not believe in a model, you should not perform operations and analyses on the data that assume the model to be true.)

Suppose now that it is believed that studying the dictionary does more good if a child already has a fair degree of verbal ability, so that the initial ability effect is multiplied by the dictionary effect and that the various random influences that affect the test scores are also dependent on the true scores, so also enter the model multiplicatively. Then an appropriate model would be

$$x_1 = \gamma\varepsilon_1$$

$$x_2 = \gamma\delta\varepsilon_2$$

Now the parameters are multiplied rather than added to give the observed scores x_1 and x_2. Here, δ might be estimated by dividing x_1 by x_2.

A further possibility is that there is a limit, λ, to improvement, and studying the dictionary improves performance on the verbal ability test by some proportion of the child's possible improvement, $\lambda - \gamma$. Here, a suitable model would be

$$x_1 = \gamma + \varepsilon_1$$

$$x_2 = \gamma + (\lambda - \gamma)\delta + \varepsilon_2$$

With this model there is no way to estimate δ from the data unless a value of λ is given or assumed. One of the principal uses of statistical models is to attempt to explain variation in measurements. This variation may be due to a variety of factors, including variation from the measurement system, variation due to environmental conditions that change over the course of a study, variation from individual to individual (or experimental unit to experimental unit), etc.

The decision about an appropriate model should be largely based on the investigator's prior knowledge of an area. In many situations, however, additive linear models are invoked by default since such models allow many powerful and informative statistical techniques to be applied to the data. Such models appear in several later chapters.

Formulating an appropriate model can be a difficult problem. The general principles of model formulation are covered in detail in books on scientific method but include the need to collaborate with appropriate experts, to incorporate as much background theory as possible, etc. Apart from those models formulated entirely on a priori theoretical grounds, most models are, to some extent at least, based on an initial examination of the data, although completely empirical models are rare. The more usual intermediate case arises when a class of models is entertained a priori, but the initial data analysis is crucial in selecting a subset of models from the class. In regression analysis, for example, the general approach is determined a priori, but a scatter diagram (see Chapter 2) will be of crucial importance in indicating the "shape" of the relationship, and residual plots (see Chapter 3) will be essential for checking assumptions such as normality, etc.

The formulation of a preliminary model from an initial examination of the data is the first step in the iterative, formulation/criticism cycle of model building. This can produce some problems since formulating a model and testing it on the same data is not generally considered good science. It is always preferable to confirm whether a derived model is sensible by testing on new data. But when data are difficult or expensive to obtain, some model modification and assessment of fit on the original data are almost inevitable. Investigators need to be aware of the possible dangers of such a process.

Perhaps the most important principle to have in mind when testing models on data is that of parsimony, that is, the "best" model is one that provides an adequate fit to data with the fewest number of parameters. This principle is often known as *Occam's razor*, which in its original form in Latin is *entia non stunt multiplicanda praeter necessitatem*, which translates roughly as "a plurality of reasons should not be posited without necessity."

One last caveat about statistical models: according to George Box, "all models are wrong, but some are useful." Statistical models are always simplifications, but some models are useful in providing insights into what is happening in complex, real-world situations.

1.6 Determining Sample Size

One of the most frequent questions faced by a statistician dealing with investigators planning a study is, "How many participants do I need to recruit?" Answering the question requires consideration of a number of factors, for example, the amount of time available for the study, the likely ease or difficulty in recruiting the type of subject required, and the possible financial constraints that may be involved. But the statistician may, initially at least, largely ignore these important aspects of the problem and apply a statistical procedure for calculating sample size. To make things simpler, we will assume that the investigation the researcher is planning is an experimental intervention with two treatment groups. To calculate the sample size, the statistician and the researcher will first need to identify the response variable of most interest and the appropriate statistical test to be used in the analysis of the chosen response; then they will need to decide on values for the following quantities:

- The size of the type I error, that is, the significance level.
- The likely variance of the response variable.
- The power they would like to achieve. (For those readers who have forgotten, or perhaps never knew, the power of a statistical test is the probability of the test rejecting the null hypothesis when the null hypothesis is false.)
- A size of treatment effect that the researcher feels is important, that is, a treatment difference that the investigators would not like to miss being able to declare to be statistically significant.

Given such information, the calculation of the corresponding sample size is often relatively straightforward, although the details will depend on the type of response variable and the type of test involved (see the following text for an example). In general terms, the sample size will increase as the variability of the response variable increases, and decrease as the chosen clinically relevant treatment effect increases. In addition, the sample size will need to be larger to achieve a greater power and a more stringent significance level.

As an example of the calculations involved in sample-size determination, consider a trial involving the comparison of two treatments for anorexia nervosa. Anorexic women are to be randomly assigned to each treatment, and the gain in weight in kilograms after three months is to be used as the outcome measure. From previous experience gained in similar trials, it is known that the standard deviation (σ) of weight gain is likely to be about 4 kg. The investigator feels that a difference in weight gain of 1 kg (Δ) would be of clinical importance and wishes to have a power of 90% when the appropriate two-sided test is used with significance level of 0.05 (α). The formula for calculating the number of women required in each treatment group (n) is

$$n = \frac{2\left(Z_{\alpha/2} + Z_\beta\right)^2 \sigma^2}{\Delta^2}$$

where β is 1 − Power, and

- $Z_{\alpha/2}$ is the value of the normal distribution that cuts off an upper tail probability of $\alpha/2$. So, for $\alpha = 0.05$, $Z_{\alpha/2} = 1.96$.
- Z_β is the value of the normal distribution that cuts off an upper tail probability of β. So, for a power of 0.90, $\beta = 0.10$ and $Z_\beta = 1.28$.

Therefore, for the anorexia trial,

$$n = \frac{2 \times (1.96 + 1.28)^2 \times 4^2}{1} = 336 \text{ women per treatment group.}$$

The foregoing example is clearly simplistic in the context of most psychiatric clinical trials in which measurements of the response variable are likely to be made at several different time points, during which time some patients may drop out of the trial (see Chapter 9 for a discussion of such longitudinal data and the drop out problem). Fortunately, there is a large volume of methodology useful in planning the size of randomized clinical trials with a variety of different types of outcome measures and with the complications outlined; some examples are to be found in Lee (1984), McHugh and Le (1984), Schoenfeld (1983), Hsieh (1987), and Wittes and Wallenstein (1987). In many cases, tables are available that enable the required sample size for chosen power, significance level, effect size, etc., to be simply read off. Increasingly, these are being replaced by computer software for determining sample size for many standard and nonstandard designs and outcome measures.

An obvious danger with the sample size determination procedure just mapped out is that investigators (and, in some cases, even their statisticians) may occasionally be led to specify an effect size that is unrealistically extreme (what Senn, 1997, has described with his usual candor as "a cynically relevant difference") so that the calculated sample size looks feasible in terms of possible pressing temporal and financial constraints. Such a possibility may be what led Senn (1997) to describe power calculations as "a guess masquerading as mathematics," and Pocock (1996) to comment that they are "a game that can produce any number you wish with manipulative juggling of the parameter values." Statisticians advising on behavioral investigations need to be active in estimating the degree of difference that can be realistically expected for a study based on previous studies of a similar type or, when such information is lacking, perhaps based on subjective opinions of investigators not involved in the putative study.

Getting the sample size right in a scientific study is generally believed to be critical; indeed, according to Simon (1991), discussing in particular clinical trials,

> An effective clinical trial must ask an important question and provide a reliable answer. A major determinant of the reliability of the answer is the sample size of the trial. Trials of inadequate size may cause contradictory and erroneous results and thereby lead to an inappropriate treatment of patients. They also divert limited resources from useful applications and cheat the patients who participated in what they thought was important clinical research. Sample size planning is, therefore, a key component of clinical trial methodology.

Studies with inadequate sample sizes risk missing important intervention differences, a risk nicely summarized in the phrase "absence of evidence is not evidence of absence." The case against studies with inadequate numbers of subjects appears strong, but as Senn (1997) points out, sometimes only a small study is possible. Also, misinterpreting a nonsignificant effect as an indication that a treatment effect is not effective, rather than as a failure to prove that it is effective, suggests trying to improve statistical education rather than totally abandoning small studies. In addition, with the growing use of systematic reviews and meta-analysis (see, for example, Everitt and Wessely, 2008), the results from small studies may prove valuable in contributing to an overview of the evidence of intervention effectiveness, a view neatly summarized by Senn in the phrase "some evidence is better than none." Perhaps size really is not always everything?

1.7 Significance Tests, *p*-Values, and Confidence Intervals

Although we are assuming that readers have had an introductory course in statistics that covered simple significance tests, *p*-values, and confidence intervals, a few more words about these topics here will hopefully not go amiss.

For many behavioral science students and researchers the still-ubiquitous *p*-value continues to be the Holy Grail of their research efforts, and many see it as the *raison d'etre* of statistics and statisticians. Despite the numerous caveats about *p*-values in the literature (e.g., Gardner and Altman, 1986), many behavioral scientists still seem determined to experience a "eureka moment" on finding a *p*-value of 0.049, and despair on finding one of 0.051. The *p*-value retains a powerful hold over the average behavioral researcher and student; there are a number of reasons why it should not.

The first is that the *p*-value is poorly understood. Although *p*-values appear in almost every account of behavioral science research findings, there is evidence that the general degree of understanding of the true meaning of the term is very low. Oakes (1986), for example, put the following test to 70 academic psychologists:

> Suppose you have a treatment which you suspect may alter performance on a certain task. You compare the means of your control and experimental groups (say 20 subjects in each sample). Further suppose you use a simple independent means t-test and your result is t = 2.7, df = 18, P = 0.01. Please mark each of the following statements as true or false.

- You have absolutely disproved the null hypothesis that there is no difference between the population means.
- You have found the probability of the null hypothesis being true.
- You have absolutely proved your experimental hypothesis.
- You can deduce the probability of the experimental hypothesis being true.
- You know, if you decided to reject the null hypothesis, the probability that you are making the wrong decision.
- You have a reliable experiment in the sense that if, hypothetically, the experiment were repeated a great number of times, you would obtain a significant result on 99% of occasions.

The subjects were all university lecturers, research fellows, or postgraduate students. The results presented in Table 1.1 are illuminating. Under a relative frequency view of probability, all six statements are in fact false. Only 3 out of the 70 subjects came to this conclusion. The correct interpretation of the probability associated with the observed t-value is:

> The probability of obtaining the observed data (or data that represent a more extreme departure from the null hypothesis) if the null hypothesis is true.

TABLE 1.1
Frequencies and Percentages of "True" Responses in Test of Knowledge about *p*-values

Statement	Frequency	Percentage
1. The null hypothesis is absolutely disproved.	1	1.4
2. The probability of the null hypothesis has been found.	25	35.7
3. The experimental hypothesis is absolutely proved.	4	5.7
4. The probability of the experimental hypothesis can be deduced.	46	65.7
5. The probability that the decision taken is wrong is known.	60	85.7
6. A replication has a 0.99 probability of being significant.	42	60.0

Clearly, the number of false statements described as true in this experiment would have been reduced if the true interpretation of a p-value had been included with the six others. Nevertheless, the exercise is extremely interesting in highlighting the misguided appreciation of p-values held by a group of behavioral researchers.

The second reason for researchers and students to be careful using p-values is that a p-value represents only limited information about the results from a study. Gardner and Altman (1986) make the point that the excessive use of p-values in hypothesis testing, simply as a means of rejecting or accepting a particular hypothesis at the expense of other ways of assessing results, has reached such a degree that levels of significance are often quoted alone in the main text and abstracts of papers with no mention of other more relevant and important quantities. The implications of hypothesis testing—that there can always be a simple "yes" or "no" answer as the fundamental result from a research study—is clearly false and, used in this way, hypothesis testing is of limited value; indeed, according to Siegfried (2010) "It's science's dirtiest secret: The 'scientific method' of testing hypotheses by statistical analysis stands on a flimsy foundation." And in 2016 in *The American Statistician* the American Statistical Association (ASA) made a lengthy statement mostly critical of the way p-values are used (see Wasserstein and Lazar, 2016).

The most common alternative to presenting results in terms of p-values in relation to a statistical null hypothesis is to estimate the magnitude of some parameter of interest along with some interval that includes the population value of the parameter with a specified probability. Such confidence intervals can be found relatively simply for many quantities of interest (see Gardner and Altman, 1986, for details), and although the underlying logic of interval estimation is essentially similar to that of significance tests, they do not carry with them the pseudoscientific hypothesis testing language of such tests. Instead, they give a plausible range of values for the unknown parameter. As Oakes (1986) rightly comments: "The significance test relates to what the population parameter is not: the confidence interval gives a plausible range for what the parameter is."

So, should the p-value be abandoned completely? Many statisticians would, grumpily, answer yes, but we think a more sensible response, at least for behavioral scientists, would be a resounding "maybe." The p-value should rarely be used in a purely confirmatory way, but in an exploratory fashion, p-values can be useful in giving some informal guidance on the possible existence of an interesting effect even when the required assumptions of whatever test is being used are known to be invalid. It is often possible to assess whether a p-value is likely to be an under- or overestimate, and whether the result is clear one way or the other. In this text, both p-values and confidence intervals will be used; purely from a pragmatic point-of-view, the former are needed by behavioral science students since they remain of central importance in the bulk of the behavioral science literature.

1.8 Summary

- Statistical principles are central to most aspects of a psychological investigation.
- Data and their associated statistical analyses form the evidential parts of behavioral science arguments.
- Significance testing is far from the be-all and end-all of statistical analyses, but it does still matter because evidence that can be discounted as an artifact of sampling will not be particularly persuasive. But p-values should not be taken too seriously; confidence intervals are often more informative.
- Good statistical analysis should highlight those aspects of the data that are relevant to the substantive arguments; do so clearly and fairly, and be resistant to criticisms.
- Although randomised controlled experiments are largely considered as the gold standard in behavioral research because they are generally considered to lead to the clearest conclusions about causal relationships, Yarkoni and Westfall (2017) argue that more emphasis should be placed of methods that accurately forecast/predict behaviors that have not yet been observed.
- Variable type often determines the most appropriate method of analysis, although some degree of flexibility should be allowed.
- Sample size determination to achieve some particular power is an important exercise when designing a study, but the result of the statistical calculation involved needs to be considered in terms of what is feasible from a practical viewpoint.
- Statisticians giving advise to behavioral scientists and behavioral scientists seeking advise from a statistician might like to take a look at the small book *Talking about Statistics* (Everitt and Hay, 1992) to see ideal examples of how such consultations should develop.
- Three more small books, one by Hand (2008), one by Magnello and van Loon (2009), and one by Stigler (2016) give brief, but useful accounts of the basics of statistics, measurement, and models, along with interesting snippets of the history of these topics.

1.9 Exercises

1.1 The Pepsi-Cola Company carried out research to determine whether people tended to prefer Pepsi Cola to Coca Cola. Participants were asked to taste two glasses of cola and then state which they preferred. The two glasses were not labeled Pepsi or Coke for obvious reasons. Instead, the Coke glass was labeled Q, and the Pepsi glass was labeled M. The results showed that "more than half chose Pepsi over Coke" (Huck and Sandler, 1979, p. 11). Are there any alternative explanations for the observed difference, other than the taste of the two drinks? Explain how you would carry out a study to assess any alternative explanation you think possible.

1.2 You develop a headache while working for hours at your computer. You stop, go into another room, and take two aspirins. After about 15 min, your headache has gone and you return to work. Can you infer a definite causal relationship between taking the aspirin and curing the headache? If not, why not?

1.3 You are interested in assessing whether or not laws that ban purchases of handguns by convicted felons reduce criminal violence. What type of study would you carry out, and how would you go about the study?

1.4 Attribute the following quotations about statistics and statisticians:

a. To understand God's thoughts we must study statistics, for these are a measure of his purpose.
b. You cannot feed the hungry on statistics.
c. A single death is a tragedy, a million deaths is a statistic.
d. Thou shall not sit with statisticians nor commit a Social Science.
e. Facts speak louder than statistics.
f. I am one of the unpraised, unrewarded millions without whom statistics would be a bankrupt science. It is we who are born, marry, and who die in constant ratios.
g. A recent survey of North American males found 42% were overweight, 34% were critically obese and 8% ate the survey.
h. Life is a gamble at terrible odds – if it was a bet you wouldn't take it.
i. Misunderstanding of probability may be the greatest of all impediments to scientific literacy.
j. A fool must now and then be right by chance.
k. The stock market has forecast nine of the last five recessions.

1.5 In reading about the results of an intervention study, you find that alternate subjects have been allocated to the two treatment groups. How would you feel about the study and why?

1.6 What is the ratio of two measurements of warmth, one of which is 25°C and the other of which is 110°F?

2

Looking at Data

2.1 Introduction

According to Chambers et al. (1983), "there is no statistical tool that is as powerful as a well-chosen graph." Certainly, graphical display has a number of advantages over tabular displays of numerical results, not least in creating interest and attracting the attention of the viewer.

But just what is a graphical display? A concise description is given by Tufte (1983):

> Data graphics visually display measured quantities by means of the combined use of points, lines, a coordinate system, numbers, symbols, words, shading and color.

Graphical displays are very popular; it has been estimated that between 900 billion (9×10^{11}) and 2 trillion (2×10^{12}) images of statistical graphics are printed each year. Perhaps one of the main reasons for such popularity is that graphical presentation of data often provides the vehicle for discovering the unexpected; the human visual system is very powerful in detecting patterns, although the following caveat from the late Carl Sagan should be kept in mind:

> Humans are good at discerning subtle patterns that are really there, but equally so at imagining them when they are altogether absent.

Some of the advantages of graphical methods have been listed by Spear (1952) and Schmid (1954):

- In comparison with other types of presentation, well-designed charts are more effective in creating interest and in appealing to the attention of the reader.
- Visual relationships as portrayed by charts and graphs are more easily grasped and more easily remembered.
- The use of charts and graphs saves time since the essential meaning of large measures of statistical data can be visualized at a glance.

- Charts and graphs provide a comprehensive picture of a problem that makes for a more complete and better-balanced understanding than could be derived from tabular or textual forms of presentation.
- Charts and graphs can bring out hidden facts and relationships, and can stimulate, as well as aid, analytical thinking and investigation.

The last point is reiterated by the legendary John Tukey in his observation that "the greatest value of a picture is when it forces us to notice what we never expected to see."

The prime objective of a graphical display is to communicate to ourselves and others. Graphic design must do everything it can to help people understand. In some cases a graphic is required to give an overview of the data and perhaps to tell a story about the data. In other cases a researcher may want a graphical display to suggest possible hypotheses for testing on new data and, after some model has been fitted to the data, a graphic that criticizes the model may be what is needed. In this chapter we will consider graphics primarily from the story-telling angle; graphics that help check model assumption will be discussed in later chapters.

Since the 1970s, a wide variety of new methods for displaying data graphically have been developed; these will hunt for special effects in data, indicate outliers, identify patterns, diagnose models, and generally search for novel and perhaps unexpected phenomena. Large numbers of graphs may be required, so computers are needed to supply them for the same reasons they are used for numerical analyses, namely, they are fast and they are accurate.

So, because the machine is doing the work, the question is no longer "shall we plot?" but rather "what shall we plot?" There are many exciting possibilities, including *dynamic graphics* (see Cleveland and McGill, 1988), and the related area of *infographics* (also called information visualization or *infovis*, see Gelman and Unwin, 2013 and Vehkalahti et al., 2018), but graphical exploration of data usually begins with some simpler, well-known methods, and it is these that we deal with in the next section.

2.2 Simple Graphics—Pie Charts, Bar Charts, Histograms, and Boxplots

2.2.1 Categorical Data

Newspapers, television, and the media in general are very fond of two very simple graphics for displaying categorical data, namely, the pie chart and the bar chart. Both can be illustrated using the data shown in Table 2.1, which show the percentage of people convicted of five different types of crime. In the pie charts for drinkers and abstainers (see Figure 2.1), the sections of the circle have areas proportional to the observed percentages. In the correspond-

TABLE 2.1
Crime Rates for Drinkers and Abstainers

Crime	Drinkers	Abstainers
Arson	6.6	6.4
Rape	11.7	9.2
Violence	20.6	16.3
Stealing	50.3	44.6
Fraud	10.8	23.5

Note: Figures are percentages.

ing bar charts (see Figure 2.2), percentages are represented by rectangles of appropriate size placed along a horizontal axis.

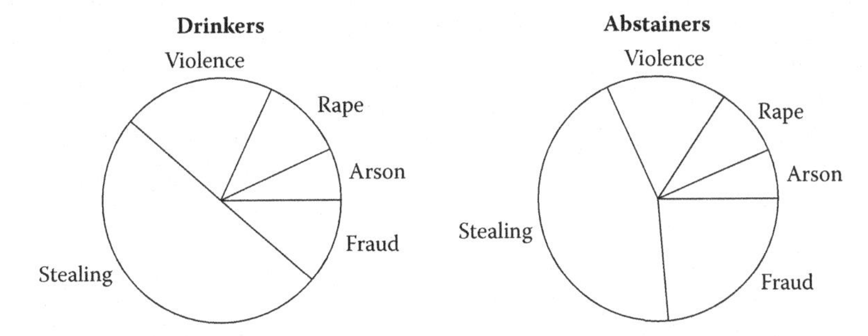

FIGURE 2.1
Pie charts for drinkers' and abstainers' crime percentages.

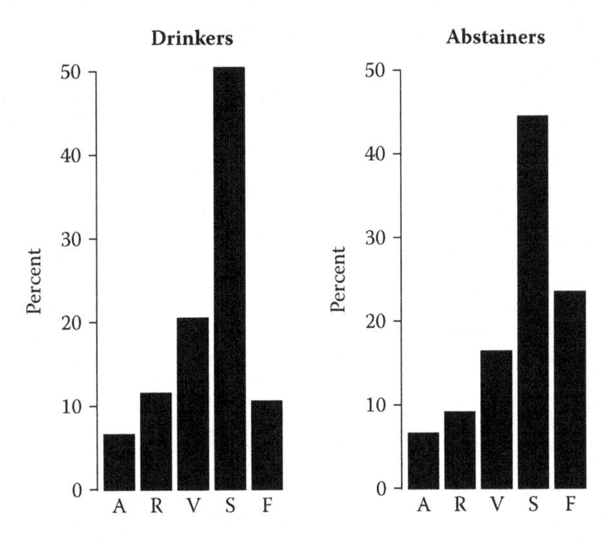

FIGURE 2.2
Bar charts for drinkers' and abstainers' crime percentages.

Despite their widespread popularity, both the general and, in particular, the scientific use of pie charts has been severely criticized. Tufte (1983), for example, comments that "tables are preferable to graphics for many small data sets. A table is nearly always better than a dumb pie chart; the only worse design than a pie chart is several of them ... pie charts should never be used." A similar lack of affection is shown by Bertin (1981), who declares that "pie charts are completely useless," and by Wainer (1997), who claims that "pie charts are the least useful of all graphical forms." Certainly in regard to the data in Table 2.1, the numerical data in the table are as informative, or perhaps even more informative, than the associated pie charts in Figure 2.1, and the bar chart in Figure 2.2 seems no more necessary than the pie chart for these data.

Two examples that illustrate why both pie charts and bar charts are often (but not always—as will be seen later) of little more help in understanding categorical data than the numerical data themselves and how other graphics are frequently more useful are given in Cleveland (1994). The first example compares the pie chart of 10 percentages with an alternative graphic, the dot plot. The plots are shown in Figures 2.3 and 2.4. The 10 percentages represented by the two graphics have a bimodal distribution; odd-numbered observations cluster around 8%, and even-numbered observations cluster around 12%. Furthermore, each even value is shifted with respect to the preceding odd value by about 4%. This pattern is far easier to spot in the dot plot than in the pie chart.

Dot plots for the crime data in Table 2.1 are shown in Figure 2.5, and these are also more informative than the corresponding pie charts in Figure 2.1.

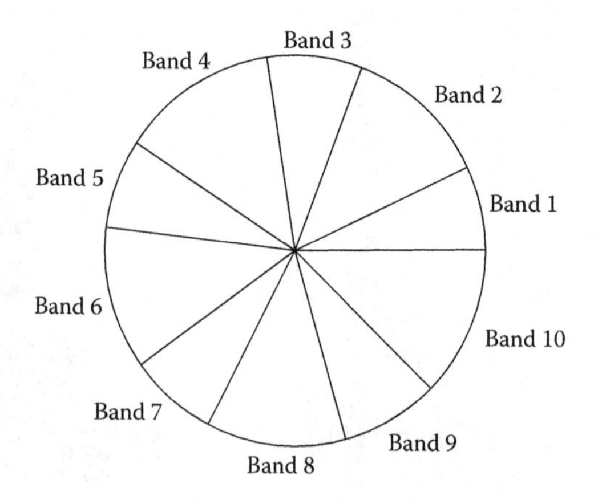

FIGURE 2.3
Pie chart for 10 percentages. (Suggested by Cleveland, 1994. Used with permission from Hobart Press.)

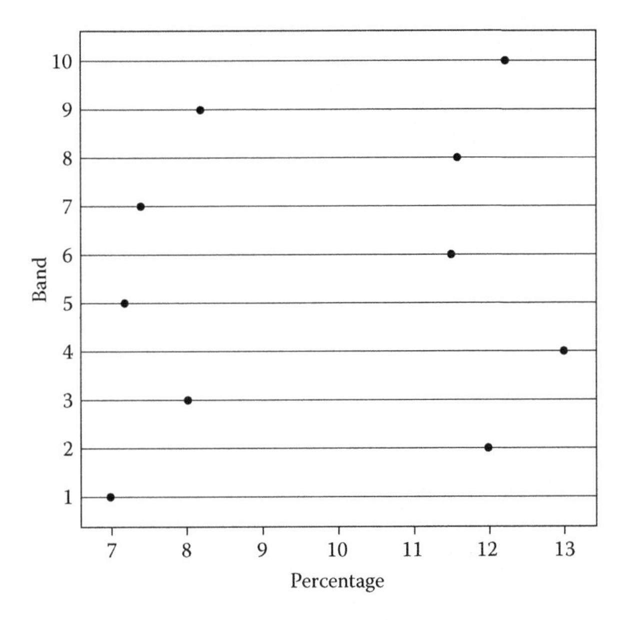

FIGURE 2.4
Dot plot for 10 percentages.

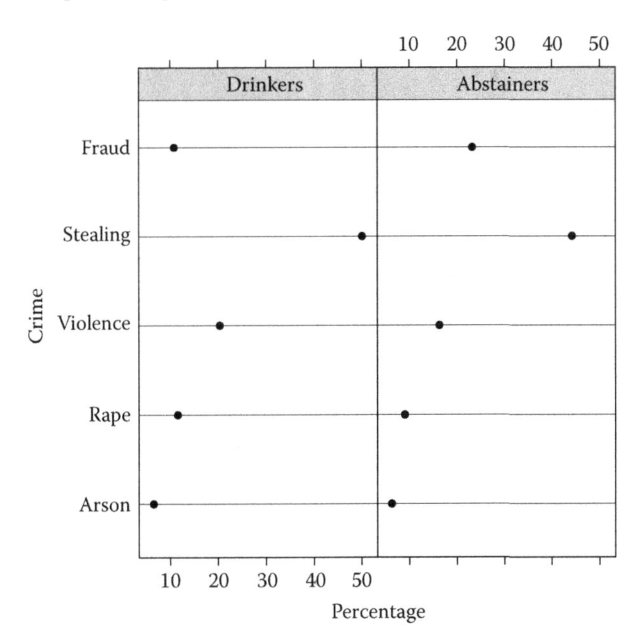

FIGURE 2.5
Dot plots for drinkers' and abstainers' crime percentages.

The second example given by Cleveland begins with the diagram shown in Figure 2.6, which originally appeared in Vetter (1980). The aim of the diagram is to display the percentages of degrees awarded to women in several disciplines of science and technology during three time periods. At first glance the labels on the diagram suggest that the graph is a standard divided bar chart with the length of the bottom division of each bar showing the percentage for doctorates, the length of the middle division showing the percentage for master's degrees, and the top division showing the percentage for bachelor's degrees. A little reflection shows that this interpretation is not correct since it would imply that, in most cases, the percentage of bachelor's degrees given to women is lower than the percentage of doctorates. Closer examination of the diagram reveals that the three values of the data for each discipline during each time period are determined by the three adjacent vertical dotted lines. The top of the left-hand line indicates the value for doctorates, the top end of the middle line indicates the value for master's degrees, and the top end of the right-hand line indicates the value for bachelor's degrees.

Cleveland (1994) discusses other problems with the diagram in Figure 2.6; in particular, he points out that the manner of the diagram's construction makes it hard to connect visually the three values of a particular type of degree for a specific discipline, thus making it difficult to see changes over time.

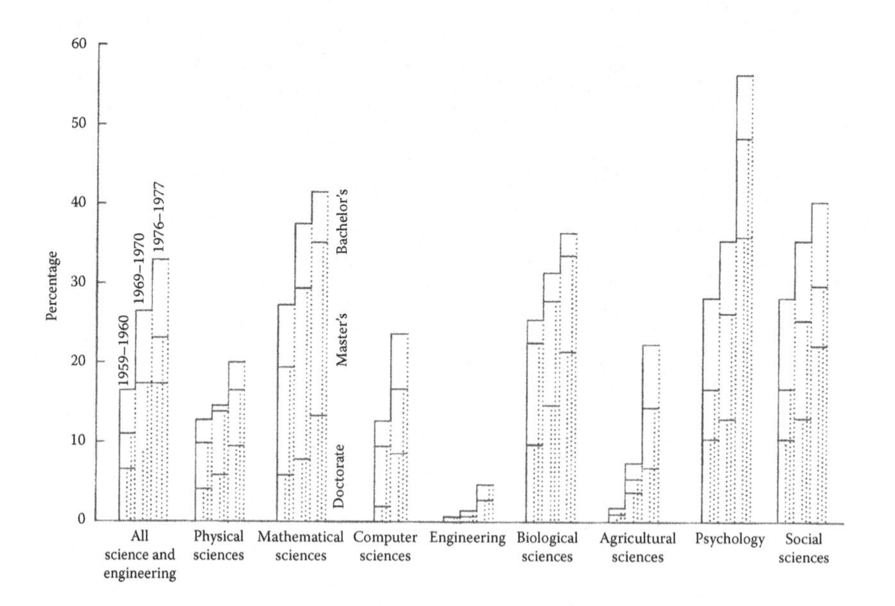

FIGURE 2.6
Proportion of degrees in science and technology earned by women in the periods 1959–1960, 1969–1970, and 1976–1977. (Reproduced with permission from Vetter, B. M., 1980, *Science*, 207, 28–34. ©1980 American Association for the Advancement of Society.)

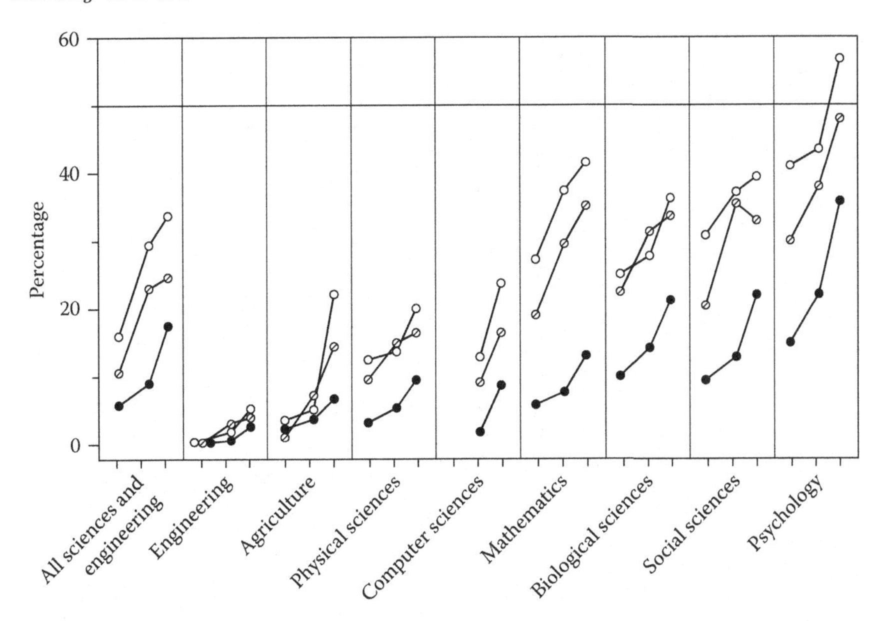

FIGURE 2.7
Percentage of degrees earned by women for three degrees (○ bachelor's degree; ϕ master's degree; • doctorate), three time periods, and nine disciplines. The three points for each discipline and degree indicate the periods 1959–1960, 1969–1970, and 1976–1977.

Figure 2.7 shows the data represented by Figure 2.6 replotted by Cleveland in a bid to achieve greater clarity. It is now clear how the data are represented, and this diagram allows viewers to see easily the percentages corresponding to each degree, in each discipline, over time. Finally the figure caption explains the content of the diagram in a comprehensive and clear fashion. All in all Cleveland appears to have produced a graphic that would satisfy even that doyen of graphical presentation, Edward R. Tufte, in his demand that "excellence in statistical graphics consists of complex ideas communicated with clarity, precision and efficiency."

Wainer (1997) gives a further demonstration of how displaying data as a bar chart can disguise rather than reveal important points about data. Figure 2.8 shows a bar chart of life expectancies in the middle 1970s, divided by sex, for ten industrialized nations. The order of presentation is alphabetical (with the U.S.S.R. positioned as Russia). The message we get from this diagram is that there is little variation and women live longer than men. But by displaying the data in the form of a simple stem-and-leaf plot (see Figure 2.9), the magnitude of the sex difference (7 years) is immediately clear as is the unusually short life expectancy for men in the U.S.S.R., whereas Russian women have life expectancy similar to women in other countries.

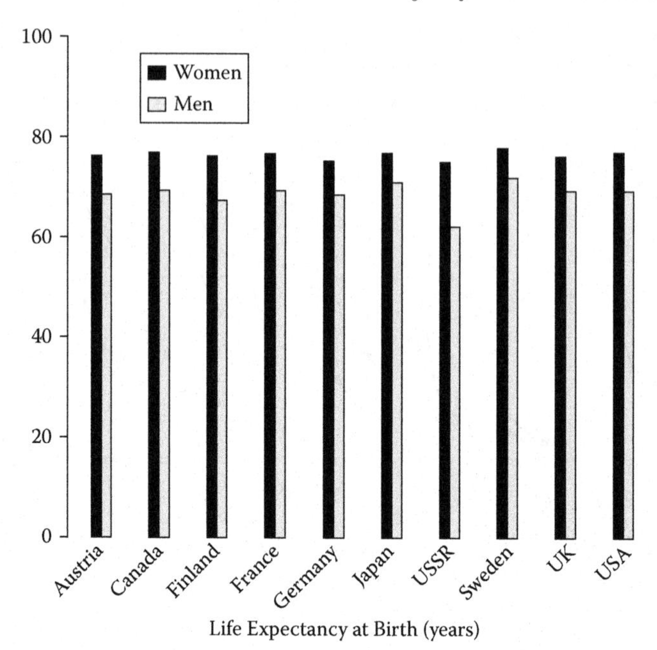

FIGURE 2.8
Bar chart showing life expectancies at birth by sex and by country.

Women	Age	Men
Sweden	78	
France, US, Japan, Canada	77	
Finland, Austria, UK	76	
USSR, Germany	75	
	74	
	73	
	72	Sweden
	71	Japan
	70	
	69	Canada, UK, US, France
	68	Germany, Austria
	67	Finland
	66	
	65	
	64	
	63	
	62	USSR

FIGURE 2.9
An alternative display of life expectancies at birth by sex and by country.

To be fair to the poor old bar chart, we will end this subsection by illustrating how a sophisticated adaptation of the graphic can become an extremely effective tool for displaying a complex set of categorical data. The example is taken from Sarkar (2008) and uses data summarizing the fates of the 2201 passengers on the Titanic. The data are categorized by economic status (class of ticket: 1st, 2nd, or 3rd, or crew), sex (male or female), age (adult or child), and whether they survived or not. The first diagram produced by Sarkar is shown in Figure 2.10. This plot looks impressive but is dominated by the third "panel" (adult males) as heights of bars represent counts, and all panels have the same limits. So, sadly, all the plot tells us is that there were many more males than females aboard (particularly among the crew, which is the largest group), and that there were even fewer children. The plot becomes more illuminating about what really happened to the passengers if the proportion of survivors is plotted and by allowing independent horizontal scales

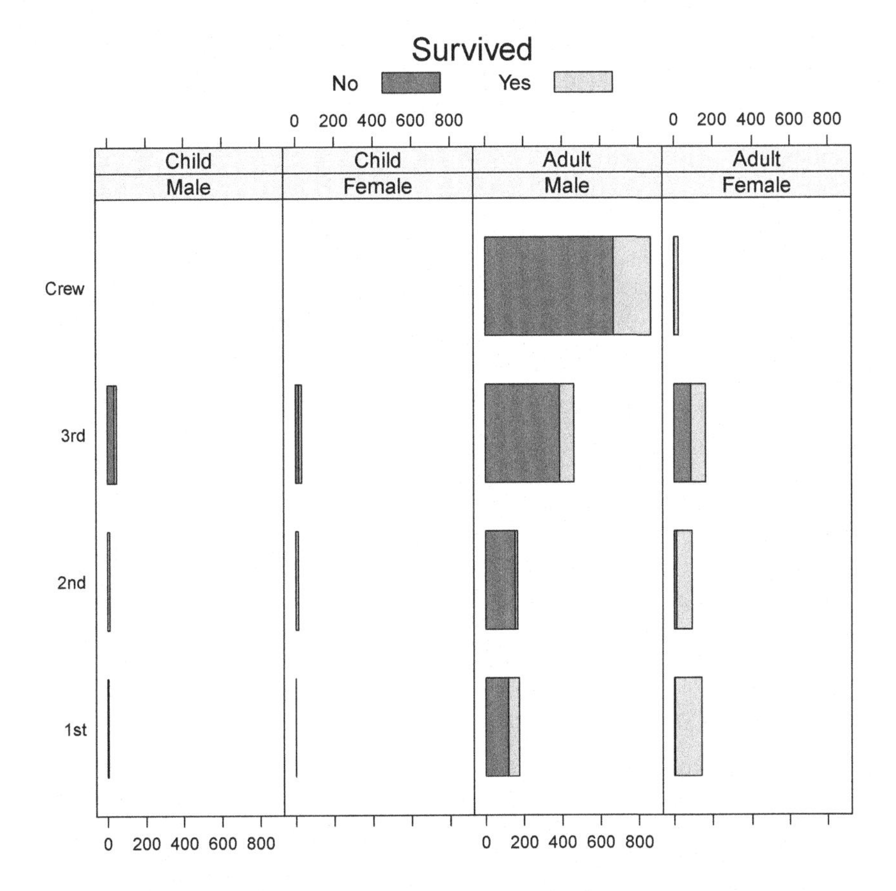

FIGURE 2.10
A bar chart summarizing the fate of passengers of the Titanic, classified by sex, age, and whether they survived or not.

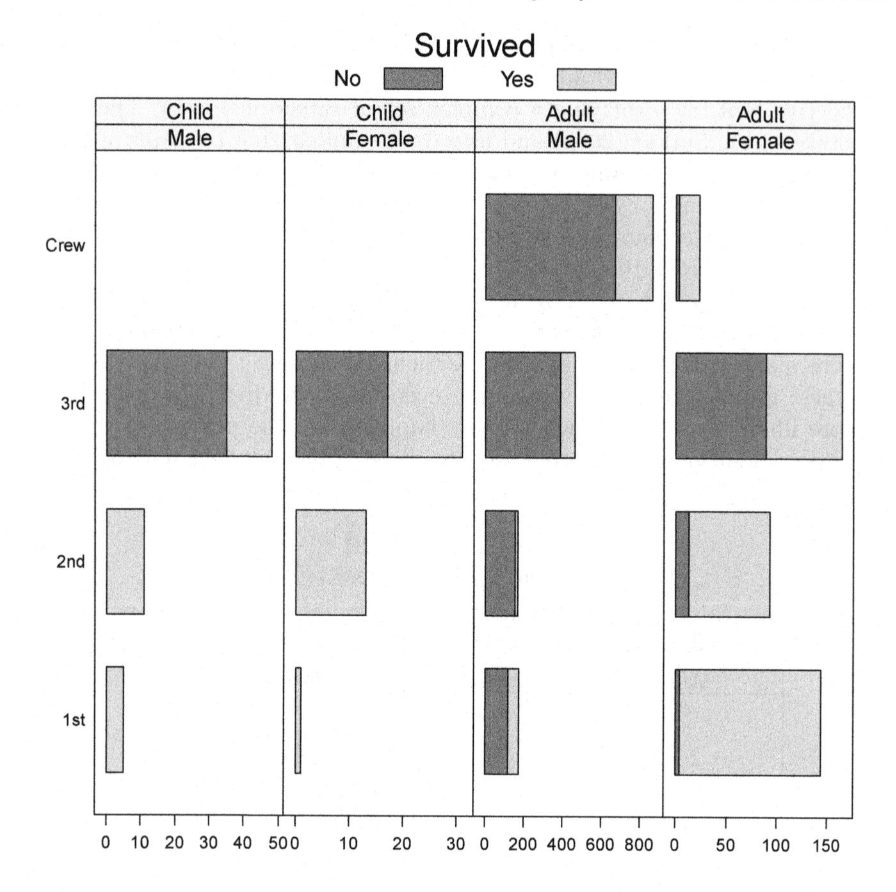

FIGURE 2.11
Survival among different subgroups of passengers on the Titanic, with a different horizontal scale in each panel.

for the different "panels" in the plot; this plot is shown in Figure 2.11. This plot emphasizes the proportion of survivors within each subgroup rather than the absolute numbers. The proportion of survivors is lowest among third-class passengers, and the diagram makes it very clear that the "women and children first" policy did not work very well for this class of passengers.

2.2.2 Interval/Quasi-Interval Data

The data shown in Table 2.2 come from an observational study described in Howell and Huessy (1981, 1985), in which a researcher asked 50 children to tell her about a given movie. For each child the researcher recorded the number of "and then ..." statements.

Let us begin by constructing that old favorite, the histogram, for these data; it is shown in Figure 2.12. Here the histogram is based on a relatively

TABLE 2.2
Number of "and Then..." Statements Made by 50 Children Recalling the Story of a Movie They Had Just Seen

18	15	22	19	18	17	18	20	17	12	16	16	17	21	23	18	20	21	20	20	15	18	17	19	20
23	22	10	17	19	19	21	20	18	18	24	11	19	31	16	17	15	19	20	18	18	40	18	19	16

small number of observations and tells us little about the data except that there is some degree of skewness perhaps and possibly two "outliers."

The histogram is a widely used graphical method that is at least 100 years old. But Cleveland (1994) makes the point that "maturity and ubiquity do not guarantee the efficacy of a tool." The histogram is generally used for two purposes: counting and displaying the distribution of a variable. However, according to Wilkinson (1992), "it is effective for neither." Histograms can often be misleading about a variable's distribution because of their dependence on the number of classes chosen, and simple tallies of the observations to give a numerical frequency distribution table are usually preferable for counting. Finally, the histogram is a poor method for comparing groups of univariate measurements (Cleveland, 1994).

A more useful graphical display of a variable's distributional properties is the boxplot. This is obtained from the five-number summary of a data set, the five numbers in question being the minimum, the lower quartile (LQ), the median, the upper quartile (UQ), and the maximum. The distance between the upper and lower quartiles, the interquartile range (IQR), is a measure of the spread of a variable's distribution. The relative distances from the

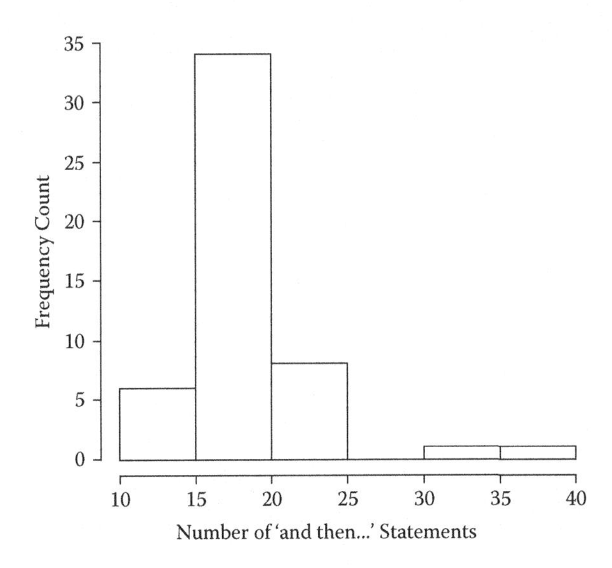

FIGURE 2.12
Histogram of count of "and then ... " statements by 50 children.

median of the upper and lower quartiles give information about the shape of a variable's distribution; for example, if one distance is much greater than the other, the distribution is skewed. In addition, the median and the upper and lower quartiles can be used to define arbitrary but often useful limits, L and U, that maybe helpful in identifying possible outliers. The two limits are calculated as follows:

$$\text{U} = \text{UQ} + 1.5\,\text{IQR}$$

$$\text{L} = \text{LQ} - 1.5\,\text{IQR}$$

Observations outside the limits can be regarded as potential outliers (they are sometimes referred to specifically as outside observations), and such observations may merit careful attention before undertaking any analysis of a data set because there is always the possibility that they can have undue influence on the results of the analysis.

The construction of a boxplot is described in Figure 2.13.

The boxplot of the data in Table 2.2 is shown in Figure 2.14. The diagram indicates a number of possible outliers and also highlights the skewness in the data.

In Table 2.3, there is a similar data set as in Table 2.2, but here the observations were collected from adults. A question of interest is whether children and adults recall stories in the same way. At some stage this may require a formal procedure such as the construction of a confidence interval for, say, the mean difference in the number of "and then ..." statements between children and adults. But here we will see how far we can get with a

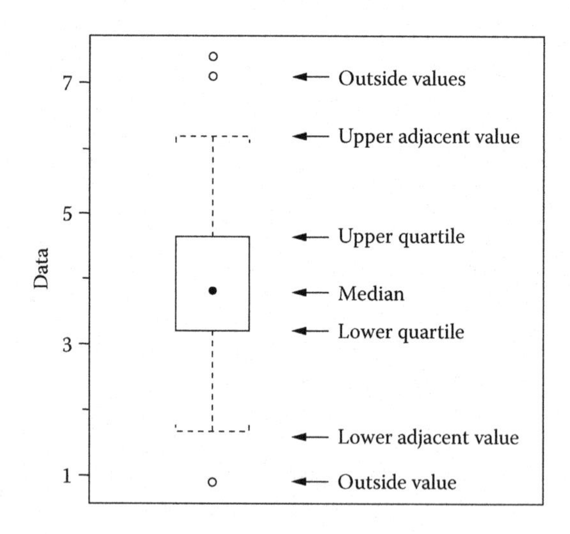

FIGURE 2.13
The construction of a boxplot.

TABLE 2.3
Number of "and Then..." Statements Made by 50 Adults Recalling the Story of a Movie They Had Just Seen

10	12	5	8	13	10	12	8	7	11	11	10	9	9	11	15	12	17	14	10	9	8	15	16	10
14	7	16	9	1	4	11	12	7	9	10	3	11	14	8	12	5	10	9	7	11	14	10	15	9

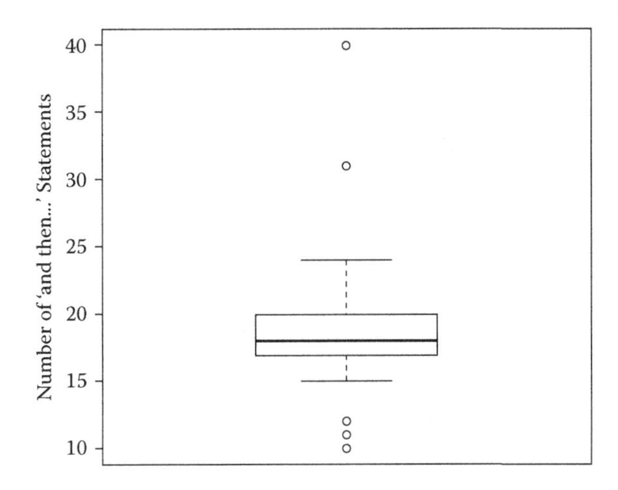

FIGURE 2.14
Boxplot of count of "and then ... " statements by children.

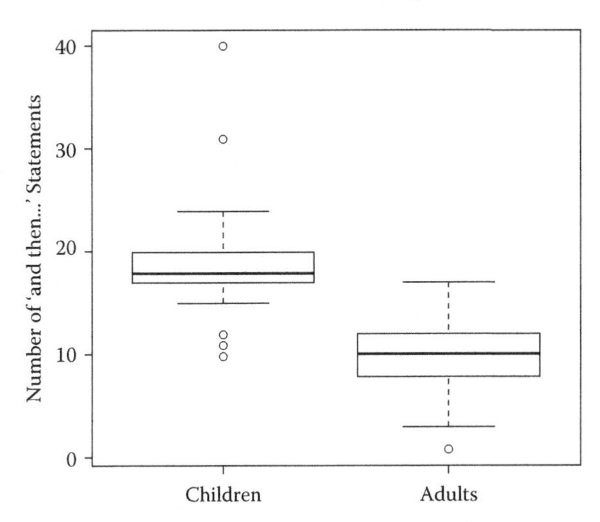

FIGURE 2.15
Side-by-side boxplots of counts of "and then ... " statements by children and adults.

graphical approach namely comparing the boxplots of each data set. The side-by-side boxplots are shown in Figure 2.15. The diagram clearly demonstrates that the adults generally use less "and then ... " statements than children and also suggests that the distribution of the adults' observations is closer to being symmetric that that of the children.

Although the boxplots in Figure 2.15 give some information about the distributions of the observations in each group, it may be useful to delve a little further and use probability plots to assess the normality of the observations for both children and adults. Probability plots are described in Technical Section 2.1.

Technical Section 2.1: Probability Plots

The classic example of a probability plot is that for investigating the assumption that a set of data is from a normal distribution; here the ordered sample values, $y_{(1)} \leq y_{(2)} \leq \cdots \leq y_{(n)}$ are plotted against the quantiles of a standard normal distribution, that is, $\Phi^{-1}(p_i)$, where

$$p_i = \frac{i - \frac{1}{2}}{n}, \quad \text{and} \quad \Phi(x) = \int_{-\infty}^{x} \frac{1}{\sqrt{2\pi}} e^{-\frac{1}{2}u^2} du$$

Departures from linearity in the plot indicate that the data do not have a normal distribution.

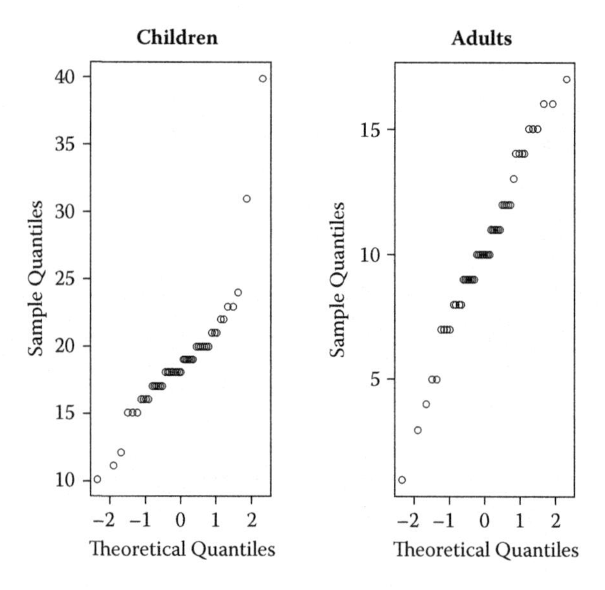

FIGURE 2.16
Probability plots of counts of "and then ... " statements by children and adults.

The normal probability plots for the children and the adults are shown in Figure 2.16. The plot for the children's observations shows a marked departure from linearity but the plot for the adults' data looks linear. These findings might need to be considered before any formal test or procedure is applied to the data set, although here constructing the usual normality based confidence interval is unlikely to be misleading.

Probability plots have been around for a long time, but they remain a useful technique for assessing distributional assumptions in some cases as here for the raw data, but also for the residuals that are used to assess the assumptions when fitting models to data, as we shall see in Chapter 3.

2.3 The Scatterplot and beyond

The simple xy scatterplot has been in use since at least the 18th century and has many advantages for an initial exploration of data. Indeed, according to Tufte (1983):

> The relational graphic—in its barest form the scatterplot and its variants—is the greatest of all graphical designs. It links at least two variables encouraging and even imploring the viewer to assess the possible causal relationship between the plotted variables. It confronts causal theories that x causes y with empirical evidence as to the actual relationship between x and y.

Let us begin by looking at a straightforward use of the scatterplot using some of the data in Table 2.4. These data were collected from a sample of 24 primary school children in Sydney, Australia. Part of the data is given in Table 2.4. Each child completed the Embedded Figures Test (EFT), which measures "field dependence," that is, the extent to which a person can abstract the logical structure of a problem from its context. Then the children were allocated to one of two experimental groups, and they were timed as they constructed a 3×3 pattern from nine colored blocks, taken from the Wechsler Intelligence Scale for Children (WISC). The two groups differed in the instructions

TABLE 2.4
Field Dependence Measure and Time to Complete a Task from the WISC for Children in Two Experimental Groups

Row Group			Corner Group		
Child	**Time**	**EFT**	**Child**	**Time**	**EFT**
1	317	59	1	342	43
2	464	33	2	222	23
3	525	49	3	219	9
4	298	69	4	513	128
5	491	65	5	295	44

they were given for the task: the "row" group was told to start with a row of three blocks, and the "corner" group was told to start with a corner of three blocks. The experimenter was interested in whether the different instructions produced any change in the average time to complete the picture and whether this time was affected by field dependence.

So, to begin, Figure 2.17 shows the scatterplots of completion time against EFT for both the row and the corner groups. The first thing to notice about the two plots is the obvious outlier in the plot for the row experimental group, and the relationship between time to completion and EFT appears to be stronger in the row group than in the corner group although the outlier in the row group may be entirely responsible for this apparent difference. Other than this, the plots are not particularly informative, and we perhaps cannot expect too much from them given the rather small samples involved. It should be remembered that calculating a correlation coefficient between two variables without looking at the corresponding scatterplot is very poor data analysis practice because a correlation coefficient can, on occasions, be badly affected by outliers, and the scatterplot is needed to spot the offending observations.

To make the scatterplots a little more informative, we can add the linear regression fit (see Chapter 3) of time to completion against EFT to each plot. The result is Figure 2.18. The plots now demonstrate more clearly that completion time appears to have a stronger linear relationship to EFT in the row group than in the corner group (but remember that outlier).

Another possible way to plot the data in Table 2.4 is to simply combine all the data in one scatterplot, identifying the row and corner group observations in some way. This is what we have done in Figure 2.19.

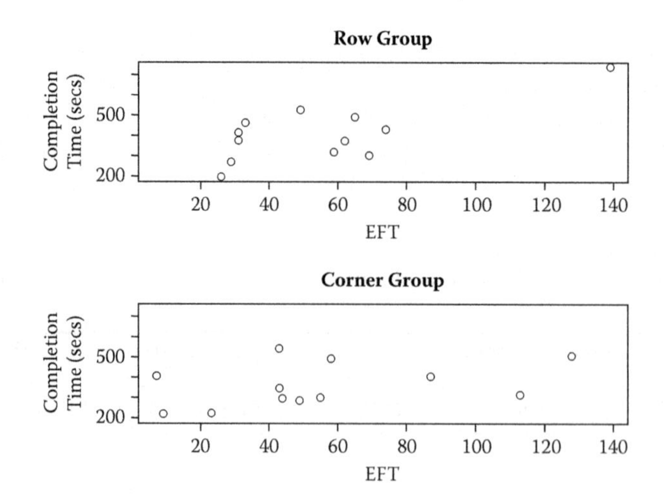

FIGURE 2.17
Scatterplots of time to completions against EFT score for row and corner groups.

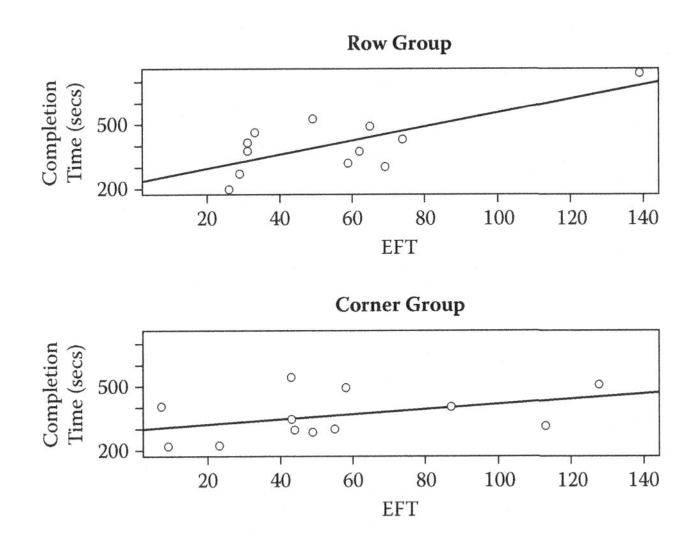

FIGURE 2.18
Scatterplots of time to completion against EFT for row and corner groups with added linear regression fit.

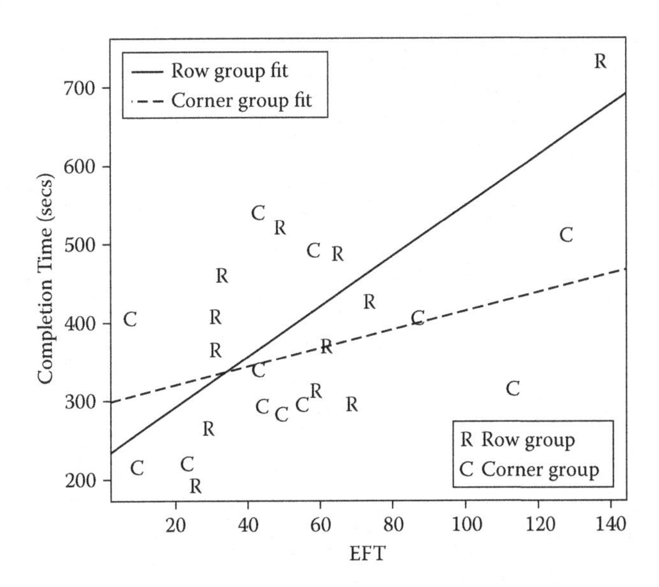

FIGURE 2.19
Scatterplot of completion time against EFT with observations labeled as row or corner group and linear regression fits for each group shown.

TABLE 2.5
Data Collected About Time Spent Looking After Car for First Five Subjects Out of 40

Subject	Sex	Age	Extro	Time
1	Male	55	40	46
2	Male	43	45	79
3	Female	57	52	33
4	Male	26	62	63
5	Female	22	31	20

Now let us move on to consider a larger set of data, part of which is given in Table 2.5. These data are taken from Miles and Shevlin (2001), and give the sex, age, extroversion score, and the average number of minutes per week a person spends looking after his or her car, for 40 people. People may project their self-image through themselves or through the objects they own, such as their cars. Therefore, a theory could be developed that predicts that people who score higher on a measure of extroversion are likely to spend more time looking after their cars. This possibility will be examined in Chapter 4; here we will see how much information about the data we can derive from some scatterplots. Any information about the data collected at this point may be very helpful in fitting formal models to the data.

To begin, we shall construct scatterplots of time spent by people looking after their cars, against age and extroversion score. Often when using scatterplots to look at data, it is helpful to add something about the marginal distributions of the two variables, and this we will do here. Further, we will add to each plot the appropriate linear regression fits separately for men and women. The two plots are shown in Figures 2.20 and 2.21. The plot in Figure 2.20 shows that the relationship between time spent looking after car and age is approximately the same for men and women and time increases a little with age. Figure 2.21 shows that time spent looking after car increases with an increase in extroversion score for both men and women, but that the increase appears greater for men. This has implications for how these data might be modeled as we shall see in Chapter 4.

2.3.1 The Bubbleplot

The scatterplot can only display two variables. However, there have been a number of suggestions as to how extra variables may be included. In this subsection we shall illustrate one of these, the bubbleplot, in which three variables are displayed; two are used to form the scatterplot itself, and then the values of the third variable are represented by circles with radii proportional to these values and centered on the appropriate point in the scatterplot.

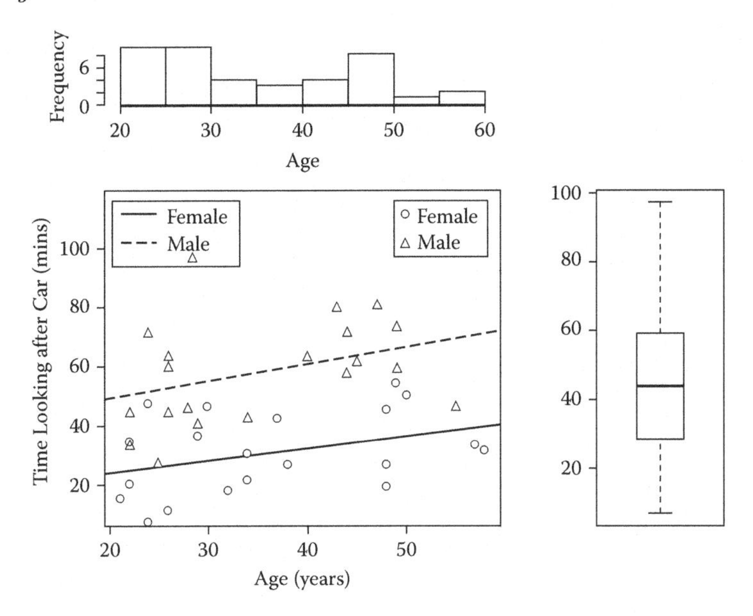

FIGURE 2.20
Scatterplot of time spent looking after car, against age, showing marginal distributions of the two variables and fitted linear regressions for men and women.

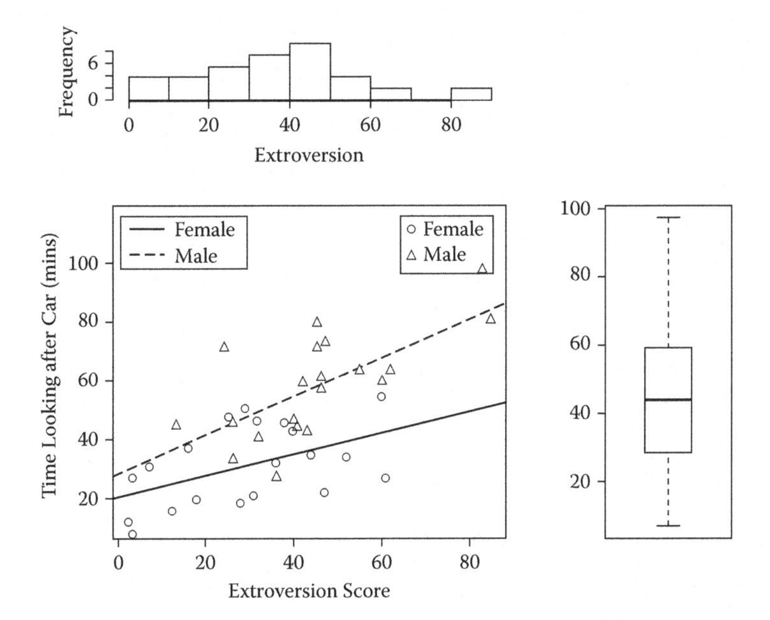

FIGURE 2.21
Scatterplot of time spent looking after car, against extroversion, showing marginal distributions of the two variables and fitted linear regressions for men and women.

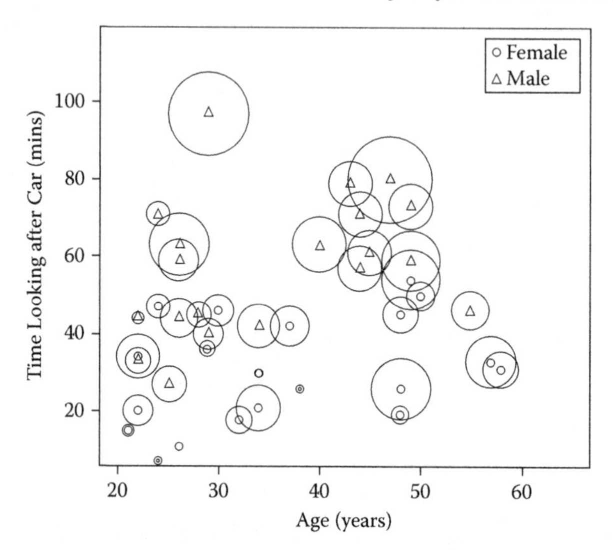

FIGURE 2.22
Bubbleplot of time spent looking after car, against age, with extroversion represented as circles.

For the data in Table 2.5, Figure 2.22 shows a bubbleplot with time spent looking after the car, against age as the scatterplot, and the extroversion scores represented by circles with appropriate radii. Gender is also displayed on the plot, so essentially, Figure 2.22 displays all four variables in the data set. Whether more information can be gleaned from this than from the plots given earlier is perhaps a moot point. But one observation does stand out: an approximately 30-year-old, extroverted man who spends almost 100 min per week looking after his car. Perhaps some counseling might be in order!

A plot a little like a bubbleplot is used by Bickel et al. (1975) to analyze the relationship between admission rate and the proportion of women applying to various academic departments at the University of California at Berkeley. The scatterplot of percentage of women applicants against percentage of applicants admitted is shown in Figure 2.23; the plots are enhanced by "boxes," the sizes of which indicate the relative number of applicants. The negative correlation indicated by the scatterplot is due almost exclusively to a trend for the large departments. If only a simple scatterplot had been used here, vital information about the relationship would have been lost.

2.3.2 The Bivariate Boxplot

A further helpful enhancement to the scatterplot is often provided by the two-dimensional analog of the boxplot for univariate data, known as the bivariate boxplot (Goldberg and Iglewicz, 1992). This type of boxplot may be useful in indicating the distributional properties of the data and in identi-

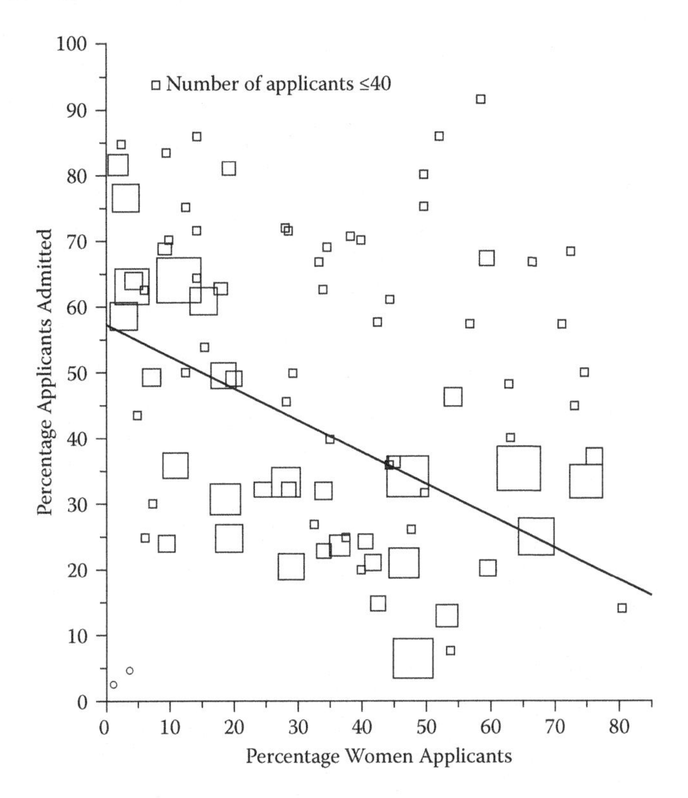

FIGURE 2.23
Scatterplot of the percentage of female applicants versus percentage of applicants admitted to 85 departments at the University of California at Berkeley. (Reproduced with permission from Bickel, P. J. et al., 1975, *Science*, 187, 398–404.)

fying possible outliers. The bivariate boxplot is based on calculating robust measures of location, scale, and correlation; it consists essentially of a pair of concentric ellipses, one of which (the "hinge") includes 50% of the data, and the other (called the "fence") which delineates potential troublesome outliers. In addition, resistant regression lines of both y on x and x on y are shown, with their intersection showing the bivariate locations estimator. The acute angle between the regression lines will be small for a large absolute value of correlations and large for a small one.

Figure 2.24 shows the bivariate boxplot of time spent looking after car and age, and Figure 2.25 shows the corresponding diagram for time and extroversion. Neither diagram shows any clear outliers, that is, observations that fall outside the dotted ellipse. But in both diagrams there is one observation that lies on the dotted ellipse.

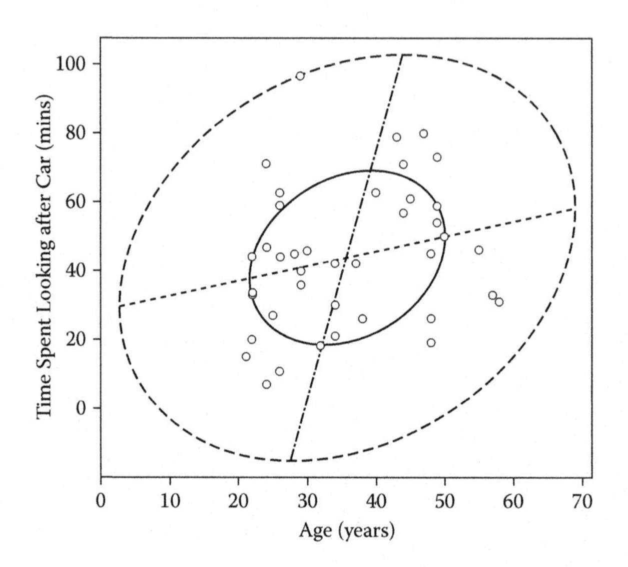

FIGURE 2.24
Bivariate boxplot of time spent looking after car and age.

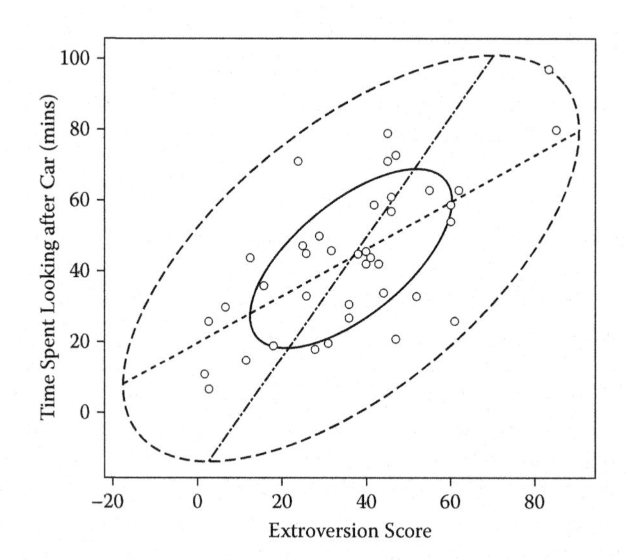

FIGURE 2.25
Bivariate boxplot of time spent looking after car and extroversion.

2.4 Scatterplot Matrices

When there are many variables measured on all the individuals in a study, an initial examination of all the separate pairwise scatterplots becomes difficult. For example, if 10 variables are available, there are 45 possible scatterplots. But all these scatterplots can be conveniently arranged into a scatterplot matrix that then aids in the overall comprehension and understanding of the data.

A scatterplot matrix is defined as a square, symmetric grid of bivariate scatterplots. The grid has q rows and columns, each one corresponding to a different variable. Each of the grid's cells shows a scatterplot of two variables. Variable j is plotted against variable i in the ijth cell, and the same variables appear in cell ji with the x- and y-axes of the scatterplots interchanged. The reason for including both the upper and lower triangles of the grid, despite the seeming redundancy, is that it enables a row and a column to be visually scanned to see one variable against all others, with the scales for the one variable lined up along the horizontal or the vertical.

To illustrate the use of a scatterplot matrix, we shall use the data shown in Table 2.6. These data arise from an experiment in which five different types of electrode were applied to the arms of 16 subjects and the resistance measured (in kilohms). The experiment was designed to see whether all electrode types performed similarly. The scatterplot matrix for the data is shown in Figure 2.26; each of the scatterplots in the diagram has been enhanced by the addition of the linear fit of the y variable on the x variable. The diagram suggests the presence of several outliers, the most extreme of which is subject 15; the reason for the two extreme readings on this subject was that he had very hairy arms. Figure 2.26 also indicates that the readings on particular pairs of electrodes, for example, electrode 1 and electrode 4, are hardly related at all.

We can use the plot of results for the first and second electrodes to demonstrate how the bivariate boxplot looks when there are probable outliers in the data (see Figure 2.27). Three outliers are identified by the bivariate boxplot. If we calculate the correlation coefficient between the two variables using all the

TABLE 2.6
Measure of Resistance (Kilohms) Made on Five Different Types of Electrode for Five of the 16 Subjects

Subject	E1	E2	E3	E4	E5
1	500	400	98	200	250
2	660	600	600	75	310
3	250	370	220	250	220
4	72	140	240	33	54
5	135	300	450	430	70

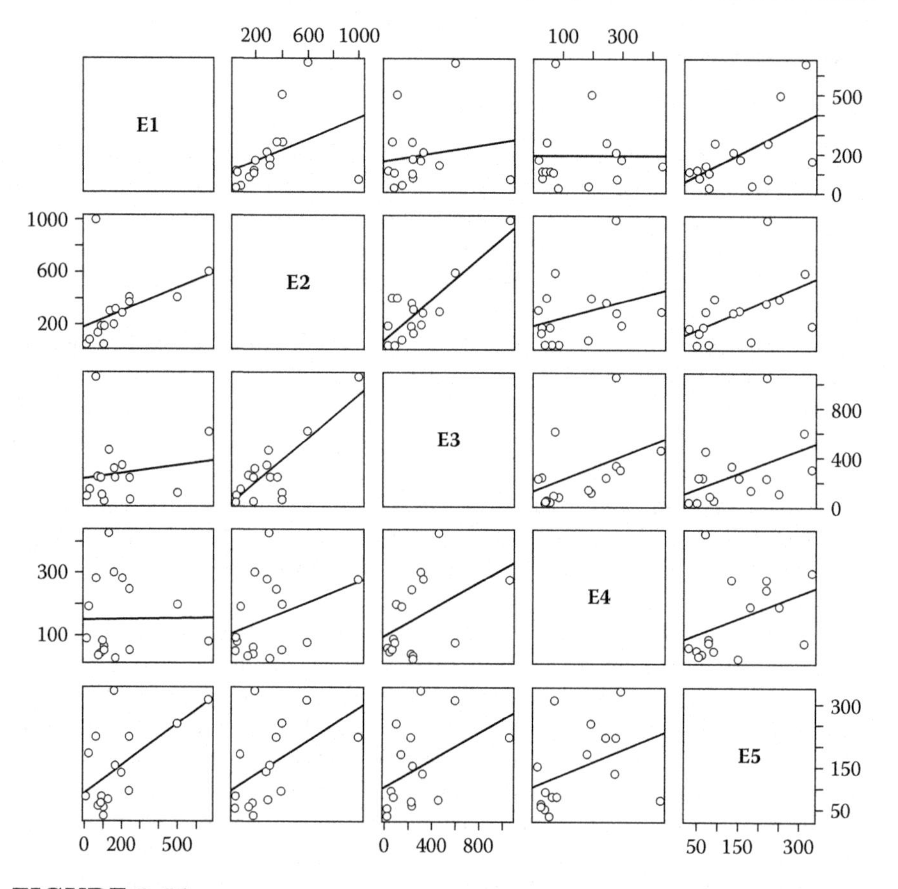

FIGURE 2.26
Scatterplot matrix for data on measurements of skin resistance made with five different types of electrodes.

data, we get a value of 0.41; if we recalculate the correlation after removing subjects 1, 2, and 15, we get a value of 0.88—more than double the previous value. This example underlines how useful the bivariate boxplot can be, and also underlines the danger of simply calculating a correlation coefficient without examining the relevant scatterplot.

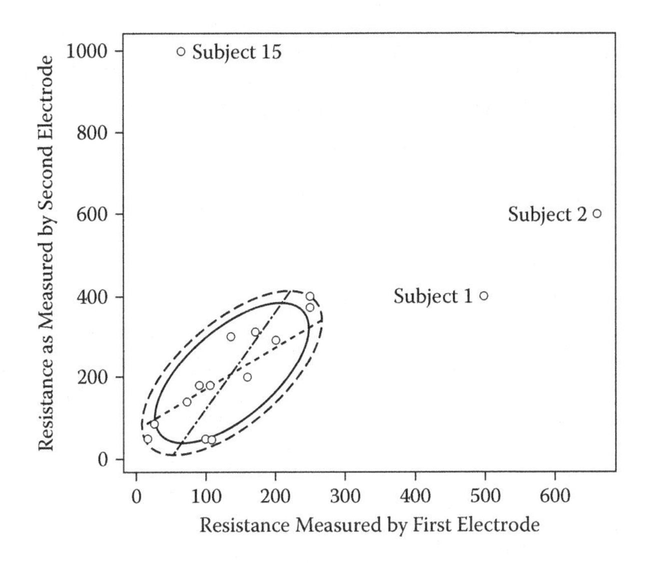

FIGURE 2.27
Bivariate boxplot for data on electrodes one and two.

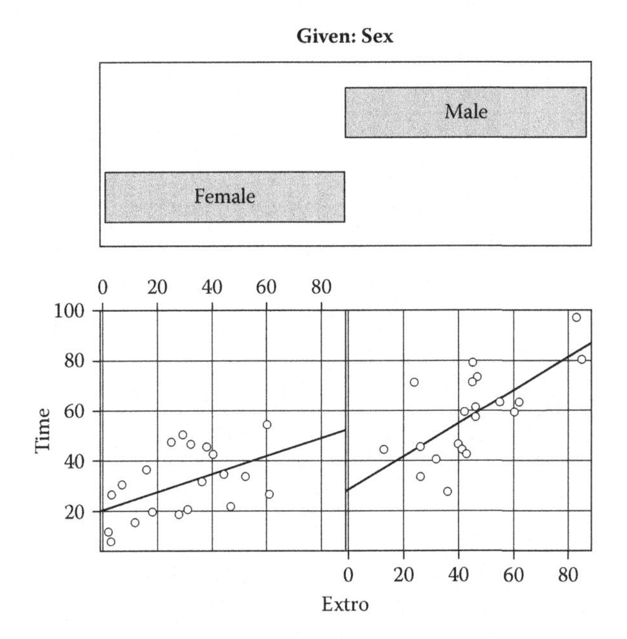

FIGURE 2.28
Coplot of time spent looking after car against extroversion conditioned on sex.

2.5 Conditioning Plots and Trellis Graphics

The conditioning plot or coplot is a potentially powerful visualization tool for studying how, say, a response variable depends on two or more explanatory variables. In essence, such plots display the bivariate relationship between two variables while holding constant (or "conditioning upon") the values of one or more other variables. If the conditioning variable is categorical, then the coplot is no more than, say, a scatterplot of two variables for each level of the categorical variable. As an example of this type of simple coplot, Figure 2.28 shows plots of time spent looking after car against extroversion score conditioned on sex; each scatterplot is enhanced by a linear regression fit. The plot highlights what was found in an earlier plot (Figure 2.21)—that the relationship between time spent looking after car and extroversion is different for men and women.

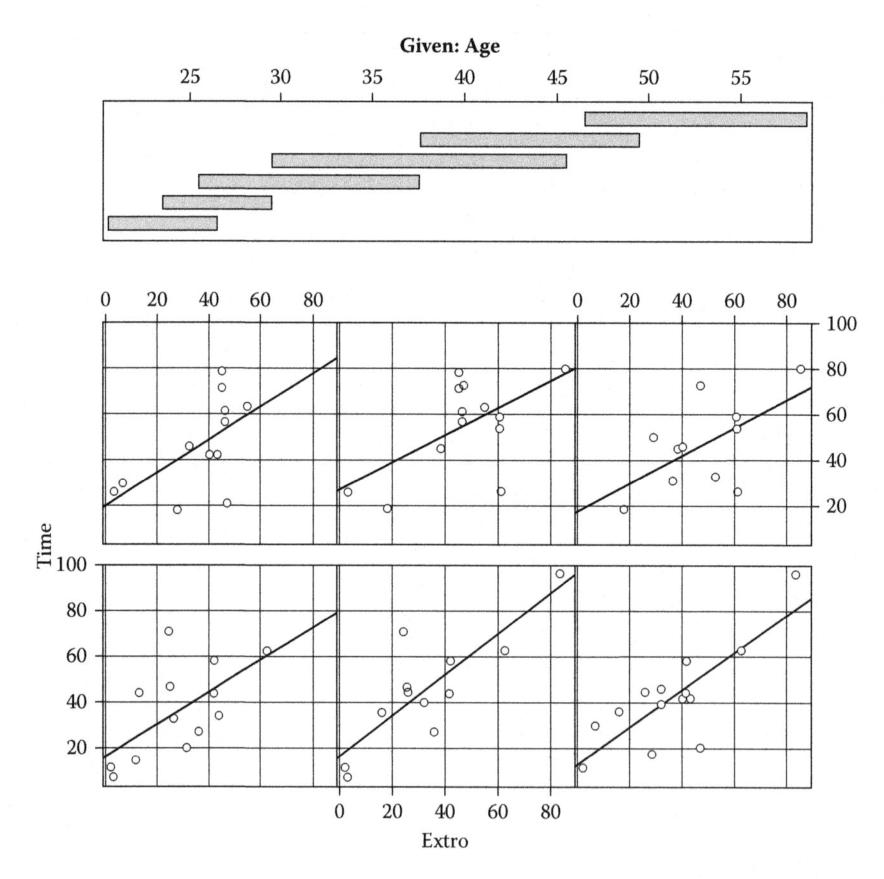

FIGURE 2.29
Coplot of time against extroversion conditioned on age.

As a more complicated coplot, Figure 2.29 shows time spent looking after car against extroversion conditioned on age. In this diagram, the panel at the top of the figure is known as the given panel; the panels below are dependence panels. Each rectangle in the given panel specifies a range of values of population size. On a corresponding dependence panel, time is plotted against age for those people whose ages lie in the particular interval. To match age intervals to dependence panels, the latter are examined in order from left to right in the bottom row and, then again, from left to right in subsequent rows. The plot suggests that the relationship between time and extroversion is much the same over the age range observed in the data set.

Conditional graphical displays are simple examples of a more general scheme known as *trellis graphics* (Cleveland, 1993). This is an approach to examining high-dimensional structure in data by means of one-, two-, and three-dimensional graphs. The problem addressed is how observations of one or more variables depend on the observations of the other variables. The essential feature of this approach is the multiple conditioning that allows some type of plot to be displayed for different values of a given variable (or variables). The aim is to help in understanding both the structure of the data and how well proposed models describe the structure. Excellent examples of the application of trellis graphics are found in Sarkar (2008).

As a relatively simple example of what can be done with trellis graphics, we will again use the data on time spent looking after car and produce a three-dimensional scatter plot for time, age, and extroversion conditioned on sex (see Figure 2.30). This diagram makes the generally longer times spent looking after their cars by men very apparent, although whether it adds anything to earlier plots is a question we leave for the reader.

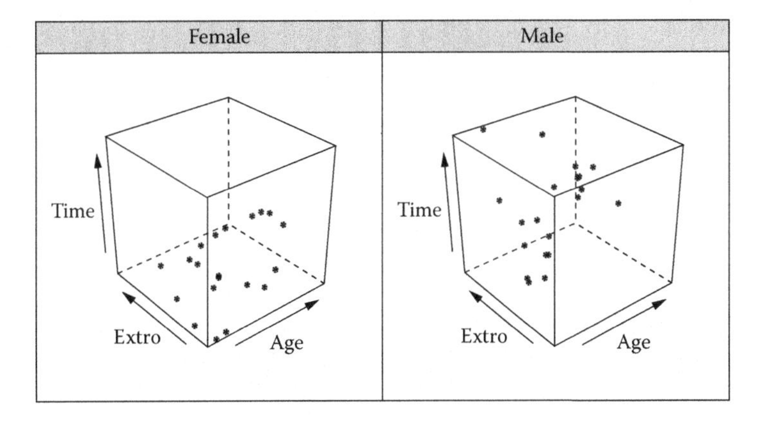

FIGURE 2.30
Three-dimensional scatterplot of time, age, and extroversion conditioned on sex.

Let us now look at two more complex examples of trellis graphics, taken from Sarkar (2008). The first involves data collected in a survey of doctorate degree recipients in the United States. The data are shown in Table 2.7. Any graphic for the data has to involve the proportions of reasons across fields of study rather than the counts because the latter do not tell us very much, except, for example, that the "Biological Sciences" subject area contributes the majority of postdocs. A stacked bar chart of the data based on the proportions rather than the counts in Table 2.7 is shown in Figure 2.31. An alternative display for the proportions, a multipanel dot plot, is shown in Figure 2.32. For comparing the proportions of reasons across areas of study, the dot plot seems preferable because it is more easily judged by eye. The multipanel dot plot becomes even more informative if the proportions are ordered from low to high within each panel, as shown in Figure 2.33. We see that the most popular reason for choosing a postdoctoral position is "Expected or Additional Training," and that this applies to all areas of study. For "Earth, Atmospheric, and Ocean Sciences," postdocs appear to mostly take a job because other employment is not available. Figure 2.33 provides an easy-to-use and informative display of the data.

TABLE 2.7
Reasons for Choosing a Postdoctoral Position After Graduating from U.S. Universities by Area of Study

Subject	Expected or Additional Training	Work with Specific Person	Training Outside PhD Field	Other Employment not Available	Other
Biological sciences	6404	2427	1950	1779	602
Chemistry	865	308	292	551	168
Earth, Atmospheric, and Ocean sciences	343	75	75	238	80
Engineering	586	464	288	517	401
Medical sciences	205	137	82	68	74
Physics and astronomy	1010	347	175	399	162
Social and behavioral sciences	1368	564	412	514	305
All postdoctorates	11197	4687	3403	4406	1914

The last example is also taken from Sarkar (2008) and is shown in Figure 2.34. The diagram gives the scatterplot matrix of violent crime rates in the 50 states of the United States in 1973, conditioned on geographical region. Each scatterplot in the diagram is enhanced by a locally weighted regression fit, an alternative to linear regression, to be discussed in Chapter 3. The relationship between each pair of crimes appears to be pretty similar in all four regions.

Trellis graphics is a potentially very exciting and powerful tool for the exploration of data from behavioral studies. However, a word of caution is perhaps in order. With small or moderately sized data sets, the number of

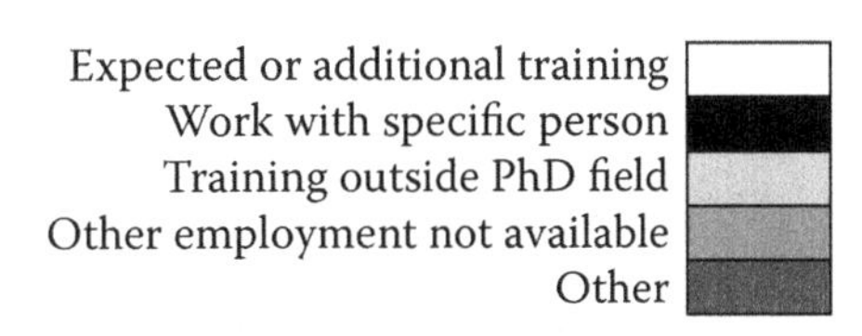

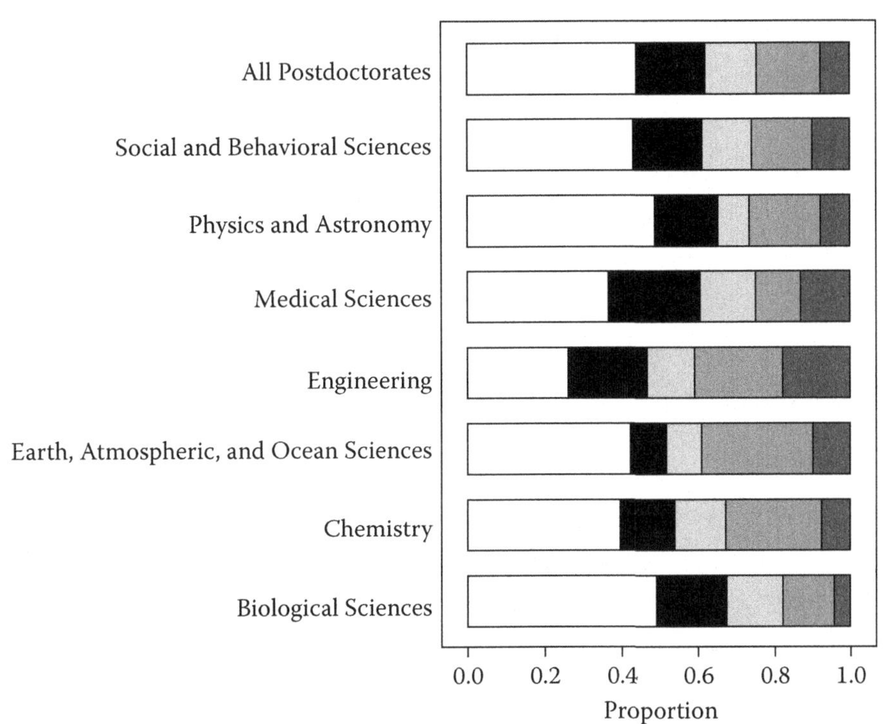

FIGURE 2.31
A stacked bar chart showing the proportion of reasons for choosing a postdoc by field of study.

observations in each panel may be too few to make the panel graphically acceptable. A further caveat is that trellis graphics can be seductive with the result that simpler graphics, which in many cases may be equally informative about a data set, may be ignored.

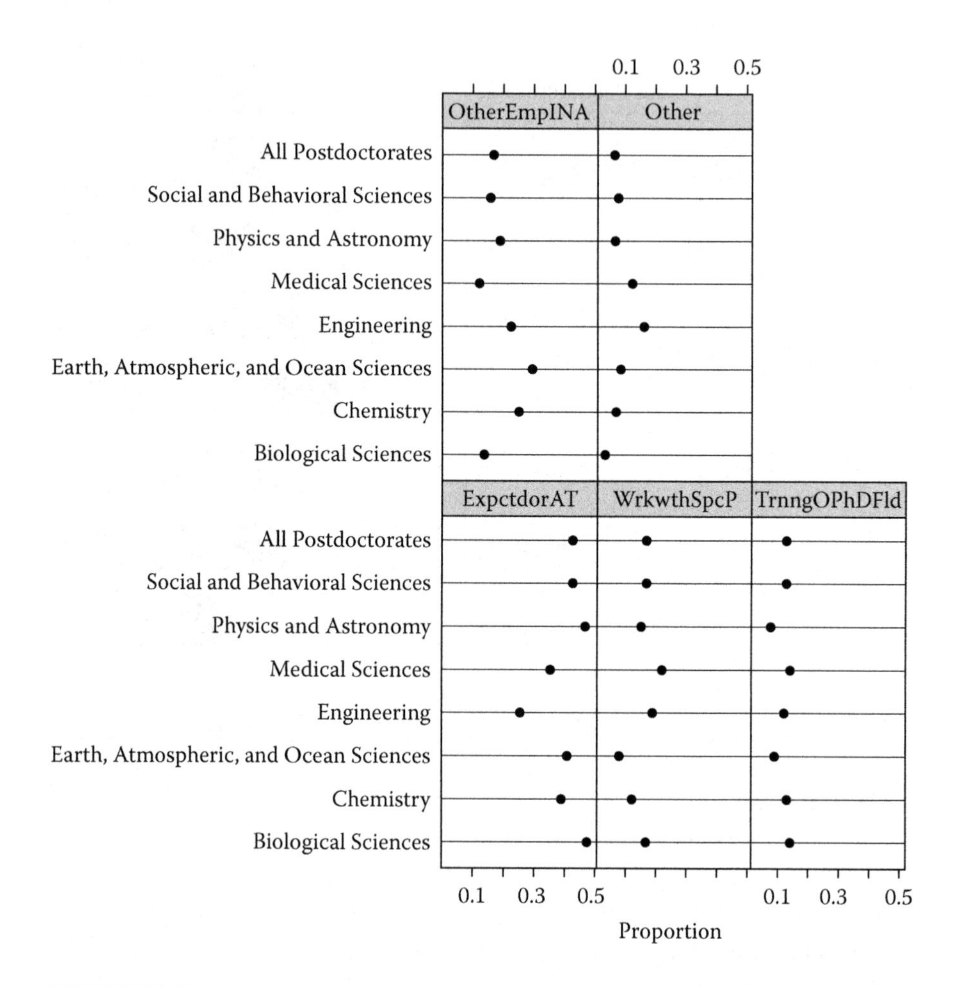

FIGURE 2.32
A multipanel dot plot showing the proportion of reasons for choosing a postdoc by field of study.

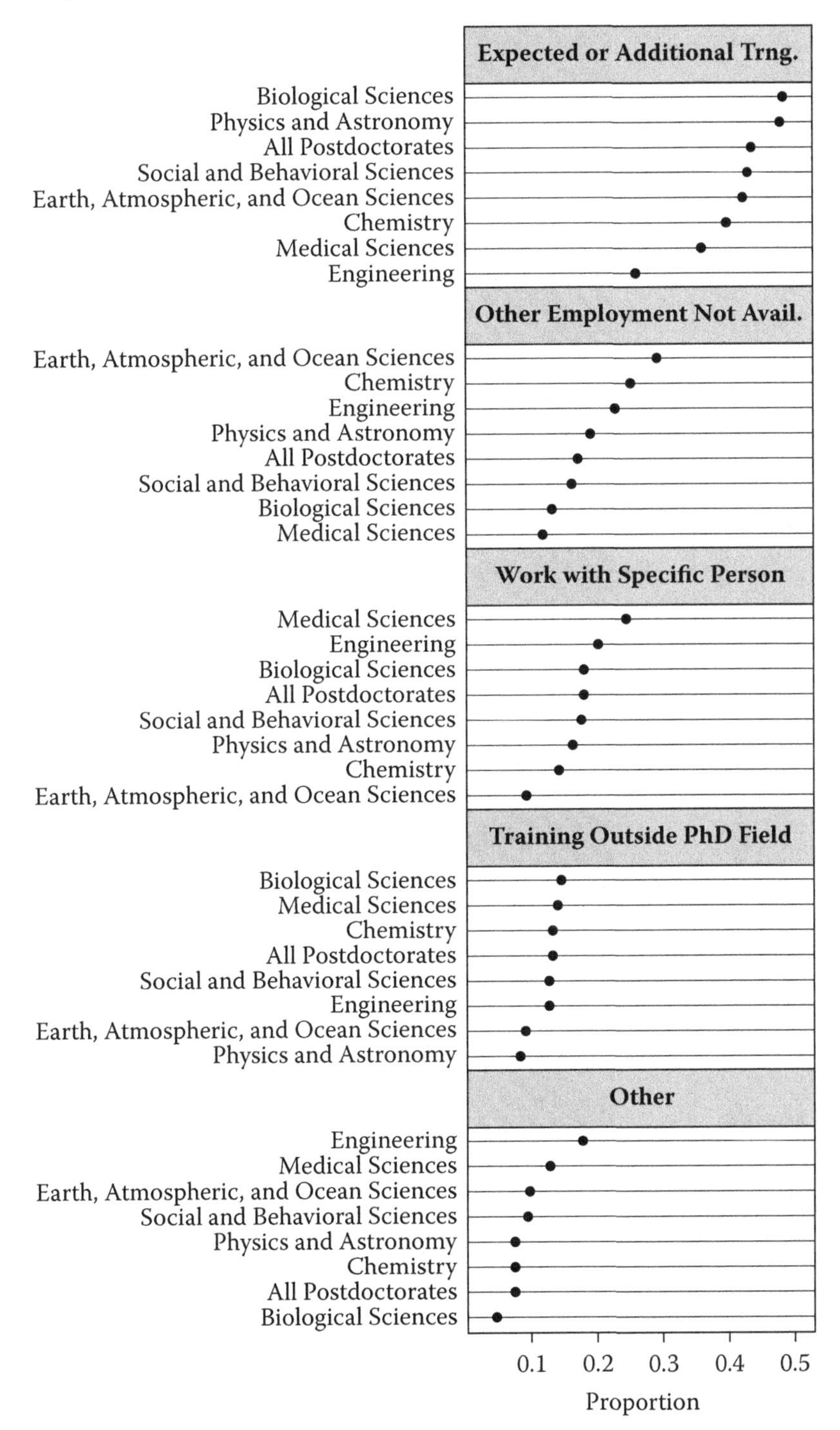

FIGURE 2.33
Reasons for choosing a postdoc position.

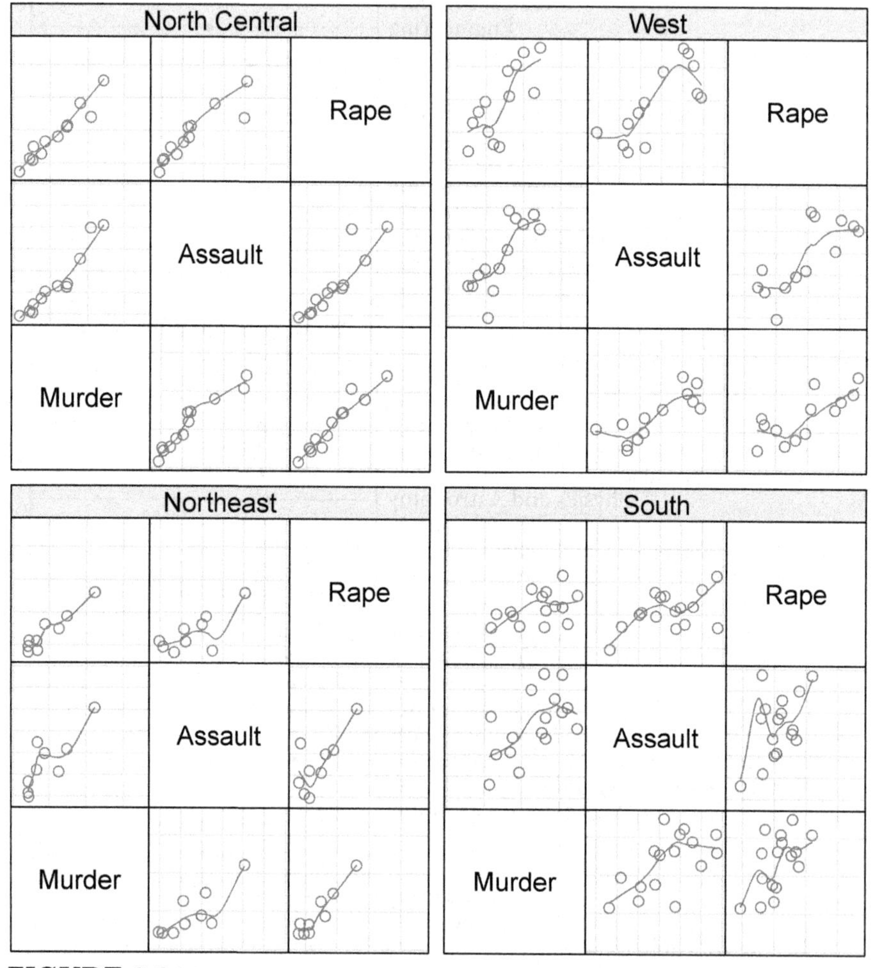

FIGURE 2.34
Scatterplot matrices of violent crime rates conditioned on geographical region.

2.6 Graphical Deception

In general, graphical displays of the kind described in previous sections are extremely useful in the examination of data; indeed, they are almost essential both in the initial phase of data exploration and in the interpretation of results from more formal statistical procedures, as will be seen in later chapters. Unfortunately, it is relatively easy to mislead the unwary with graphical material, and not all graphical displays are as honest as they should be. For example, consider the plot of the death rate per million from breast cancer for several periods over three decades, shown in Figure 2.35a. The rate appears to show a rather alarming increase. However, when the data are replotted with the vertical scale beginning at zero, as shown in Figure 2.35b, the increase in the breast cancer death rate is altogether less startling. This example illustrates that undue exaggeration or compression of the scales is best avoided when drawing graphs (unless, of course, you are actually in the business of deceiving your audience).

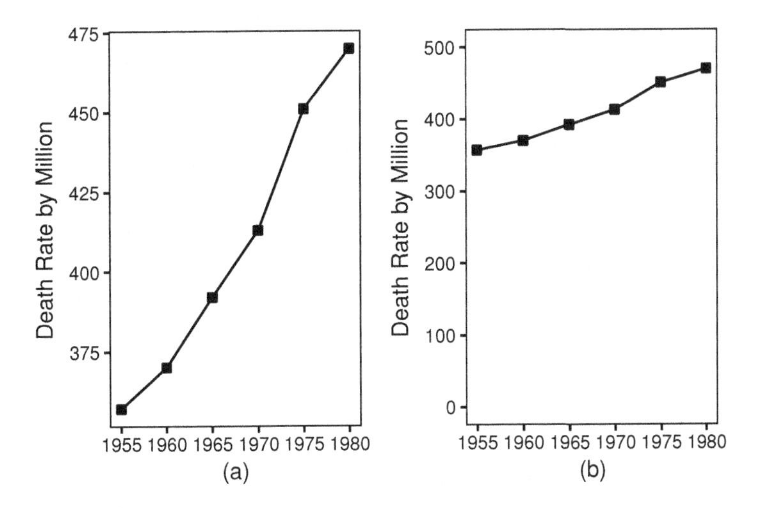

FIGURE 2.35
Death rates from breast cancer where (a) the y-axis does not include the origin and (b) the y-axis does include the origin.

A very common distortion introduced into the graphics most popular with newspapers, television, and the media in general is when both dimensions of a two-dimensional figure or icon are varied simultaneously in response to changes in a single variable. The examples shown in Figure 2.36, both taken from Tufte (1983), illustrate this point. Tufte quantifies the distortion with

FIGURE 2.36
Graphics exhibiting lie factors of (a) 9.4 and (b) 2.8.

what he calls the *lie factor* of a graphical display, which is defined as the size of the effect shown in the graph divided by the size of the effect in the data. Lie factor values close to unity show that the graphic is probably representing the underlying numbers reasonably accurately. The lie factor for the "oil barrels" is 9.4 since a 454% increase is depicted as 4280%. The lie factor for the "shrinking doctors" is 2.8.

A further example given by Cleveland (1994) and reproduced here in Figure 2.37 demonstrates that even the manner in which a simple scatterplot is drawn can lead to misperceptions about data. The example concerns the way in which judgment about the correlation of two variables made on the basis of looking at their scatterplot can be distorted by enlarging the area in which the points are plotted. The coefficient of correlation in the diagram on the right in Figure 2.37 appears greater.

Some suggestions for avoiding graphical distortion, taken from Tufte (1983), are

- The representation of numbers, as physically measured on the surface of the graphic itself, should be directly proportional to the numerical quantities represented.

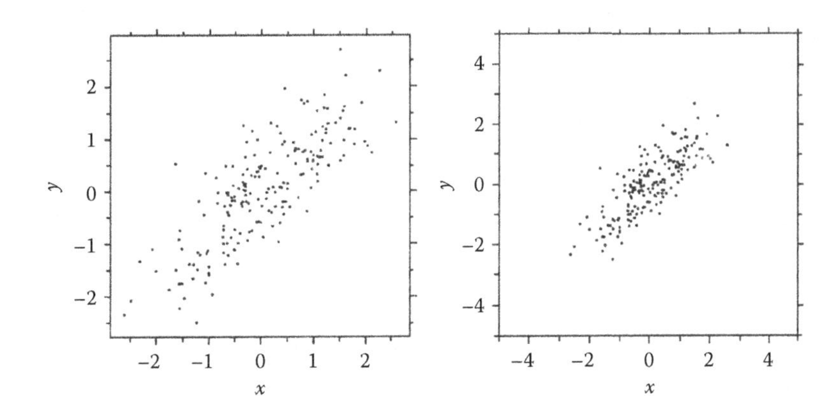

FIGURE 2.37
Misjudgment of size of correlation caused by enlarging the plot area.

- Clear, detailed, and thorough labeling should be used to defeat graphical distortion and ambiguity. Write out explanations of the data on the graphic itself. Label important events in the data.
- To be truthful and revealing, data graphics must bear on the heart of quantitative thinking: "compared to what?" Graphics must not quote data out of context.
- Above all else, show the data.

Being misled by graphical displays is usually a sobering but not a life-threatening experience. However, Cleveland (1994) gives an example, where using the wrong graph contributed to a major disaster in the American space program, namely, the explosion of the Challenger space shuttle and the deaths of the seven people on board. To assess the suggestion that low temperature might affect the performance of the O-rings that sealed the joints of the rocket motor, engineers studied the graph of the data shown in Figure 2.38. Each data point was from a shuttle flight in which the O-rings had experienced thermal distress. The horizontal axis shows the O-ring temperature, and the vertical scale shows the number of O-rings that had experienced thermal distress. On the basis of these data, Challenger was allowed to take off when the temperature was 31°F, with tragic consequences.

The data for "no failures" are not plotted in Figure 2.38 because the engineers involved believed that these data were irrelevant to the issue of dependence. They were mistaken, as shown by the plot in Figure 2.39, which includes all the data. Here a pattern does emerge, and a dependence of failure on temperature is revealed.

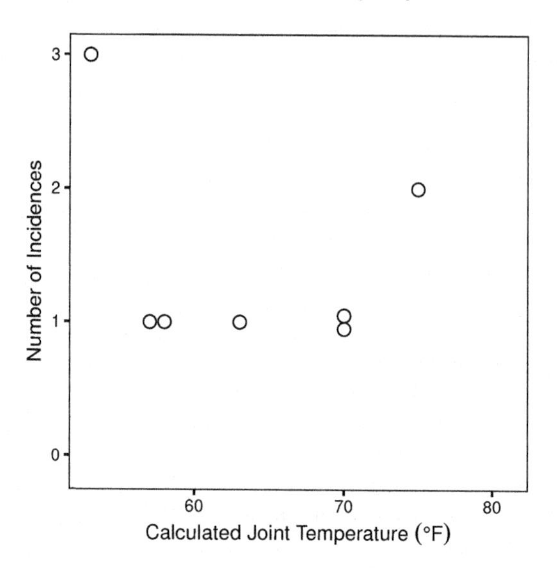

FIGURE 2.38
Data taken from Siddhartha et al. (1989) plotted by space shuttle engineers the evening before the Challenger accident to determine the dependence of O-ring failure on temperature.

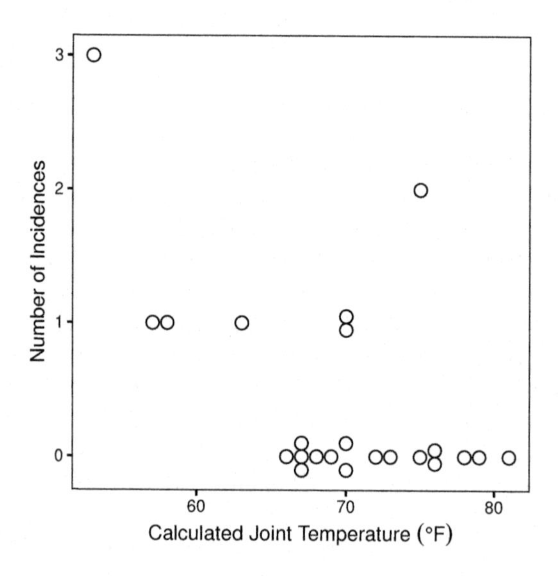

FIGURE 2.39
A plot of the complete O-ring data.

To end on a less somber note and to show that misperception and miscommunication are certainly not confined to statistical graphics, see Figure 2.40!

FIGURE 2.40
Misperception and miscommunication are sometimes a way of life! (Drawing by Charles E. Martin, ©1961, 1969, *The New Yorker Magazine.* Used with permission.)

2.7 Summary

- Graphical displays are an essential feature in the analysis of empirical data. The prime objective is to communicate to ourselves and others.
- Graphic design must do everything it can to help people understand the subject.
- In some cases, a graphical "analysis" may be all that is required (or merited) for a data set.
- Pie charts and bar plots are rarely more informative than a numerical tabulation of the data.
- Boxplots are more useful than histograms for displaying most data sets and are very useful for comparing groups. In addition, they are useful for identifying possible outliers.
- Scatterplots are the fundamental tool for examining relationships between variables. They can be enhanced in a variety of ways to provide

extra information. Scatterplots are always needed when considering numerical measures of correlation between pairs of variables.

- Scatterplot matrices are a useful first step in examining data with more than two variables.
- Trellis graphs can look very enticing and may in many, but not all, cases give greater insights into patterns in the data than simpler plots.
- Beware of graphical deception.
- Unless graphs are relatively simple, they are not likely to survive the first glance.

2.8 Exercises

2.1 According to Cleveland (1994), "The histogram is a widely used graphical method that is at least a century old. But maturity and ubiquity do not guarantee the efficiency of a tool. The histogram is a poor method."

Do you agree with Cleveland? Give your reasons.

2.2 Shortly after metric units of length were officially introduced in Australia, each of a group of 44 students was asked to guess, to the nearest meter, the width of the lecture hall in which they were sitting. Another group of 69 students in the same room were asked to guess the width in feet, to the nearest foot. (The true width of the hall was 13.1 m or 43.0 ft).

The data contains the students' guesses. Construct suitable graphical displays for both sets of guesses with the aim of answering the question "which set of guesses is most accurate?"

2.3 The data set contains values of seven variables for 10 states in the United States. The seven variables are

1. Population size divided by 1000
2. Average per capita income
3. Illiteracy rate (% population)
4. Life expectancy (years)
5. Homicide rate (per 1000)
6. Percentage of high school graduates
7. Average number of days per year below freezing

- Construct a scatterplot matrix of the data, labeling the points by state name.

- On each panel of the scatterplot matrix show the corresponding bivariate boxplot.
- Construct a coplot of life expectancy and homicide rate conditional on average per capita income.

2.4 Figure 2.41 shows the traffic deaths in a particular area before and after stricter enforcement of the speed limit by the police. Does the graph convince you that the efforts of the police have had the desired effect of reducing road traffic deaths? If not, why not?

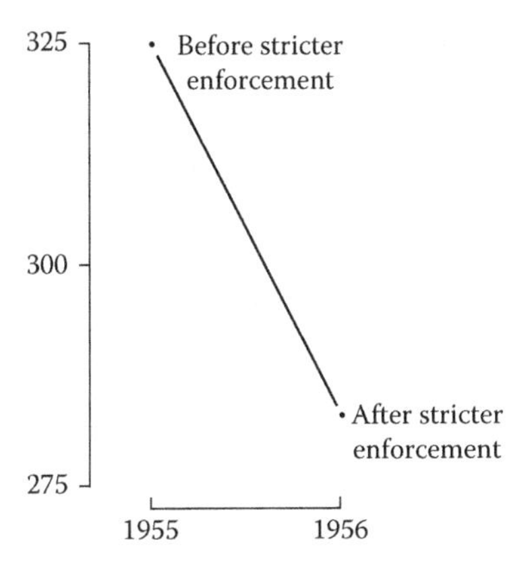

FIGURE 2.41
Traffic deaths before and after introduction of stricter enforcement of speed limit.

2.5 Mortality rates per 100,000 from male suicides for a number of age groups and a number of countries are given. Construct side-by-side boxplots for the data from different age groups, and comment on what the graphics tell us about the data.

3

Simple Linear and Locally Weighted Regression

3.1 Introduction

Table 3.1 shows the heights (in centimeters) and the resting pulse rates (beats per minute) for 5 of a sample of 50 hospital patients (data sets with two variables are often referred to as bivariate data). Is it possible to use these data to construct a model for predicting pulse rate from height, and what type of model might be used? Such questions serve to introduce one of the most widely used statistical techniques: regression analysis. In very general terms, regression analysis involves the development and use of statistical techniques designed to reflect the way in which variation in an observed random variable changes with changing circumstances. More specifically, the aim of a regression analysis is to derive an equation relating a dependent and an explanatory variable or, more commonly, several explanatory variables. The derived equation may sometimes be used solely for prediction, but more often its primary purpose is as a way of establishing the relative importance of the explanatory variables in determining the response variable, that is, in establishing a useful model to describe the data. (Incidentally, the term regression was first introduced by Galton in the 19th century to characterize a tendency toward mediocrity, that is, more average, observed in the offspring of parents.)

In this chapter, we shall concern ourselves with regression models for a response variable that is continuous and for which there is a single explanatory variable. In Chapter 4, we will extend the model to deal with the situation in

TABLE 3.1
Pulse Rates and Heights Data

Subject	Heights (cm)	Pulse Rates (beats/min)
1	160	68
2	167	80
3	162	84
4	175	80
5	185	80

which there are several explanatory variables, and then, in Chapters 5 and 6, we shall consider various generalizations to this model, for example, allowing the response variable to be categorical.

No doubt most readers will have covered simple linear regression for a response variable and a single explanatory variable in their introductory statistics course. Despite this, it may be worthwhile reading Section 3.2 both as an aide-memoire and as an initial step in dealing with the more complex procedures needed when several explanatory variables are considered, a situation to be discussed in Chapter 4. It is less likely that readers will have been exposed to locally weighted regression, which will also be covered in this chapter and which can often serve as a useful antidote to the (frequently unthinking) acceptance of the simple linear model per se.

3.2 Simple Linear Regression

The technical details of the simple linear regression model are given in Technical Section 3.1.

Technical Section 3.1: Simple Linear Regression

Assume that y_i represents the value of what is generally known as the response variable on the ith individual and x_i represents the individual's values on what is most often called an explanatory variable; the simple linear regression model is

$$y_i = \beta_0 + \beta_1 x_i + \varepsilon_i$$

where β_0 is the intercept, and β_1 is the slope of the linear relationship assumed between the response variable y and the explanatory variable x, and ε_i is an error term measuring the amount by which the observed value of the response differs from the value predicted by the fitted model. ("Simple" here means that the model contains only a single explanatory variable; we shall deal with the situation where there are several explanatory variables in Chapter 4.) The error terms are assumed to be statistically independent random variables having a normal distribution with mean 0 and the same variance σ^2 at every value of the explanatory variable. The parameter β_1 measures the change in the response variable produced by a change of one unit in the explanatory variable.

The regression coefficients β_0 and β_1 may be estimated as $\hat{\beta}_0$ and $\hat{\beta}_1$ using least-squares estimation in which the sum of squared differences between the observed values of the response variable y_i and the values "predicted" by the regression equation $\hat{y}_i = \hat{\beta}_0 + \hat{\beta}_1 x_i$ is minimized, leading to the following estimates:

$$\hat{\beta}_0 = \bar{y} - \hat{\beta}_1 \bar{x}$$

$$\hat{\beta}_1 = \frac{\sum (y_i - \bar{y})(x_i - \bar{x})}{\sum (x_i - \bar{x})^2}$$

The predicted values of y_i from the model are

$$\hat{y}_i = \hat{\beta}_0 + \hat{\beta}_1 x_i$$

This fitted equation could be used to predict the value of the response variable for some particular value of the explanatory variable, but it is very important to note that trying to predict values of the response variable outside the observed range of the explanatory variable is a potentially dangerous business.

The variability of the response variable can be partitioned into a part that is due to regression on the explanatory variable, the regression mean square (RGMS) given by RGMS $= \sum_{i=1}^{n} (\hat{y}_i - \bar{y})^2$, and a residual mean square(RMS) given by RMS $= \sum_{i=1}^{n} (y_i - \hat{y}_i)^2/(n-2)$. The RMS gives an estimate of σ^2, and the F-statistic given by F = RGMS/RMS with 1 and $n-2$ degrees of freedom (DF) gives a test that the slope parameter β_1 is 0. (This is of course equivalent to testing that the correlation of the two variables is 0.)

The estimated variance of the slope parameter estimate is

$$\text{Var}\left(\hat{\beta}_1\right) = \frac{s^2}{\sum_{i=1}^{n} (x_i - \bar{x})^2}$$

The estimated variance of a predicted value y_{pred} at a given value of x, say x_0, is

$$\text{Var}\left(y_{\text{pred}}\right) = s^2 \left[\frac{(x_0 - \bar{x})^2}{\sum_{i=1}^{n} (x_i - \bar{x})^2} + \frac{1}{n} + 1 \right]$$

where s^2 is the RMS value defined earlier. (Note that the variance of the prediction increases as x_0 gets further away from $\bar{x}$.)

A confidence interval for the slope parameter can be constructed in the usual way from the estimated standard error of the parameter, and the variance of a predicted value can be used to construct confidence bounds for the fitted line.

In some applications of simple linear regression, a model without an intercept is required (when the data is such that the line must go through the origin), that is, a model of the form

$$y_i = \beta x_i + \varepsilon_i$$

In this case, application of least squares gives the following estimator for β:

$$\hat{\beta} = \sum_{i=1}^{n} x_i y_i / \sum_{i=1}^{n} x_i^2$$

3.2.1 Fitting the Simple Linear Regression Model to the Pulse Rates and Heights Data

Fitting the simple linear regression model to the data in Table 3.1 gives the results shown in Table 3.2. Figure 3.1 shows the fitted line on a scatterplot of the data and a 95% confidence interval for the predicted values calculated from the relevant variance term given in Technical Section 3.1; the diagram also contains some graphical displays giving information about the marginal distributions of each of the two variables. The results in Table 3.2 show that there is no evidence of any linear relationship between pulse rate and height. The multiple R-squared, which in this example with a single explanatory variable is simply the square of the correlation coefficient between pulse rate and height, is 0.0476, so that less than 5% of the variation in pulse rate is explained by the variation in height. Figure 3.1 shows that the fitted line is almost horizontal and that a horizontal line could easily be placed within the two dotted lines indicating the confidence interval for predicted values. Clearly, the fitted linear regression would be very poor if used to predict pulse rate from height.

Figure 3.1 also shows that pulse rate has a very skewed distribution. Because of the latter, it may be of interest to repeat the plotting and fitting process after some transformation of pulse rate (see Exercise 3.1).

TABLE 3.2
Results from Fitting a Simple Linear Regression to the Pulse and Heights Data

	Coefficients			
	Estimate	**Standard Error**	**t-Value**	**Pr(>|t|)**
Intercept	46.9069	22.8793	2.050	0.0458
Heights	0.2098	0.1354	1.549	0.1279

Note: Residual standard error: 8.811 on 48 DF; multiple R-squared: 0.04762; F-statistic: 2.4 on 1 and 48 DF; *p*-value: 0.1279.

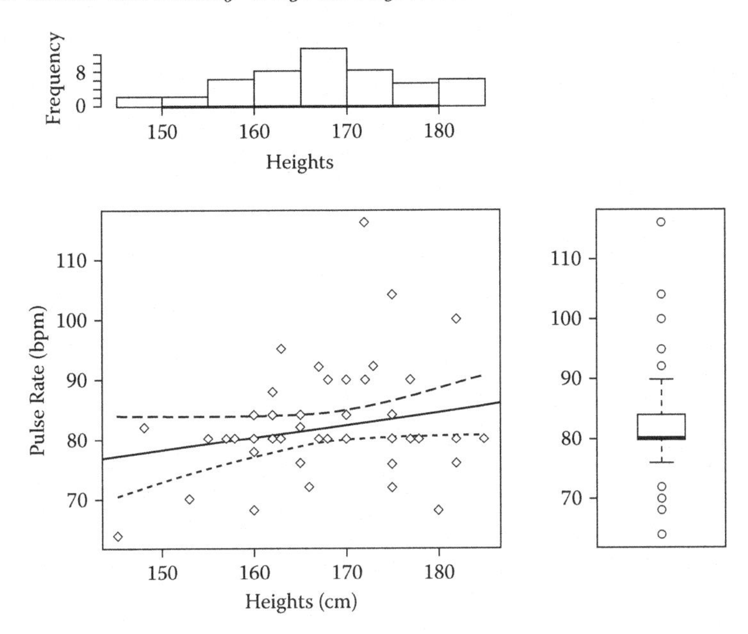

FIGURE 3.1
Scatterplot and fitted linear regression for pulse and heights data.

3.2.2 An Example from Kinesiology

For our second example of simple linear regression, we will use some data from an experiment in kinesiology, a natural care system that uses gentle muscle testing to evaluate many functions of the body in the structural, chemical, neurological, and biological realms. A subject performed a standard exercise at a gradually increasing level. Two variables were measured: (1) oxygen uptake and (2) expired ventilation, which is related to the exchange of gases in the lungs. Part of the data is shown in Table 3.3 (there are 53 subjects in the full data set), and the researcher was interested in assessing the relationship between the two variables. A scatterplot of the data along with the fitted simple linear regression is shown in Figure 3.2. The estimated regression coef-

TABLE 3.3
Data on Oxygen Uptake and Expired Volume (in Liters)

Subject	Oxygen Uptake	Expired Ventilation
1	574	21.9
2	592	18.6
3	664	18.6
4	667	19.1
5	718	19.2

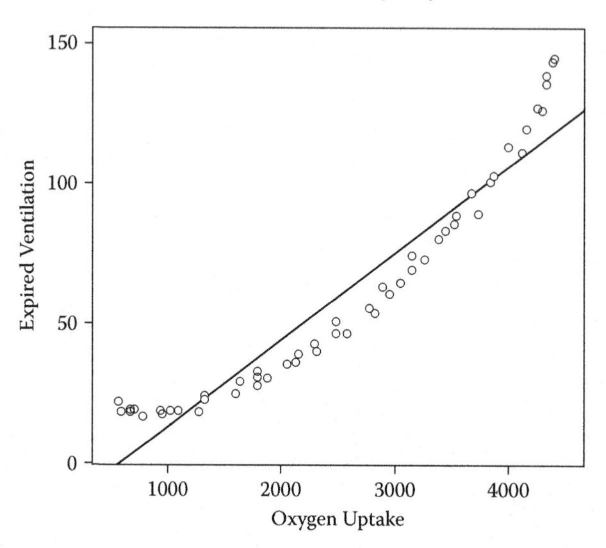

FIGURE 3.2
Scatterplot of expired ventilation against oxygen uptake with fitted simple linear regression.

ficient in Table 3.4 is highly significant, but Figure 3.2 makes it very clear that the simple linear model is not appropriate for these data; we need to consider a more complicated model. The obvious choice here is to consider a model that, in addition to the linear effect of oxygen uptake, includes a quadratic term in this variable, that is, a model of the form

$$y_i = \beta_0 + \beta_1 x_i + \beta_2 x_i^2 + \varepsilon_i$$

Such a model can easily be fitted by least squares to give estimates of its three parameters. One point to note about this model that may seem confusing is that it remains, similar to the simple model outlined in Technical Section 3.1, a linear model despite the presence of the quadratic term. The reason for this

TABLE 3.4
Results from Fitting a Simple Linear Regression to the Kinesiology Data

	Coefficients			
	Estimate	**Standard Error**	**t-Value**	**Pr(>|t|)**
(Intercept)	−18.448734	3.815196	−4.836	<0.001
Oxygen	0.031141	0.001355	22.987	<0.001

Note: Residual standard error: 11.96 on 51 DF; multiple R-squared: 0.912; adjusted R-squared: 0.9103; F-statistic: 528.4 on 1 and 51 DF; p-value: <0.001.

is that "linear" in linear regression models refers to the parameters rather than the explanatory variables. An example of a nonlinear model would be

$$y_i = \beta_1 x_i + \exp(\beta_2 x_i) + \varepsilon_i$$

We will not deal with such models in this book. It is worth mentioning here that including polynomial terms, for example, x and x^2, in a linear regression model can sometimes lead to a problem known as collinearity, which will be discussed in Chapter 4. This can often be overcome by what is known as centering the explanatory variable, that is, using the original variable with its mean subtracted as the explanatory variable. Kleinbaum et al. (2013) provide an example of the effectiveness of such an approach for correcting collinearity.

Fitting the model containing the quadratic term in oxygen uptake gives the numerical results shown in Table 3.5 and shows that the regression coefficient for the quadratic term is highly significant. The numerical results are summarized in Figure 3.3, which shows a scatterplot of the data with the addition of the fitted quadratic curve. Clearly, the new model provides an excellent fit.

Note that the numerical results in Table 3.5 are written in scientific notation where, for example, 1.5e–3 means $1.5 \times 10^{-3} = 0.0015$. The reason for this is that values of oxygen squared are very large, so the corresponding estimated regression coefficient and its standard error are very small.

3.3 Regression Diagnostics

Having fitted a simple regression model to our data and estimated and interpreted the regression coefficients, there still remains work to be done. We need to assess the model to see whether, for example, the assumption that the variance of the response does not change with the values of the explanatory

TABLE 3.5
Results from Fitting a Linear Regression Model to the Kinesiology Data with Linear and Quadratic Terms for Oxygen Uptake

	Coefficients			
	Estimate	**Standard Error**	**t-Value**	**Pr(>|t|)**
Intercept	2.427e+01	1.940e+00	12.509	<2e–16
Oxygen	−1.344e–02	1.762e–03	−7.628	6.27e–10
Oxygen^2	8.902e–06	3.444e–07	25.850	<2e–16

Note: Residual standard error: 3.186 on 50 DF; multiple R-squared: 0.9939; F-statistic: 4055 on 2 and 50 DF; *p*-value: <2.2e–16.

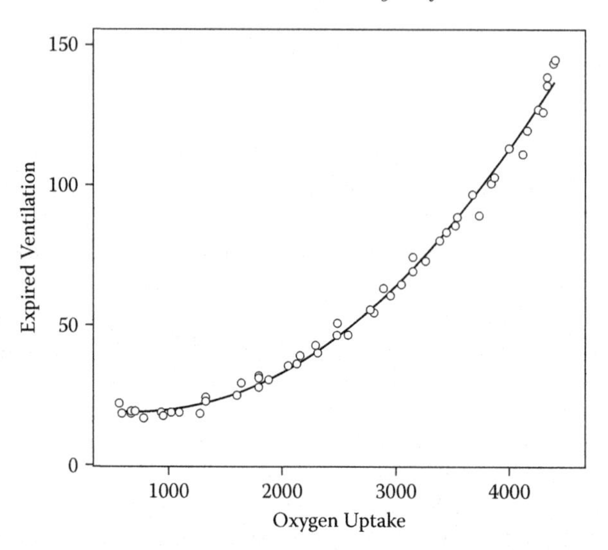

FIGURE 3.3
Kinesiology data showing a fitted linear regression model that includes both oxygen uptake and oxygen uptake squared as explanatory variables.

variable, the constant variance assumption, is reasonable. Further, we need to discover if the model we have used is sensible for the data at hand. Not checking assumptions or assessing if, say, a more complex model is needed and fitting a model that is, in one way or another, unsuitable for the data, are likely to lead to incorrect inferences and conclusions. One way of investigating both the assumptions made and the possible failings of a model is an examination of residuals, that is, the difference between an observed value of the response variable y_i and the fitted value $\hat{y}_i$. (The residuals essentially estimate the error terms in the model.)

In regression analysis, there are various ways of plotting residual values that can be helpful in assessing particular components of the regression model. The most useful plots are as follows:

- A boxplot or probability plot of the residuals can be useful in checking for symmetry and specifically the normality of the error terms in the regression model.

- Plotting the residuals against the corresponding values of the explanatory variable. Any sign of curvature in the plot might suggest that, say, a quadratic term in the explanatory variable should be included in the model.

- Plotting the residuals against the fitted values of the response variable (not the response values themselves for reasons spelt out in Rawlings et al., 1998). If the variability of the residuals appears to increase with the size of the fitted values, a transformation of the response variable prior to fitting is indicated.

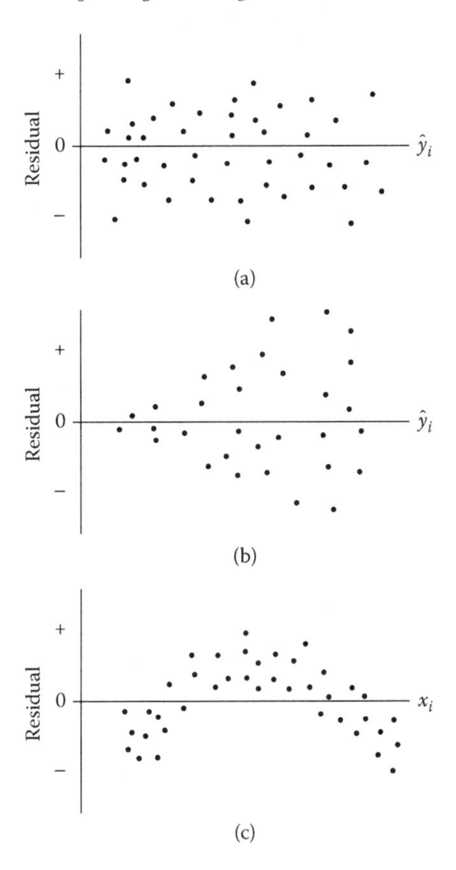

FIGURE 3.4
Idealized residual plots.

Figure 3.4 shows some idealized residual plots that indicate particular points about models:

- Figure 3.4a is what is looked for to confirm that the fitted model is appropriate.
- Figure 3.4b suggests that the assumption of constant variance is not justified so that some transformation of the response variable before fitting might be sensible.
- Figure 3.4c implies that the model needs a quadratic term in the explanatory variable.

(In practice, of course, the residual plots obtained might be somewhat more difficult to interpret than these idealized plots!)

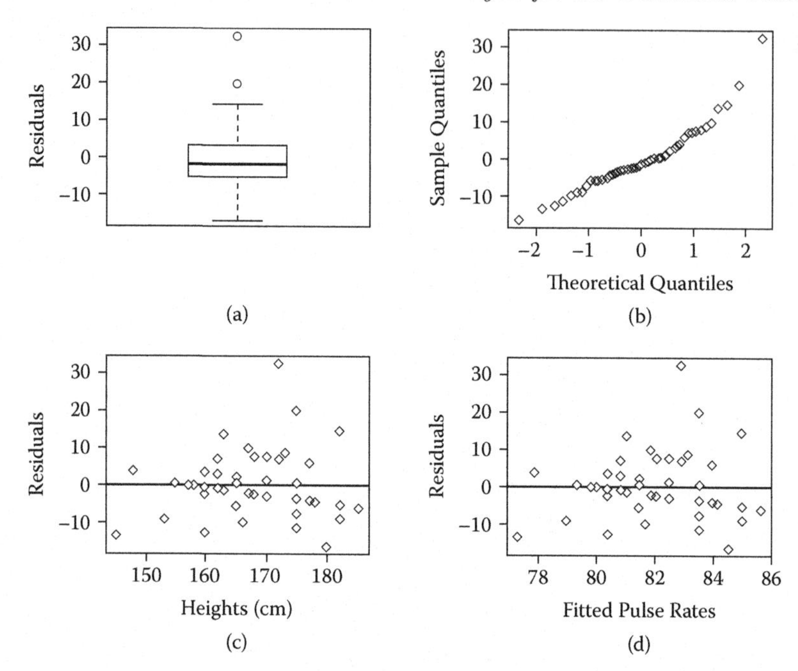

FIGURE 3.5
Residual plots for the pulse rates and heights data form fitting a simple linear regression model: (a) boxplot of residuals, (b) probability plot of residuals, (c) plot of residuals against height, and (d) plot of residuals against fitted pulse rates.

Let us now look at some residual plots for the two examples considered earlier. For the pulse rate and heights data, Figure 3.5 shows four residual plots. The boxplot indicates two very large residuals, but the probability shows little evidence of a departure from linearity, so there is no evidence of a departure from normality. Both Figure 3.5c and 3.5d suggest that the variance of the residuals increases both with height and the fitted values of the pulse rate; the constant variance assumption seems questionable for these data, and a transformation of the response may be helpful (again, see Exercise 3.1).

Moving on to the kinesiology data, Figure 3.6 shows the same four residual plots for a simple linear regression fitted to these data as the plots in Figure 3.5. The probability plot indicates that the residuals do not have a normal distribution, and the plots of residuals against oxygen uptake and fitted expired ventilation show very clearly that a model with a quadratic term in oxygen uptake is needed. For these data, the need for a quadratic term was clear by looking at the scatterplot of expired ventilation against oxygen uptake, but this will not always be the case, and in many cases, the residual plots will uncover problems or the need for other terms in the current model that are not apparent in the scatterplot of the data.

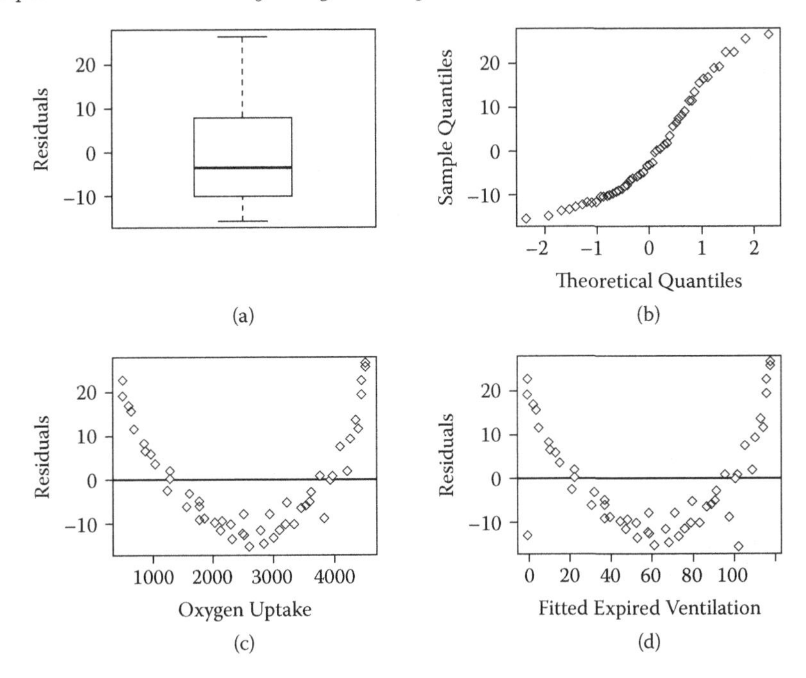

FIGURE 3.6
Residual plots for the oxygen uptake and expired ventilation data from fitting a simple linear regression model: (a) boxplot of residuals, (b) probability plot of residuals, (c) plot of residuals against oxygen uptake, and (d) plot of residuals against fitted expired ventilation.

In Figure 3.7, the same four residual plots are given for the kinesiology data after fitting the model with both a linear and a quadratic term for oxygen uptake. We see that now the residuals are far better behaved than in Figure 3.6; clearly, this more complicated model is a far better fit than the simple linear regression model.

The "raw" residuals used here suffer from certain problems that make them less helpful in investigating fitted models than they might be with some relatively simple adjustments. These adjustments, along with a number of other diagnostic tools for regression models, will be discussed in Chapter 4.

3.4 Locally Weighted Regression

When investigating the relationship between two variables, the first stop is the simple linear regression model. A scatterplot of the data that also shows the fitted line provides an excellent first graphic for studying the dependence of two variables. After looking at this graph and also viewing the residual plots, we may perhaps decide to add, say, a quadratic explanatory variable

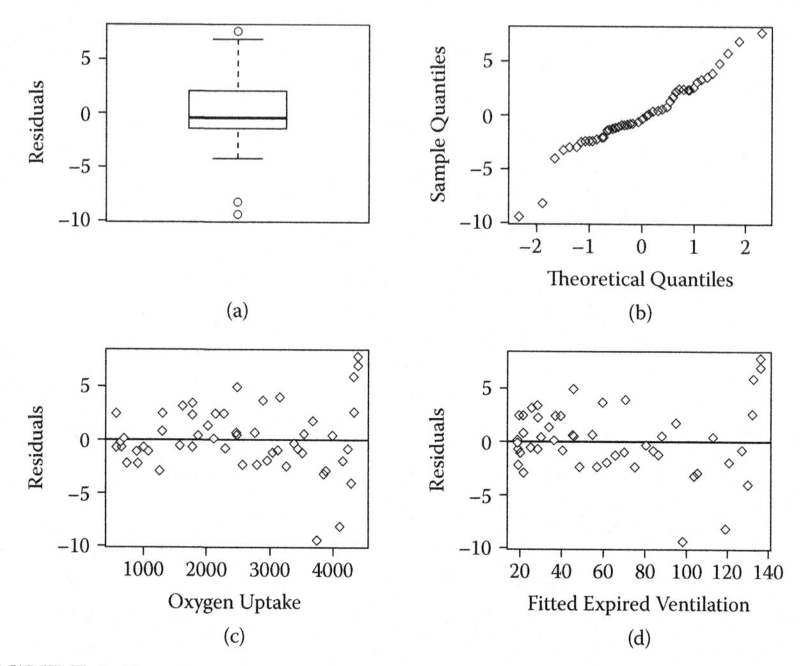

FIGURE 3.7
Residual plots for the oxygen uptake and expired ventilation data from fitting a linear regression model that includes both a linear and quadratic term for oxygen uptake: (a) boxplot of residuals, (b) probability plot of residuals, (c) plot of residuals against oxygen uptake, and (d) plot of residuals against fitted expired ventilation.

term. But, instead of assuming we know the functional form for a regression model, is there any way to allow the data themselves to suggest the appropriate functional form? The secret is to replace the global estimates from the regression models considered earlier in this chapter with local estimates in which the statistical dependency between two variables is described not with a single parameter such as a regression coefficient but with a series of local estimates. For example, a regression might be estimated between the two variables for some restricted range of values for each variable and the process repeated across the range of each variable. The series of local estimates is then aggregated by drawing a line to summarize the relationship between the two variables. In this way, no particular functional form is imposed on the relationship. Such an approach is particularly useful when

- The relationship between the variables is expected to be of a complex form not easily fitted by standard linear or nonlinear models.
- There is no a priori reason for using a particular model.
- We would like the data themselves to suggest the appropriate functional form.

This approach is essentially an example of *exploratory data analysis* (Tukey 1977), in which the form of any functional relationship emerges from a set of data rather than from, say, a theoretical construct. The starting point for a local estimation approach to fitting relationships between variables is the scatterplot smoother, which will now be described.

3.4.1 Scatterplot Smoothers

The local estimation procedures to be discussed here are essentially *nonparametric* because, unlike a parametric technique, for example, linear regression, they do not summarize the relationship between two variables with a parameter such as a regression or correlation coefficient. Instead, nonparametric smoothers summarize the relationship between two variables using a line drawing. The simplest of this collection of nonparametric smoothers is a *locally weighted regression* or *lowess* fit, first suggested by Cleveland (1979). Technical details are given in Technical Section 3.2.

Technical Section 3.2: Lowess Fit

The locally weighted regression approach assumes that the explanatory variable x and the response variable y are related in the following way:

$$y_i = g(x_i) + \varepsilon_i$$

where g is a p-degree polynomial function in the predictor variable x, and ε_i are random variables with mean 0 and constant scale. Values of y_i are used to estimate y_i at each x_{ij} and are found by fitting the polynomials using weighted least squares with large weights for points near to x_i and small otherwise.

Two parameters control the shape of a lowess curve. The first is a smoothing parameter, α (often known as the *span*), with larger values leading to smoother curves—typical values are 1/4 to 1. In essence, the span decides the amount of the trade-off between reduction in bias and increase in variance. If the span is too large, the nonparametric regression estimate will be biased, but if the span is too small, the estimate will be overfitted with inflated variance. Keele (2008) gives an extended discussion of the influence of the choice of span on nonparametric regression.

The second parameter, γ, is the degree of the polynomials that are fitted by the method; γ can be 1 or 2. In any specific application, the change of the two parameters α and γ must be based on a combination of judgment and trial and error. Residual plots may be helpful in judging a particular combination of values.

Our first example of using a locally weighted regression approach involves again the data on pulse rates and heights given in Table 3.1. A scatterplot of the data that shows both the simple linear regression fit and the locally weighted regression fit is shown in Figure 3.8. The lowess fit shows some degree of curvature explained perhaps by the locally weighted approach being less influenced by the observations with large pulse rates. This possible curvature may be worth investigating by fitting a linear model with both a linear and a quadratic term for height (see Exercise 3.1). This demonstrates one way of using locally weighted regression fits—they may indicate a possible parametric model for the data.

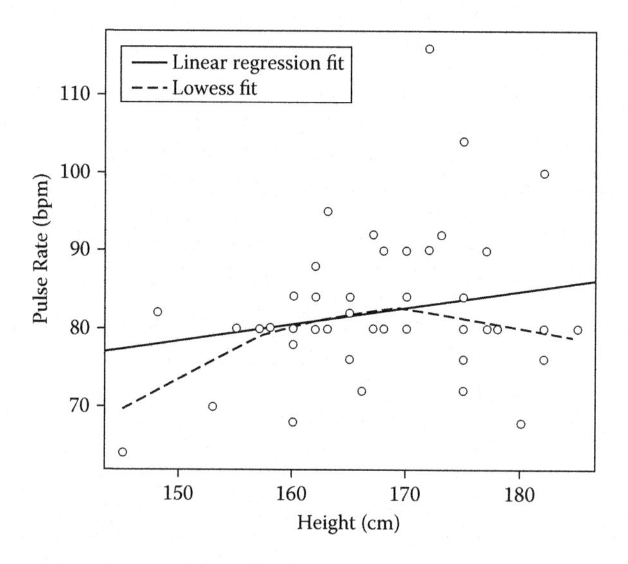

FIGURE 3.8
Scatterplot of pulse rate against height showing fitted linear and locally weighted regression fits.

An alternative smoother that can often be usefully applied to bivariate data is some form of spline function. (A spline is a term for a flexible strip of metal or rubber used by a draftsman to draw curves.) Such functions are described in Technical Section 3.3.

Technical Section 3.3: Spline Smoothers

Spline functions are polynomials within intervals of the x variable that are connected across different values of x. Figure 3.9, for example, shows a linear spline function, that is, a piecewise linear function, of the form

$$f(x) = \beta_0 + \beta_1 x + \beta_2 (x - a)_+ + \beta_3 (x - b)_+ + \beta_4 (x - c)_+$$

where $(u)_+ = u$, when $u > 0$, and 0 otherwise.

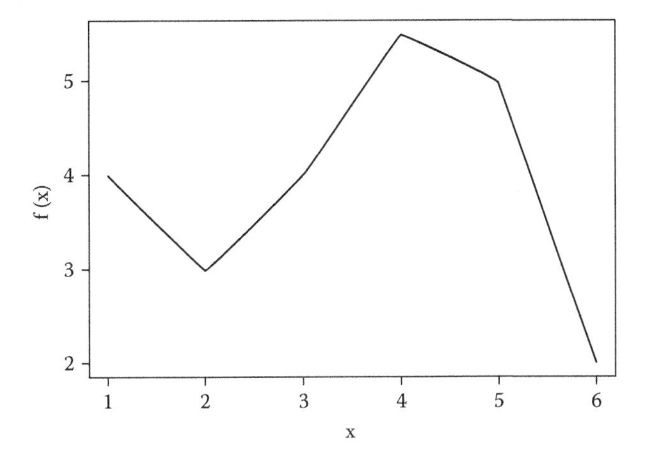

FIGURE 3.9
A linear spline function with knots at $a = 1, b = 3$, and $c = 5$.

The interval endpoints a, b, and c are called *knots.* The number of knots can vary according to the amount of data available for fitting the function.

The linear spline is simple and can approximate some relationships, but it is not smooth and so will not fit highly curved functions well. This problem is overcome by using piecewise polynomials—in particular, cubics, which have been found to have beneficial properties with good ability to fit a variety of complex relationships. The result is a *cubic spline.*

Again, we wish to fit a smooth curve, $g(x)$, that summarizes the dependence of y on x. A natural first attempt might be to try to determine g by least squares as the curve that minimizes $\sum \left[y_i - g\left(x_i\right)\right]^2$, but this would simply result in an interpolating curve and would not be smooth at all. Instead, an amended least-squares criterion can be used to determine g, namely,

$$\sum \left[y_i - g\left(x_i\right)\right]^2 + \lambda \int g''\left(x\right)^2 dx$$

where $g''(x)$ represents the second derivation of $g(x)$ with respect to x. Although when written formally this criterion looks a little formidable, it is really nothing more than an effort to govern the trade-off between the goodness of fit of the data (as measured by $\sum[y_i - g(x_i)]^2$) and the "wiggliness" or departure of linearity of g (as measured by $\int g''(x)^2 dx$); for a linear function, this part of the fitting criterion would be zero. The parameter λ governs the smoothness of g, with larger values resulting in a smoother curve.

The function that minimizes the amended least-squares fitting criterion is known as a cubic spline and is essentially a series of cubic polynomials joined at the unique observed values of the explanatory variables x_i (for more details, see Keele, 2008).

The spline smoother does have a number of technical advantages over the lowess smoother, such as providing the best mean square error and avoiding overfitting, which can cause smoothers to display unimportant variation between x and y, which is of no real interest. But, in practice, the lowess smoother and the cubic spline smoother will give very similar results on many examples.

As an example of using a spline smoother, Figure 3.10 shows the pulse rate and height data yet again with, in this case, added linear, lowess, and spline fits. The latter two fits are very similar for these data.

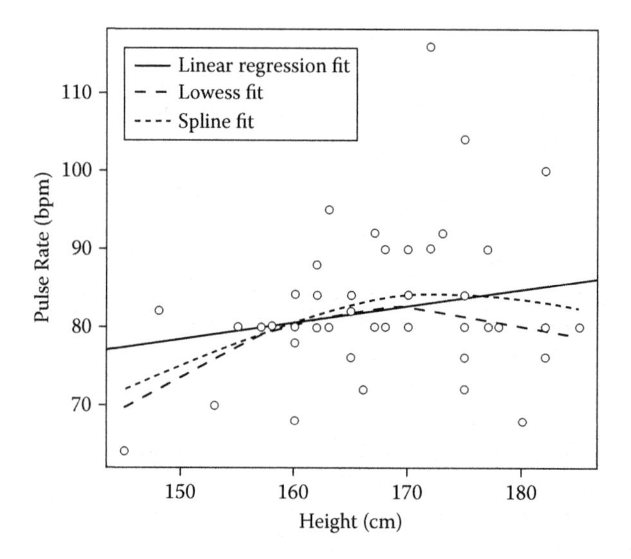

FIGURE 3.10
Scatterplot of pulse rate against height showing linear, lowess, and spline fits.

As a further example of the use of scatterplot smoothers, we will use data from Jacobson and Dimock's (1994) study of the 1992 U.S. House elections, an example also used in Keele (2008). In the 1992 House elections, many incumbents were defeated, and Jacobson and Dimock investigated the factors that contributed to the unusually high number of incumbents who were beaten that year. In 1992, dissatisfaction with Congress was high because of a weak economy and a number of congressional scandals. Jacobson and Dimock suggest that one possible indicator of such dissatisfaction was the percentage of vote for H. Ross Perot in the 1992 presidential election. The district-level vote between the president and members of the house is highly correlated, and Jacobson and Dimock explored whether the level of support for Perot in the

TABLE 3.6
First Five Observations of Challenger Vote and Perot Vote

District	Challenger Vote (%)	Perot Vote (%)
1	37.9	11.7
2	38.2	10.8
3	29.8	11.9
4	32.8	14.8
5	53.8	9.2

presidential election increased the support for the challengers in the House elections. The first five observations (out of a total of 312) of the challenger's percentage vote and the percentage vote for Perot in each congressional district in the 1992 election are shown in Table 3.6.

A scatterplot of the Jacobson and Dimock data is shown in Figure 3.11. Jacobson and Dimock (1994) assumed the relationship between the two variables to be linear, and the fitted simple linear regression of challenger vote on Perot vote is also shown in Figure 3.11. We can investigate the fitted model by looking at the four residual plots used previously; they are shown in Figure 3.12. These plots give no cause for concern about the simple linear regression model; the probability plots give no evidence that the residuals depart from normality, and Figures 3.12c and 3.12d give no indication that higher-order terms in the explanatory variable are needed in the model. Nevertheless, we shall fit both a lowess and a spline smoother to the data, with the result shown

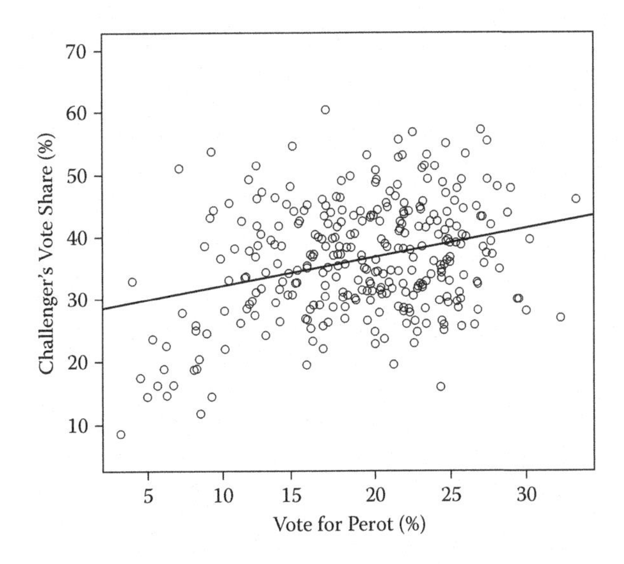

FIGURE 3.11
Scatterplot of challenger vote (%) and Perot vote (%) showing fitted simple linear regression.

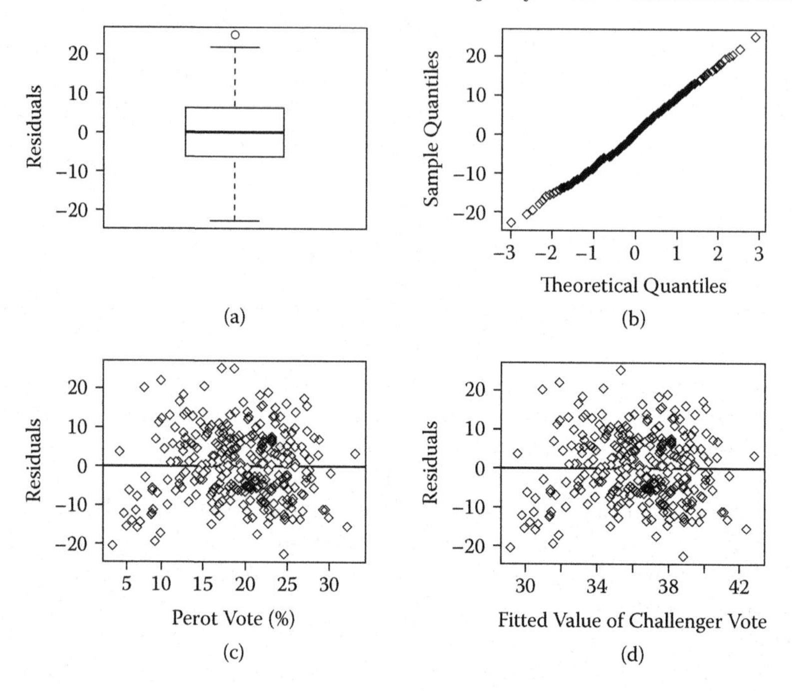

FIGURE 3.12
Residual plots from the fitting of a simple linear regression model to the congressional vote data.

in Figure 3.13. Both smoothers suggest some leveling off of the relationship between challenger vote and Perot vote; given the evidence from the residual plots in Figure 3.12, we might conclude that the lowess and spline fits are misleading in this case. But Keele (2008) shows this is not true by carrying out some inferential tests, which clearly demonstrate that there is significant curvature in the relationship between challenger vote and Perot vote. So, here is an example in which use of a locally weighted regression approach leads to a discovery not apparent with the use of simple linear regression and associated residual plots. It appears that increases in the challenger's vote share does not occur uniformly as support for Perot increases (which is implied by the simple linear regression model); rather, support for Perot has a diminishing effect on the challenger's vote share.

In this section we have given only a brief, relatively informal account of scatterplot smoothers that hopefully demonstrates how they might be useful in practice. We have not covered issues of inference, overfitting, etc., and for these and other details of the approach, readers are referred to the excellent book by Keele (2008).

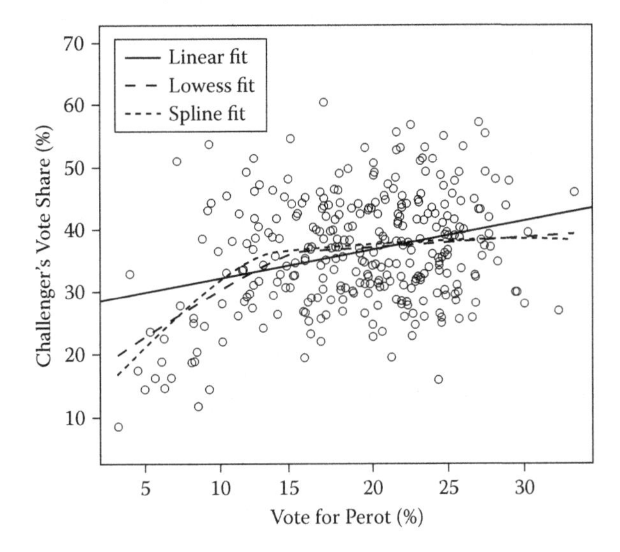

FIGURE 3.13
Scatterplot of challenger's vote share and Perot vote 1992, showing linear, lowess, and spline fits.

3.5 Summary

- The scatterplot and a fit of a simple linear regression model are the first points to be considered when dealing with a set of bivariate data.
- The assumptions of the fitted model must be checked by looking at a variety of residual plots. In this way the assumptions of normality and constant variance can be assessed.
- Residual plots may also be helpful in indicating the possible need for higher-order terms in the explanatory variable to be included in the model.
- Scatterplot smoothers can be a useful additional tool in the exploration of a bivariate data set. They often provide a helpful antidote to the unthinking application of simple linear regression.

3.6 Exercises

3.1 Reanalyze the pulse rates and heights data after taking a log transformation of pulse rate. Contrast and compare the results with those described in the text, remembering that using a log transformation changes the scale of this variable.

3.2 The data gives the final examination scores (out of 75) and corresponding exam completion times (seconds) of 134 individuals. Construct a scatterplot of the data that shows the simple linear regression fit of exam score on time and also gives suitable graphics for the marginal distributions of each variable. Use residual plots to check the assumptions of the model and whether a more complex model might be needed for these data.

3.3 The data gives the average vocabulary size of children at various ages. Construct the scatterplot of the data and use the scatterplot and knowledge of the data to fit a suitable model.

3.4 The data gives marriage and divorce rates (per 1000 population per year) for 14 countries. Derive the linear regression equation of divorce rate on marriage rate and show the fitted line on a scatterplot of the data. On the basis of the regression line, predict the divorce rate for a country with a marriage rate of 8 per 1000 and also for a country with a marriage rate of 14 per 1000. How much conviction do you have in each prediction?

3.5 The data gives the average percentage memory retention measured against passing time (minutes). The measurements were taken five times during the first hour after subjects memorized a list of disconnected items, and then at various times up to a week later. Plot the data (after a suitable transformation if necessary) and investigate the relationship between retention and time using a suitable regression model.

4

Multiple Linear Regression

4.1 Introduction

Multiple linear regression represents a generalization, to more than a single explanatory variable, of the simple linear regression procedure described in Chapter 3. It is now that the relationship between a response variable and several explanatory variables becomes interesting. The adjective "multiple" indicates that at least two explanatory variables are involved in the modeling exercise. At the onset, it is important to note that the explanatory variables are strictly assumed to be fixed and under the control of the investigator, that is, they are not considered to be random variables; only the response variable is considered to be a random variable. In practice, of course, this assumption is unlikely to be true, in which case the results from a multiple linear regression are interpreted as being conditional on the observed values of the explanatory variables, and the inherent variation in the explanatory variables is ignored. Because there are no distributional assumptions about the explanatory variables, they may be categorical with more than two categories (such nominal variables need to be coded in an appropriate way as we shall see later), ordered categorical, or interval. The goals of a multiple regression may be to determine whether the response variable and one or more explanatory variables are associated in some systematic way or to predict values of the response variables from values of the explanatory variables, or both.

Details of the model, including the estimation of its parameters by least squares and the calculation of standard errors, are given in Technical Section 4.1.

Technical Section 4.1: Multiple Linear Regression

The multiple linear regression model for a response variable y with observed values $y_1, y_2, \ldots, y_n$ and q explanatory variables $x_1, x_2, \ldots, x_q$ with observed values $x_{i1}, x_{i2}, \ldots, x_{iq}$ for $i = 1, 2, \ldots, n$ is

$$y_i = \beta_0 + \beta_1 x_{i1} + \beta_2 x_{i2} + \cdots + \beta_q x_{iq} + \varepsilon_i$$

The regression coefficient β_i measures the change in the mean response associated with a unit change in the corresponding explanatory variable, provided the values of all the other explanatory variables do not change. This is often referred to as *partialling out* or *controlling for* other variables, although such terms are probably best avoided. The "linear" in multiple linear regression model refers to the parameters rather than to the explanatory variables, as discussed in Chapter 3.

The error terms in the model $\varepsilon_1, \varepsilon_2, \ldots, \varepsilon_n$ are assumed to have a normal distribution with zero mean and the same variance σ^2 for all values of the explanatory variables. This assumption implies that, for given values of the explanatory variables, the response variable is normally distributed with a mean that is a linear function of the explanatory variables and a variance that is not dependent on them.

The least-squares estimation process is used to estimate the parameters in the multiple linear regression model, and the resulting estimators are most conveniently described with the use of a matrix and vector notation. So we introduce a vector $\mathbf{y}' = [y_1, y_2, \ldots, y_n]$ and an $n \times (q+1)$ matrix $\mathbf{X}$ given by

$$\mathbf{X} = \begin{bmatrix} 1 & x_{11} & x_{12} & \cdots & x_{1q} \\ 1 & x_{21} & x_{22} & \cdots & x_{2q} \\ \vdots & \vdots & \vdots & \ddots & \vdots \\ 1 & x_{n1} & x_{n2} & \cdots & x_{nq} \end{bmatrix}$$

Now we can write the multiple linear regression model for all n observations as

$$\mathbf{y} = \mathbf{X}\boldsymbol{\beta} + \boldsymbol{\varepsilon}$$

where $\boldsymbol{\varepsilon}' = [\varepsilon_1, \varepsilon_2, \ldots, \varepsilon_n]$ and $\boldsymbol{\beta}' = [\beta_0, \beta_1, \beta_2, \ldots, \beta_q]$. The least-squares estimators of the parameters in the multiple linear regression model are given by the set of equations

$$\hat{\boldsymbol{\beta}} = (\mathbf{X}'\mathbf{X})^{-1}\mathbf{X}'\mathbf{y}$$

These matrix manipulations are easily performed on a computer, but you must ensure that there are no linear relationships between the explanatory variables, such as one variable is the sum of several others; otherwise, your regression software will complain because the inverse of the matrix $\mathbf{X}'\mathbf{X}$ will be singular. (More details of the model in matrix form and the least-squares estimation process are given in Rawlings et al., 1998.) The estimated regression coefficients have the same interpretation as given earlier for the population values of these parameters. Of course, each estimated coefficient and its interpretation are only applicable within the range of values of the corresponding explanatory variable that has been used in fitting the multiple linear regression model.

The variation in the response variable can be partitioned into a part due to regression on the explanatory variables and a residual as in the case of simple linear regression. These can be arranged in an analysis of variance table as follows:

Source	Df	Sum of Squares	Mean Square	F-statistic
Regression	q	RGSS	RGMS = RGSS/q	RGMS/RSMS
Residual	$n-q-1$	RSS	RSMS = RSS/ $(n-q-1)$	

The F-statistic gives a test of the omnibus null hypothesis that all the regression coefficients are zero, that is, none of the explanatory variables are associated with the response variable; in most practical situations, this is a relatively uninteresting hypothesis. The residual mean square s^2 is an estimator of σ^2, and the estimator of the covariance matrix (see Chapter 12) of the parameters is

$$\mathbf{S}_{\hat{\beta}} = s^2(\mathbf{X}'\mathbf{X})^{-1}$$

The diagonal elements of this matrix give estimates of the variances of the estimated regression coefficients, and the off-diagonal elements give estimates of the estimated covariances. The estimated variances are used to assess the statistical significance of the regression coefficients and to construct confidence intervals for them.

A measure of the fit of the model is provided by the multiple correlation coefficient R, which is defined as the correlation between the observed values of the response variable y_i and the values predicted by the fitted model $\hat{y}_i$, which are given by

$$\hat{y}_i = \hat{\beta}_0 + \hat{\beta}_1 x_{i1} + \hat{\beta}_2 x_{i2} + \cdots + \hat{\beta}_q x_{iq}$$

The value of R^2 gives the proportion of variability in the response variable accounted for by the explanatory variables.

4.2 An Example of Multiple Linear Regression

As our first example of fitting a multiple linear regression model, we will return to the data introduced in Chapter 2 (see Table 2.5) concerned with how long each week people spend looking after their cars. The interest lies in investigating which of the three explanatory variables—age, a measure of extroversion, and gender—are most important in determining the amount of time people spend looking after their cars. Note that, in this example, one of the explanatory variables, gender, is a binary variable, but as explained earlier, since no distributional assumptions are made about explanatory variables, such a variable causes no problems when applying multiple linear regression.

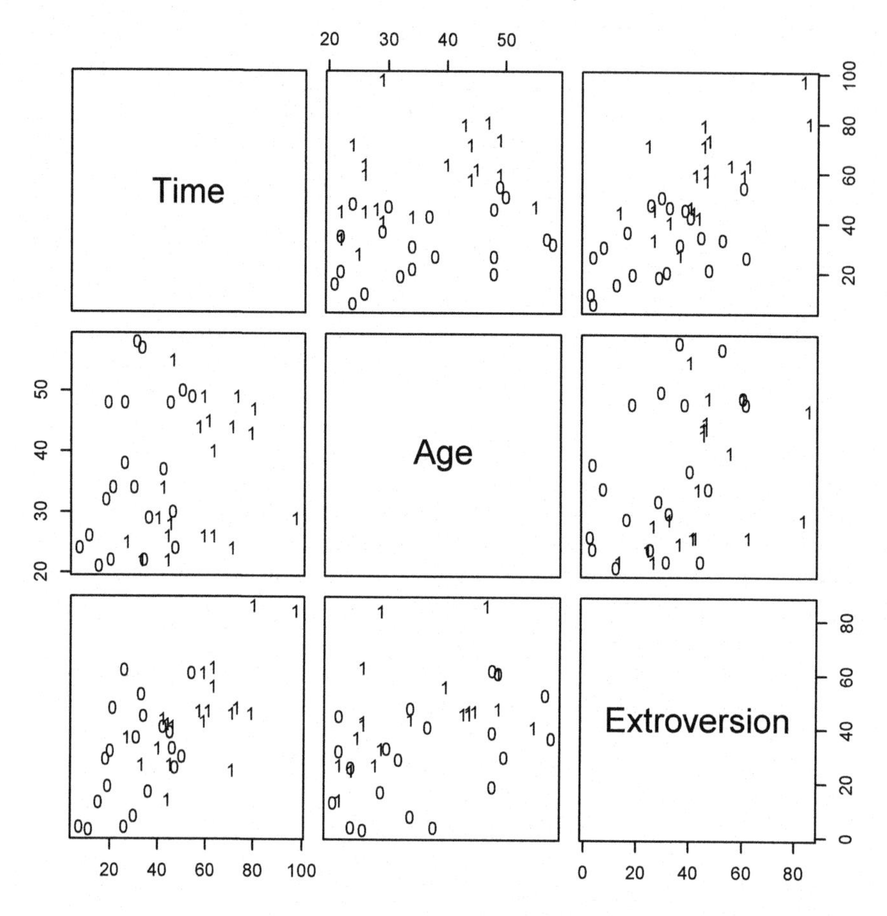

FIGURE 4.1
Scatterplot matrix for the time spent looking after car data, with gender labeled 1 = male, 0 = female.

(Categorical explanatory variables with more than two categories also cause no problems, but they have to be coded appropriately as we shall see in later examples.)

Let us begin with constructing a scatterplot matrix for the three variables—time, age, and extroversion—labeling on each panel of the plot the gender of a subject. The resulting diagram is shown in Figure 4.1. It is clear from this diagram that men spend more time looking after their cars and that time is strongly related to extroversion; perhaps it is less strongly related with age. In addition, the diagram suggests that age and extroversion scores are related, and that there may be two potentially troublesome outliers.

The results from fitting the multiple linear regression model are shown in Table 4.1. The omnibus F-test for testing the hypothesis that all three

TABLE 4.1
Results from Fitting Multiple Linear Regression Model to Time Spent Looking After the Car Data with Age, Extroversion Score, and Gender as Explanatory Variables

	Estimate	Standard Error	t-Value	Pr(>|t|)
Intercept	11.306	7.315	1.546	0.130956
Age	0.156	0.206	0.754	0.455469
Extroversion	0.464	0.130	3.564	0.001053
Gender	20.071	4.651	4.315	0.000119

Note: Residual standard error: 13.02 on 36 degrees of freedom (DF); multiple R-squared: 0.638; F-statistic: 21.13 on 3 and 36 DF; *p*-value: 4.569e-08.

regression coefficients are zero has a very low associated *p*-value; there is strong evidence that not all three coefficients are zero. The square of the multiple correlation coefficient is 0.638; the three explanatory variables together account for about 64% of the variation in time spent looking after the car.

The size of the "raw" regression coefficients in Table 4.1 should not be used to judge the relative importance of the explanatory variables in predicting the response variable, although what are known as standardized values of these coefficients can, partially at least, be used in this way. The standardized values might be obtained by applying the regression model to the values of the response variable and explanatory variables, standardized by (divided by) their respective standard deviations. In such an analysis, each regression coefficient represents the change in the standardized response variable associated with a change of one standard deviation unit in the explanatory variable, again conditional on the other explanatory variables remaining constant. The standardized regression coefficients can, however, be found without undertaking this further analysis, by simply multiplying the raw regression coefficient by the standard deviation of the appropriate explanatory variable and dividing by the standard deviation of the response variable. For the time spent looking after car data, the relevant standard deviations are time (20.79), age (11.39), extroversion (19.67), and gender (0.51); so, the required standardized regression coefficients are

$$\begin{array}{ll} \text{Age:} & 0.16 \times 11.39/20.79 = 0.09 \\ \text{Extroversion:} & 0.46 \times 19.67/20.79 = 0.44 \\ \text{Gender:} & 20.1 \times \ \ 0.51/20.79 = 0.49 \end{array}$$

In any case, it looks like extroversion and gender are more important than age in predicting the time spent looking after the car. For binary explanatory variables such as gender in this example, the unstandardized regression coefficients are more directly interpretable than the standardized versions. This is because the unstandardized coefficients for such explanatory variables simply estimate the difference in the average value of the response between the two categories defined by the variable, holding the other explanatory variables in the model constant.

The t-values associated with each explanatory variable are obtained by simply dividing the estimated regression coefficient by the standard error of the estimate, and it might be thought that the associated significance levels would indicate the importance of the explanatory variables. Here, the values of these associated p-values appear to imply that extroversion and gender are strongly associated with time spent looking after car, whereas age seems to not be associated with the response variable. But, this rather simplistic interpretation of the t-statistics is not always appropriate as we shall make clear later in the chapter.

The estimated regression coefficients give the changes in the value of the response variable when the corresponding explanatory variable changes by *one unit*; for a binary variable such as gender in this example, this statement means a change from one category to the other, so the regression coefficient gives an estimated difference between the two categories conditional on the other variables. Here, the estimated difference in time spent looking after the car between men and women, conditional on age and extroversion staying constant, is 20 min longer for men than for women, with a 95% confidence interval of $[20 - 2.04 \times 4.65, 20 + 2.04 \times 4.65]$, that is, $[10.5, 29.5]$ (2.04 is the value of a t-statistic with 36 degrees of freedom for a 0.05 significance level).

If here we accept, for the moment, the results from the t-statistics at face value, then we might conclude that a model that includes only the explanatory variables, extroversion score and gender, will be adequate for these data, thus providing a more parsimonious model for the data. If extroversion score and gender were both independent of age, then their regression coefficients in the new model would be the same as they are in Table 4.1. But because age is certainly related to extroversion (see Figure 4.1), the model with only extroversion and gender as explanatory variables needs to be fitted anew to get the correct regression coefficients for the gender and extroversion score explanatory variables. The results of fitting this simpler model are shown in Table 4.2. The regression coefficients for gender and extroversion have changed a little from those given in Table 4.1, but the t-statistics for both variables remain highly significant. The square of the multiple correlation is now 0.632, implying that the two explanatory variables in this model account for 63% of the variation in time spent looking after the car, only a very small reduction from the model with three explanatory variables.

The fitted model with gender and extroversion as explanatory variables is

$$\text{time} = 15.68 + 19.18 \times \text{gender} + 0.51 \times \text{extroversion}$$

So, for men (gender = 1) this becomes

$$\text{time} = 15.68 + 19.18 + 0.51 \times \text{extroversion}$$

TABLE 4.2
Results from Fitting Multiple Linear Regression Model to Time Spent Looking After Car Data, with Extroversion Score and Gender as Explanatory Variables

	Estimate	Standard Error	t-Value	Pr(>|t|)
Intercept	15.680	4.437	3.534	0.001118
Extroversion	0.509	0.115	4.423	8.24e-05
Gender	19.180	4.473	4.288	0.000124

Note: Residual standard error: 12.95 on 37 DF; multiple R-squared: 0.632; F-statistic: 31.77 on 2 and 37 DF; *p*-value: 9.284e-09.

and for women (gender = 0) the model is

$$\text{time} = 15.68 + 0.51 \times \text{extroversion}$$

The fitted model is seen to be equivalent to two simple regression fits, each with the same slope but with a different intercept for men and women. The model is conveniently summarized in Figure 4.2.

Another model we might consider is one in which an extroversion × gender interaction is allowed. The results from fitting such a model are shown in Table 4.3. The fitted model is now

$$\begin{aligned}\text{time} = {} & 20.02 + 0.36 \times \text{extroversion} + 7.82 \times \text{gender} \\ & + 0.31 \times (\text{extroversion} \times \text{gender})\end{aligned}$$

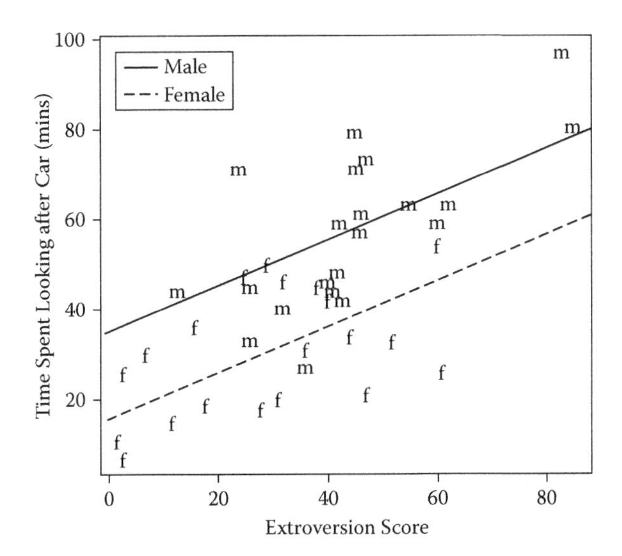

FIGURE 4.2
Plot illustrating the multiple linear regression model fitted to time spent looking after car, with extroversion and gender as the explanatory variables.

TABLE 4.3
Results from Fitting a Multiple Linear Model to the Time Spent Looking after Car Data, with Explanatory Variables, Extroversion, Gender, and Extroversion × Gender Interaction

	Estimate	Standard Error	t-Value	Pr(>|t|)
Intercept	20.018	5.456	3.669	0.000782
Extroversion	0.361	0.159	2.268	0.029430
Gender	7.818	9.571	0.817	0.419379
Extroversion × Gender	0.305	0.228	1.339	0.188970

Note: Residual standard error: 12.81 on 36 DF; multiple R-squared: 0.650; F-statistic: 22.23 on 3 and 36 DF; *p*-value: 2.548e-08.

So, for males (gender = 1) this becomes

$$\text{time} = 20.02 + 0.36 \times \text{extroversion} + 7.82 + 0.31 \times \text{extroversion}$$

and for females (gender = 0)

$$\text{time} = 20.02 + 0.36 \times \text{extroversion}$$

In this case, the model allows the fitted simple linear regression fits for men and women to have both different slopes and different intercepts. Figure 4.3 illustrates the results of fitting this model. Of course, Table 4.3 shows that the interaction term is not significant and so is not needed for these data; the simple model with parallel fits is to be preferred. However, the more complex model is illustrated simply as a useful teaching aid.

4.3 Choosing the Most Parsimonious Model When Applying Multiple Linear Regression

Now we introduce some data taken from Howell (2012), which arise from an evaluation of several hundred courses taught at a large university during the preceding semester. Students in each course had completed a questionnaire in which they rated a number of different aspects of the course on a five-point scale (1 = failure, very bad; . . . ; 5 = excellent, exceptional). The data we will use are the mean scores on six variables for a random sample of 50 courses; the scores for the first five chosen courses are shown in Table 4.4. The six variables are:

1. overall quality of lectures (Overall),
2. teaching skills of the instructor (Teach),

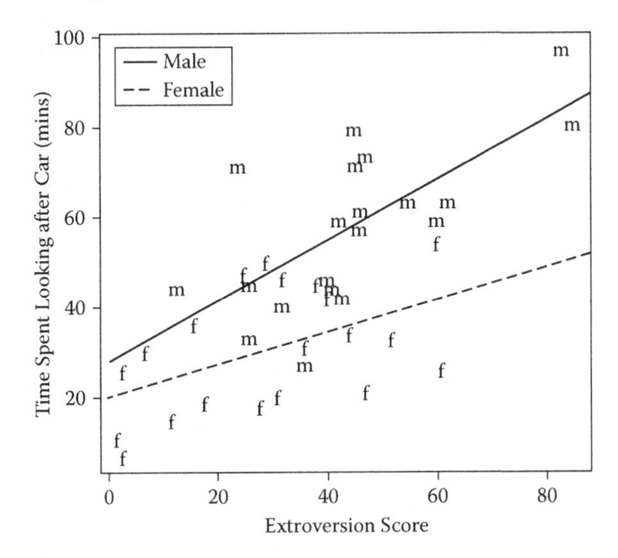

FIGURE 4.3
Diagram illustrating the results given in Table 4.3.

3. quality of the tests and exams (Exam),
4. instructor's perceived knowledge of the subject matter (Knowledge),
5. the student's expected grade in the course (Grade, where higher means better), and
6. the enrollment of the course (Enroll).

Interest lies in how variables 2 to 5 associate with or predict the Overall variable.

Before we begin the model-fitting exercise, we should examine the data graphically, and Figure 4.4 shows a scatterplot matrix of the six variables, each individual scatterplot being enhanced with both a linear and a lowess fit. The plot indicates that the overall rating is related to Teach, Exam, Knowledge, and Grade, and that these explanatory variables are also related to each other. For all the scatterplots in Figure 4.4, the fitted linear and lowess regressions are very similar, suggesting that for none of these explanatory variables is it

TABLE 4.4
Course Evaluation Data

Course	Overall	Teach	Exam	Knowledge	Grade	Enroll
1	3.4	3.8	3.8	4.5	3.5	21
2	2.9	2.8	3.2	3.8	3.2	50
3	2.6	2.2	1.9	3.9	2.8	800
4	3.8	3.5	3.5	4.1	3.3	221
5	3.0	3.2	2.8	3.5	3.2	7

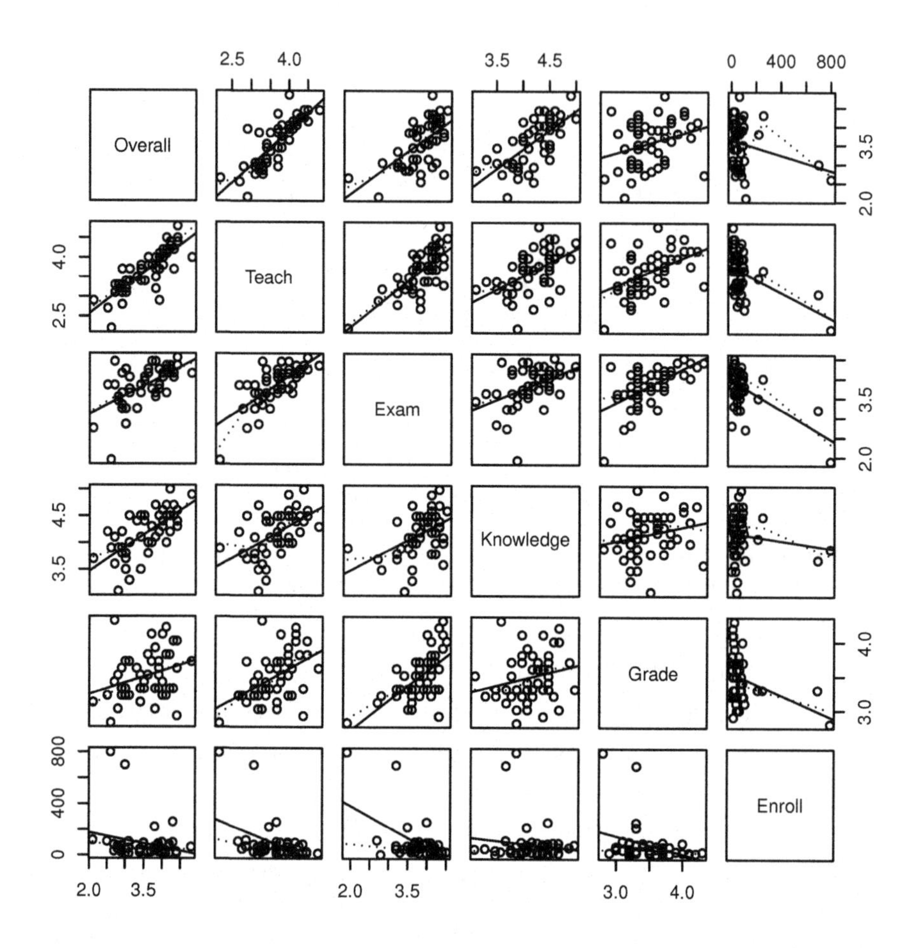

FIGURE 4.4
Scatterplot matrix of course evaluation data showing both linear regression and lowess fits for each pair of variables.

necessary to consider quadratic or higher-order terms in any model. The Enroll variable is problematic because of the presence of two very obvious outliers, one of which is course number 3 with an enroll value of 800, and the other is course 45 with an enroll value of 700. For the moment, we will not remove these two observations and will consider another problem that can occur when using multiple linear regression in practice, which we have not considered up to this point, namely, *multicollinearity.* The term is used to describe situations in which there are moderate to high correlations among some or all of the explanatory variables. Multicollinearity gives rise to a number of difficulties when multiple regression is applied:

- It severely limits the size of the multiple correlation coefficient because the explanatory variables largely attempt to explain much of the same variability in the response variable (see Dizney and Gromen, 1967, for an example).
- It makes determining the importance of a given explanatory variable difficult because the effects of explanatory variables are confounded due to their intercorrelations.
- It increases the variances of the regression coefficients, making the use of the fitted model for prediction less stable. The parameter estimates become unreliable.

Spotting multicollinearity among a set of explanatory variables may not be easy. The obvious course of action is to simply examine the correlations between these variables, but while this is often helpful, it is by no means foolproof; more subtle forms of multicollinearity may be missed. An alternative and generally far more useful approach is to examine what are known as the *variance inflation factors* of the explanatory variables. The variance inflation factor VIF_j for the jth variable is given by

$$\text{VIF}_j = \frac{1}{1 - R_j^2}$$

where R_j is the multiple correlation coefficient from the regression of the jth explanatory variable on the remaining explanatory variables. The variance inflation factor of an explanatory variable indicates the strength of the linear relationship between the variable and the remaining explanatory variables. A rough rule of thumb is that variance inflation factors greater than 10 give some cause for concern. For the course evaluation data, the required variance inflation factors are as follows:

Teach: 2.38; Exam: 3.12; Knowledge: 1.49; Grade: 1.61; Enroll: 1.54

It appears that multicollinearity is not a problem for the course evaluation data. In situations where multicollinearity may be a problem, what should be done? One possibility is to combine in some way explanatory variables that are highly correlated; an alternative approach is simply to select one of the set of correlated variables. Two more complex possibilities are regression on principal components and ridge regression, both of which are described in Chatterjee and Hadi (2012).

But here, we can now go ahead and fit the multiple linear regression model to give the results shown in Table 4.5. Together, the five explanatory variables account for about 76% of the variation in the overall rating. The omnibus F-test has a very low associated p-value, so there is very strong evidence that not all the five regression coefficients are zero. From the t-statistics, it is probably a good bet that the two most important variables for predicting

TABLE 4.5
Results of Fitting a Multiple Linear Regression Model to the Course Evaluation Data

	Estimate	Standard Error	t-Value	Pr(>|t|)
Intercept	−1.195	0.6312	−1.894	0.064875
Teach	0.763	0.1329	5.742	8.06e-07
Exam	0.132	0.1628	0.811	0.421716
Knowledge	0.489	0.1365	3.581	0.000849
Grade	−0.184	0.1655	−1.113	0.271586
Enroll	0.001	0.0004	1.348	0.184555

Note: Residual standard error: 0.3202 on 44 DF; multiple R-squared: 0.756; F-statistic: 27.19 on 5 and 44 DF; p-value: 1.977e-12.

the overall rating are the teaching skills of the instructor and the instructor's perceived knowledge of the subject matter. But as mentioned earlier, the use of t-statistics in this simplistic way is not really appropriate, the reason being that if say we were to drop Exam from the model because its associated t-test has the highest p-value, we would need to refit the model with the remaining four explanatory variables before making any further statements about their importance because the estimated regression coefficients will now change. Of course, if the explanatory variables happened to be independent of one another, there would be no problem and the t-statistics could be used in selecting the most important explanatory variables. This is, however, of little consequence in most practical applications of multiple linear regression.

Before moving on, we should ponder the question of how the results in Table 4.5 are affected by removing the two outlier courses—course 3 and course 5—from the data and refitting the model. The answer is "not very much" as readers can verify by carrying out the task themselves.

So, if using the simple t-statistics identifying a more parsimonious model, that is, one with fewer explanatory variables but still providing an adequate fit, might be suspect in many practical applications of multiple linear regression, what are the alternatives? One approach is *all subsets regression* in which all possible models, or perhaps a subset of possible models, are compared using some suitable numerical criterion; when there are q explanatory variables, there are a total of $2^q - 1$ models (each explanatory variable can be in or out of a model, and the model in which they are all out is excluded). The course evaluation data has five explanatory variables, and so there are 31 possible models to consider. With larger numbers of explanatory variables, the number of models to consider rapidly becomes large; for example, for $q = 12$ there are 4095 models to consider. Special search algorithms are used to make this method feasible. We shall not consider this method any further.

4.3.1 Automatic Model Selection

Software packages frequently offer automatic methods of selecting variables for a final regression model from a list of candidate variables. There are three typical approaches:

- Forward selection
- Backward elimination
- Stepwise regression

The forward selection approach begins with an initial model that contains only a constant term, and successively adds explanatory variables to the model until the pool of candidate variables remaining contains no variables that, if added to the current model, would contribute information that is statistically important concerning the mean value of the response. The backward elimination method begins with an initial model that contains all the explanatory variables under investigation and successively removes variables until no variables among those remaining in the model can be eliminated without adversely affecting the predicted value of the mean response in a statistical sense. Various criteria have been suggested for assessing whether a variable should be added to an existing model in forward selection or removed in backward elimination—for example, the change in the residual sum of squares that results from the inclusion or exclusion of a variable.

The stepwise regression method of variable selection combines elements of both forward selection and backward elimination. The initial model of stepwise regression is one that contains only a constant term. Subsequent cycles of the approach involve first the possible addition of an explanatory variable to the current model, followed by the possible elimination of one of the variables included earlier if the presence of new variables has made its contribution to the model no longer important.

In the best of all possible worlds, the final model selected by applying each of the three procedures outlined here would be the same. Often this does happen, but it is in no way guaranteed. Certainly, none of the automatic procedures for selecting subsets of variables are foolproof. For example, if two explanatory variables are highly correlated with each other, it is highly unlikely that any of the usual automatic methods of model selection will produce a final model that includes both variables. In one way, this is good because it avoids the problem of collinearity discussed earlier. But the final model that automatic selection produces hides the fact that another line of modeling exists based on the second of the two highly correlated variables, and the end results of pursuing that direction might be equally satisfactory, statistically or scientifically—it may even be better (Matthews, 2005).

Automatic model selection methods must be used with care, and the researcher using them should approach the final model selected with a healthy degree of skepticism. Agresti (1996) nicely summarizes the problems:

> Computerized variable selection procedures should be used with caution. When one considers a large number of terms for potential inclusion in a model, one or two of them that are not really important may look impressive simply due to chance. For instance, when all the true effects are weak, the largest sample effect may substantially overestimate its true effect. In addition, it often makes sense to include variables of special interest in a model and report their estimated effects even if they are not statistically significant at some level.

(See McKay and Campbell, 1982a, 1982b, for some more thoughts on automatic selection methods in regression.)

4.3.2 Example of Application of the Backward Elimination

With all the caveats of the previous subsection in mind, we will illustrate how the backward elimination approach works on the course evaluation data using what is known as *Akaike's information criterion* (AIC) to decide whether a variable can be removed from the current candidate model. The AIC index takes into account both the statistical goodness of fit and the number of parameters that have to be estimated to achieve this degree of fit, by imposing a penalty for increasing the number of parameters. In a series of competing models, "lower" values of the AIC are preferable; in what follows, the judgment necessary will be made informally.

The AIC for a model is defined explicitly as minus twice the maximized log-likelihood of the model plus twice the number of parameters in the model; as the log-likelihood of the model gets larger, the AIC goes down, and as the number of parameters of the model increases, so does the value of the AIC.

The results of the backward elimination approach using the AIC are as follows:

Start: AIC $= -108.28$

Explanatory variables in the model are Teach, Exam, Knowledge, Grade, and Enroll.

Step 1: Removing one explanatory variable at a time and leaving the other four in the model

Remove Exam: AIC $= -109.54$

Remove Grade: AIC $= -108.89$

Remove Enroll: AIC $= -108.26$

Remove Knowledge: AIC $= -97.49$

Remove Teach: AIC = −82.32

Removing Exam leads to a model containing the other four explanatory variables, and this model has a lower AIC than the original five-explanatory-variable model. Consequently, we drop the Exam variable and start afresh with the model containing the explanatory variables Grade, Enroll, Knowledge, and Teach.

Current: AIC = −109.54

Explanatory variables in the model are now Teach, Knowledge, Grade, and Enroll.

Step 2: Removing one explanatory variable at a time and leaving the other three in the model

Remove Grade: AIC = −110.74

Remove Enroll: AIC = −110.14

Remove Knowledge: AIC = −97.22

Remove Teach: AIC = −76.54

Removing Grade leads to a model containing the other three explanatory variables, and this model has a lower AIC than the current four-explanatory-variable model. Consequently, we drop the Grade variable and start afresh with the model containing the explanatory variables Enroll, Knowledge, and Teach.

Current: AIC = −110.74

Explanatory variables in the model are now Enroll, Knowledge, and Teach.

Step 3: Removing one explanatory variable at a time and leaving the other two in the model

Remove Enroll: AIC = −110.98

Remove Knowledge: AIC = −98.58

Remove Teach: AIC = −77.38

Removing Enroll leads to a model containing the other two explanatory variables, and this model has a lower AIC than the current three-explanatory-variable model. Consequently, we drop the Enroll variable and start afresh with the model containing the variables Knowledge and Teach.

Current: AIC = −110.98

Explanatory variables in the model are now Teach and Knowledge.

TABLE 4.6
Results of Fitting the Multiple Linear Regression Model with the Two Explanatory Variables, Teach and Knowledge, to the Course Evaluation Data

	Estimate	Standard Error	t-Value	Pr(>|t|)
Intercept	−1.2984	0.4773	−2.720	0.009121
Teach	0.7097	0.1011	7.021	7.6e-09
Knowledge	0.5383	0.1319	4.082	0.000172

Note: Residual standard error: 0.3202 on 47 DF; multiple R-squared: 0.739; F-statistic: 66.47 on 2 and 47 DF; *p*-value: 1.991e-14.

Step 4: Removing one explanatory variable at a time and leaving the other one in the model

Remove Knowledge: AIC = −97.81

Remove Teach: AIC = −77.12

Removal of either one of the two variables, Teach and Knowledge, results in a model with a far higher value of the AIC than the model containing both these variables. Consequently, we accept this as our final model.

We now need to fit the chosen model to the data to get the relevant estimated regression coefficients, etc. The results are shown in Table 4.6. We see that an increase in one unit in Teach leads to an estimated increase of 0.71 overall, conditional on Knowledge, and an increase of one unit in Knowledge leads to an increase of 0.54 overall, conditional on Teach. The square of the multiple correlation coefficient for this model is 0.74, only a little less than its value of 0.76 in the five-variable model.

4.4 Regression Diagnostics

Having selected a more parsimonious model, there still remains one further important aspect of a regression analysis to consider, and that is to check the assumptions on which the model is based. We have already described in Chapter 3 the use of residuals for this purpose, but in this section we shall go into a little more detail and introduce several other useful regression diagnostics that are now available. These diagnostics provide ways for identifying and understanding the differences between a model and the data to which it is fitted. Some differences between the data and the model may be due to isolated observations; one, or a few, observations may be outliers, or may differ in some unexpected way from the rest of the data. Other differences may be systematic; for example, a term may be missing in a linear model. Technical Section 4.2 describes a number of regression diagnostics.

Technical Section 4.2: Regression Diagnostics

To begin, we need to introduce what is known as the hat matrix $\mathbf{H}$, defined as $\mathbf{H} = \mathbf{X}(\mathbf{X}'\mathbf{X})^{-1}\mathbf{X}'$, where $\mathbf{X}$ is the matrix introduced earlier in the chapter in Technical Section 4.1 dealing with estimation of the multiple linear regression model.

In a multiple linear regression, the predicted values of the response variable can be written in matrix form as $\hat{\mathbf{y}} = \mathbf{H}\mathbf{y}$ so that $\mathbf{H}$ "puts the hats" on $\mathbf{y}$. The diagonal elements of $\mathbf{H}$, h_{ii}, $i = 1, 2, \ldots, n$, are such that $0 \leq h_{ii} \leq 1$, and have an average value of q/n. Observations with large values of h_{ii} are said to have high *leverage*, and such observations have the most effect on the estimation of the model parameters. It is often helpful to produce a plot of h_{ii} against i, an *index plot*, to identify any observations that may have undue influence on fitting the model.

The raw residuals introduced in Chapter 3 are not independent of one another, nor do they have the same variance because the variance of $r_i = y_i - \hat{y}_i$ is $\sigma^2 = (1 - h_{ii})$. Both properties make the raw residuals less useful than they might be when amended a little. Two alternative residuals are the *standardized residual* and the *deletion residual*; both are based on the raw residual r_i and are defined as follows:

$$r_i^{\text{std}} = \frac{r_i}{\sqrt{s^2(1 - h_{ii})}}$$

$$r_i^{\text{del}} = \frac{r_i}{\sqrt{s^2_{(i)}(1 - h_{ii})}}$$

where $s^2_{(i)}$ is the residual mean square estimate of σ^2 after the deletion of observation i.

The deletion residuals are often particularly helpful for the identification of outliers. A further useful regression diagnostic is *Cook's distance*, D_i, defined as

$$D_i = \frac{r_i h_{ii}}{\sqrt{q s^2(1 - h_{ii})}}$$

Cook's distance measures the influence of observation i on the estimation of all the parameters in the model. Values greater than 1 suggest that the corresponding observation has undue influence on the estimation process.

A full account of regression diagnostics is given in Cook and Weisberg (1999).

We can now take a look at these regression diagnostics using the final model chosen for the course evaluation data, namely, a model containing only the two explanatory variables: Teach and Knowledge. Figure 4.5 shows boxplots and normal probability plots for both the standardized and deletion residuals.

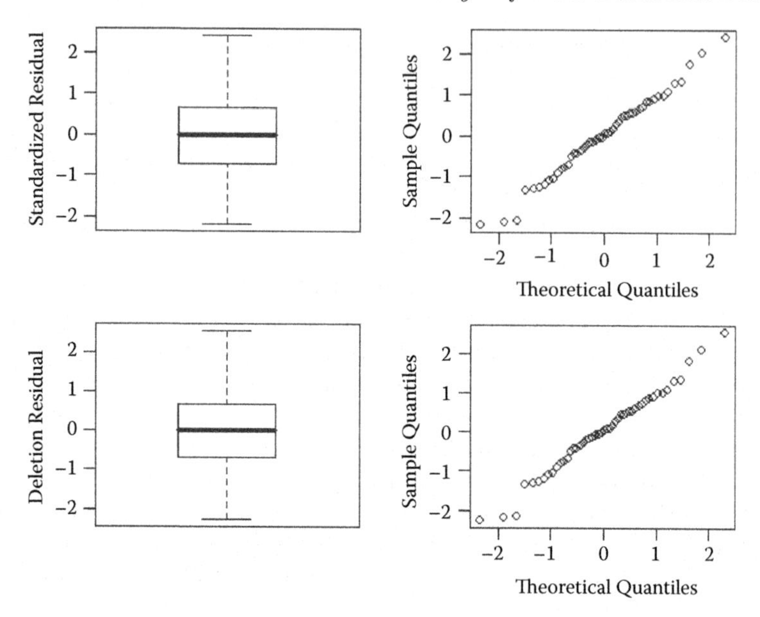

FIGURE 4.5
Boxplots and normal probability plots for both standardized and deletion residuals from the final model chosen for the course evaluation data.

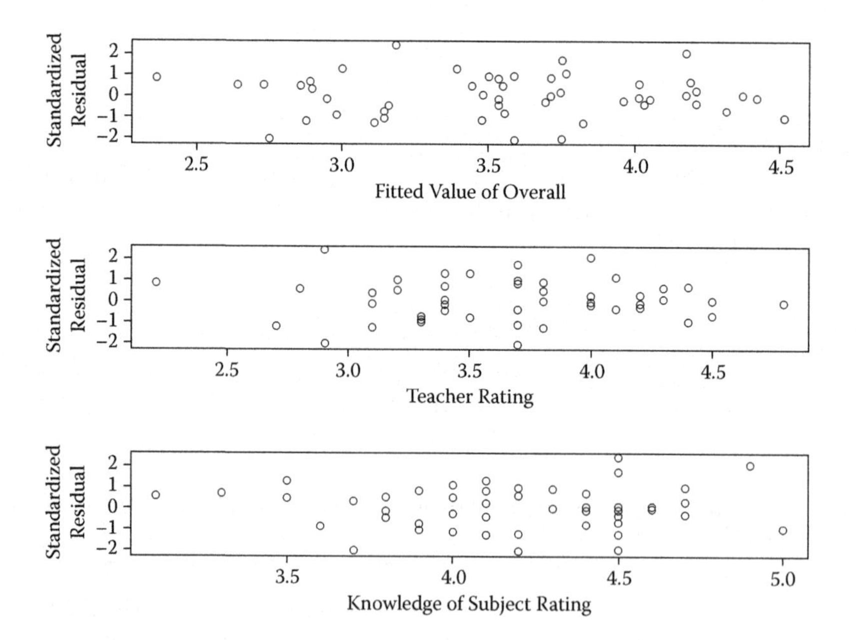

FIGURE 4.6
Plots of standardized residuals against fitted value of the overall rating—the rating of teaching ability and perceived knowledge for the course evaluation data.

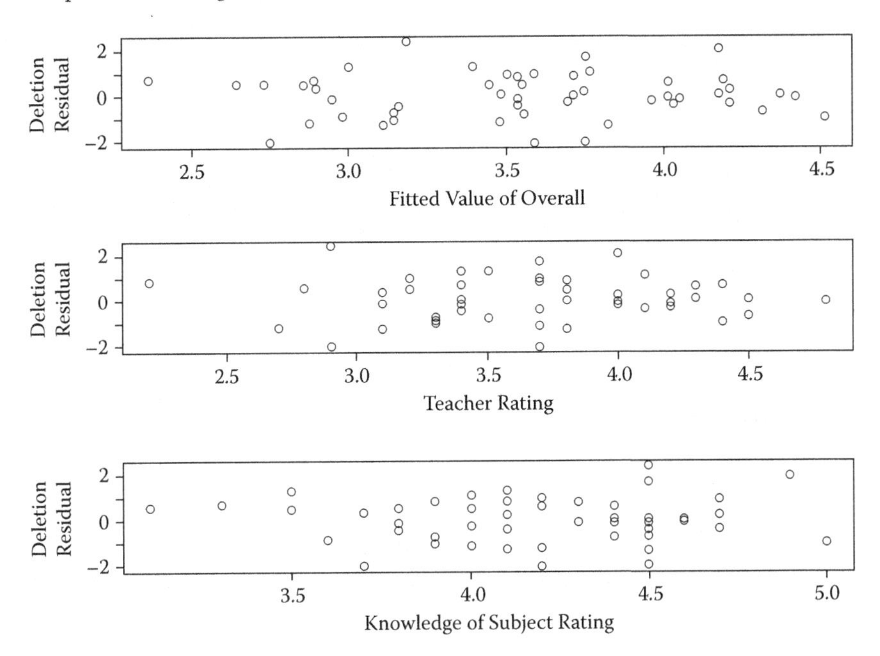

FIGURE 4.7
Plots of deletion residuals against fitted value of the overall rating—the rating of teaching ability and perceived knowledge for the course evaluation data.

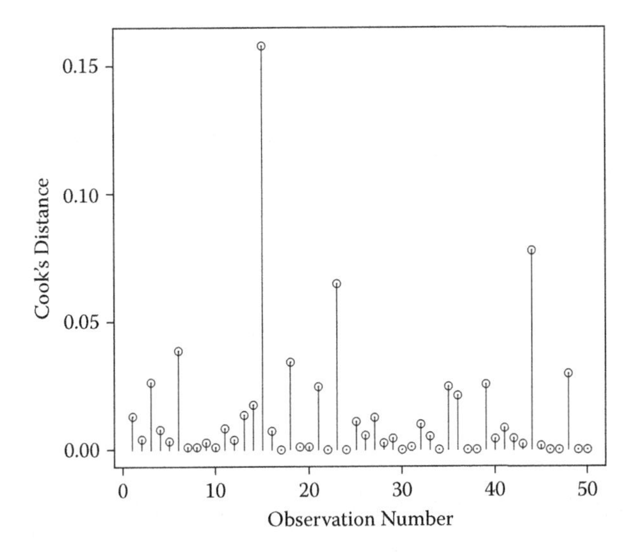

FIGURE 4.8
Index plot of Cook's distances for the final model chosen for the course evaluation data.

The corresponding plots look very similar and give no cause for concern for the model fitted. Figure 4.6 shows plots of the standardized residuals against the fitted value of the overall rating: the rating of teaching ability and perceived knowledge. Figure 4.7 shows the same three plots using deletion residuals. Again, the two sets of plots are very similar and raise no issues about the fitted model. Figure 4.8 shows an index plot of Cook's distances; none is greater than 1, and so, once again, this diagnostic, like the others, gives us some confidence that we have not violated any assumptions when fitting the chosen model.

4.5 Multiple Linear Regression and Analysis of Variance

The phrase "analysis of variance" (ANOVA) was coined by arguably the most famous statistician of the 20th century, Sir Ronald Aylmer Fisher, who defined the technique as "the separation of variance ascribable to one group of causes from the variance ascribable to other groups." ANOVA is probably the piece of statistical methodology most widely used by behavioral scientists, but there is no chapter simply entitled analysis of variance in this book. Why not? The primary reason is that the multiple linear regression model is essentially ANOVA in disguise, and so there is really no need to describe each technique separately. Instead, we will show the equivalence of ANOVA and multiple linear regression through analysis of practical examples.

4.5.1 Analyzing the Fecundity of Fruit Flies by Regression

In a study of fecundity of fruit flies, per-diem fecundity (average number of eggs laid per female per day for the first 14 days of life) for 25 females of each of three genetic lines of the fruit fly *Drosophila melanogaster* was recorded. The lines labeled RS and SS were selectively bred for resistance and susceptibility to the pesticide DDT, and the line NS as a nonselected control strain. The results for the first three fruit flies of each genetic line are shown in Table 4.7. Of interest here is whether the data give any evidence of a difference in fecundity of the three strains.

TABLE 4.7
Fecundity of Fruit Flies

Resistant (RS)	Susceptible (SS)	Nonselected (NS)
12.8	38.4	35.4
21.6	32.9	35.4
14.8	48.5	19.3

In this study, the effect of a single independent factor (genetic strain) on a response variable (per-diem fecundity) is of interest. The data arise from what is generally known as a one-way design and would usually be dealt with by analysis of variance based on the following model:

$$y_{ij} = \mu + \alpha_i + \varepsilon_{ij}$$

where y_{ij} represents the value of the jth observation in the ith genetic line, μ represents the overall mean of the response, α_i is the effect of the ith genetic line ($i = 1, 2, 3$), and ε_{ij} are random error terms assumed to have a normal distribution with mean zero and variance σ^2. The model has four parameters to describe three group means and is said to be overparameterized, which causes problems because it is impossible to find unique estimates for each parameter. This aspect of ANOVA models is discussed in detail in Maxwell and Delaney (2003), but essentially, overparameterization is overcome by imposing constraints on the parameters. In the fruit fly example, we will assume that $\alpha_1 + \alpha_2 + \alpha_3 = 0$. The usual analysis of variance table for the fruit fly data is shown in Table 4.8. The F-test is highly significant, and there is strong evidence that the average number of eggs laid per day differs among the three lines.

How can this analysis be undertaken using multiple linear regression? First, we introduce two dummy variables x_1 and x_2 defined below, which are used to label the three genetic lines:

	Genetic line		
	RS	**SS**	**NS**
x_1	1	0	−1
x_2	0	1	−1

The usual one-way ANOVA model for this situation is the one described earlier:

$$y_{ij} = \mu + \alpha_i + \varepsilon_{ij} \quad \text{with} \quad \alpha_1 + \alpha_2 + \alpha_3 = 0$$

This can be rewritten in terms of the variables x_1 and x_2 as

$$y_{ij} = \mu + \alpha_1 x_1 + \alpha_2 x_2 + \varepsilon_{ij}$$

TABLE 4.8
Analysis of Variance (ANOVA) Table for Fruit Fly Data

Source	**Sum of Squares**	**Df**	**Mean Square**	**F**	***p*-Value**
Between lines	1362.21	2	681.11	8.67	< 0.001
Within lines (error)	5659.02	72	78.60	–	–

and this is exactly the same form as a multiple linear regression model with two explanatory variables. So, applying multiple regression and regressing average number of eggs laid per day on x_1 and x_2, what do we get? The regression sum of squares is 1362.21 with 2 degrees of freedom, and the residual sum of squares is 5659.02 with 72 degrees of freedom. The results are identical to those from ANOVA, and the estimates of the regression coefficients from the regression analysis are

$$\hat{\mu} = 27.42, \quad \hat{\alpha_1} = -2.16, \quad \hat{\alpha_2} = -3.79$$

The estimates of α_1 and α_2 are simply the differences between each genetic line mean and the grand mean.

4.5.2 Multiple Linear Regression for Experimental Designs

Now let us consider a 2 × 2 factorial design with factors A and B both at two levels: A1 and A2, and B1 and B2. The usual ANOVA model for such a design is

$$y_{ijk} = \mu + \alpha_i + \beta_j + \gamma_{ij} + \varepsilon_{ijk}$$

where y_{ijk} represents the kth observation in the ijth cell of the design; α_i represents the effect of the ith level of factor A; β_j represents the effect of the jth level of factor B; γ_{ij} represents the interaction of A and B; and, as always, ε_{ijk} represents random error terms with the usual distributional assumptions. The usual constraints on the parameters to deal with overparameterization in this case are

$$\sum_{i=1}^{2} \alpha_i = 0, \quad \sum_{j=1}^{2} \beta_j = 0, \quad \sum_{i=1}^{2} \gamma_{ij} = \sum_{j=1}^{2} \gamma_{ij} = 0$$

These constraints imply that the parameters in the model are such that

$$\alpha_1 = -\alpha_2, \quad \beta_1 = -\beta_2, \quad \gamma_{1j} = -\gamma_{2j}, \quad \gamma_{i1} = -\gamma_{i2}$$

The last two equations imply that

$$\gamma_{12} = -\gamma_{11}, \quad \gamma_{21} = -\gamma_{11}, \quad \gamma_{22} = \gamma_{11}$$

showing that there is only a single interaction parameter, namely γ_{11}. The model for the observations in each of the four cells of the design can now be written explicitly as follows:

	A1	**A2**
B1	$\mu + \alpha_1 + \beta_1 + \gamma_{11}$	$\mu - \alpha_1 + \beta_1 - \gamma_{11}$
B2	$\mu + \alpha_1 - \beta_1 - \gamma_{11}$	$\mu - \alpha_1 - \beta_1 + \gamma_{11}$

We define two variables as follows:

$$x_1 = 1 \text{ if first level of A, } x_1 = -1 \text{ if second level of A.}$$

$$x_2 = 1 \text{ if first level of B, } x_2 = -1 \text{ if second level of B.}$$

The original ANOVA model for the design can now be written as

$$y_{ijk} = \mu + \alpha_1 x_1 + \beta_1 x_2 + \gamma_{11} x_3 + \varepsilon_{ijk}, \text{ where } x_3 = x_1 \times x_2$$

We can recognize this as a multiple linear regression model with three explanatory variables, and we can fit it in the usual way.

4.5.3 Analyzing a Balanced Design

Here, the fitting process can be used to illustrate the difference in analyzing a balanced 2 × 2 design (equal number of observations per cell) and an unbalanced design (unequal number of observations per cell). To begin, we will apply the multiple regression model to the balanced data in Table 4.9.

TABLE 4.9
A Balanced 2 × 2 Data Set

	A1	A2
B1	23	22
	25	23
	27	21
	29	21
B2	26	37
	32	38
	30	40
	31	35

So, for fitting the multiple regression model, all observations in cell A1,B1 have $x_1 = 1$ and $x_2 = 1$; all observations in cell A1,B2 have $x_1 = 1$, $x_2 = -1$; and so on for the remaining observations in Table 4.9. To begin, we will fit the model with the single explanatory variable x_1 to give the following results:

Source	Sum of Squares	Df	Mean Square
Regression	12.25	1	12.25
Residual	580.75	14	41.48

together with $\hat{\mu} = 28.75$ and $\hat{\alpha_1} = -0.875$.

The regression sum of squares 12.25 is what would be the between levels of A sum of squares in an ANOVA table.

Now, fit the regression with x_1 and x_2 as explanatory variables to give the following results:

Source	Sum of Squares	Df	Mean Square
Regression	392.50	2	196.25
Residual	200.50	13	15.42

together with $\hat{\mu} = 28.75$, $\hat{\alpha_1} = -0.875$, and $\hat{\beta_1} = -4.875$.

The difference between the regression sums of squares for the two-variable and one-variable models gives the sum of squares for factor B that would be obtained in an ANOVA.

Finally, we can fit a model with three explanatory variables to give the following:

Source	Sum of Squares	Df	Mean Square
Regression	536.50	3	178.83
Residual	56.50	12	4.71

together with $\hat{\mu} = 28.75$, $\hat{\alpha_1} = -0.875$, $\hat{\beta_1} = -4.875$, and $\hat{\gamma_{11}} = 3.0$.

The difference between the regression sums of squares for the three-variable and two-variable models gives the sum of squares for the A $\times$ B interaction that would be obtained in an analysis of variance. The residual sum of squares in the final step corresponds to the error sum of squares in the usual ANOVA table. (Readers might like to confirm these results by running an analysis of variance on the data.)

Note that, unlike the estimated regression coefficients in the examples considered in earlier sections, the estimated regression coefficients for the balanced 2×2 design do not change as extra explanatory variables are introduced into the regression model. The factors in a balanced design are independent; a more technical term is that they are orthogonal.

4.5.4 Analyzing an Unbalanced Design

Above we saw that when the explanatory variables are orthogonal, adding variables to the regression model in a different order than the one used earlier will alter nothing; the corresponding sums of squares and regression coefficient estimates will be the same. Is the same true of an unbalanced design? To answer this question, we shall use the data in Table 4.10.

Again, we will fit regression models first with only x_1, then with x_1 and x_2, and finally with x_1, x_2, and x_3.

TABLE 4.10
Unbalanced 2 × 2 Data Set

	A1	**A2**
B1	23	22
	25	23
	27	21
	29	21
	30	19
	27	23
	23	17
	25	—
B2	26	37
	32	38
	30	40
	31	35
	—	39
	—	35
	—	38
	—	41
	—	32
	—	36
	—	40
	—	41
	—	38

Results for x_1 model:

Source	**Sum of Squares**	**Df**	**Mean Square**
Regression	149.63	1	149.63
Residual	1505.87	30	50.19

with $\hat{\mu} = 29.567$ and $\hat{\alpha_1} = -2.233$.
The regression sum of squares gives the sum of squares for factor A.
Results for x_1 and x_2 model:

Source	**Sum of Squares**	**Df**	**Mean Square**
Regression	1180.86	2	590.42
Residual	476.55	29	16.37

with $\hat{\mu} = 29.667$, $\hat{\alpha_1} = -0.341$, and $\hat{\beta}_1 = -5.997$.

The difference in the regression sums of squares for the two-variable and one-variable models gives the sum of squares due to factor B, conditional on A already being in the model.

Results for x_1, x_2, and x_3 model:

Source	Sum of Squares	Df	Mean Square
Regression	1474.25	3	491.42
Residual	181.25	28	6.47

with $\hat{\mu} = 28.606$, $\hat{\alpha_1} = -0.667$, $\hat{\beta_1} = -5.115$, and $\hat{\gamma_{11}} = 3.302$.

The difference in the regression sums of squares for the three-variable and two-variable models gives the sum of squares due to the A × B interaction, conditional on A and B being in the model.

For an unbalanced design the factors are no longer orthogonal, and so the estimated regression parameters change as further variables are added to the model, and the sums of squares for each term in the model are now conditional on what has entered the model before them. If variable x_2 was entered before x_1, then the results would differ from those given earlier, as readers should confirm by repeating the fitting process as an exercise.

So, using the regression approach clearly demonstrates why there is a difference between analyzing a balanced design (not just a 2 × 2 design as in the example) and an unbalanced design. In the latter, the order of entering effects is important. From the need to consider order, a great deal of confusion has arisen. For example, some authors have suggested that, in a two-way unbalanced design with factors A and B, the main effects of A and B can be entered after the A × B interaction to give what are called type III sums of squares; indeed, this is the default in many software packages. However, this approach is heavily criticized by Nelder (1977) and Aitkin (1978). The arguments are relatively subtle, but they go something like this:

- When fitting models to data, the principle of parsimony is of critical importance. In choosing among possible models, we do not want to adopt complex models for which there is no empirical evidence.

- Thus, if there is no convincing evidence of an A × B interaction, we do not retain this term in the model. Thus, additivity of A and B is assumed unless there is convincing evidence to the contrary.

- So, the argument proceeds that type III sum of squares for, say, A, in which it is adjusted for the A × B interaction, makes no sense.

- First, if the interaction term is necessary in the model, then the experimenter will usually want to consider simple effects of A at each level of B separately. A test of the hypothesis of no A main effect would not usually be carried out if the A × B interaction is significant.

- If the A × B interaction is not significant, then adjusting for it is of no interest and causes a substantial loss of power in testing A and B main effects.

The arguments of Nelder and Aitkin against the use of type III sums of squares are persuasive and powerful. Their recommendation to use what are generally known as type I sums of squares in which interaction terms are considered after the main effects of the factors in the interaction term, perhaps considering main effects in a number of orders, as the most suitable way in which to identify a suitable model for a data set is also convincing and strongly endorsed.

Note that, although ANOVA models can be expressed as multiple linear regression models (as seen in a series of examples), we are not suggesting that behavioral researchers should stop using the ANOVA module in whatever software they use, because that module will conveniently take care of the conversion to a multiple linear regression model and print out the usual analysis of variance table that is required by the researcher.

In the next chapter, we will look at a general framework for linear models, which will allow the appropriate models to be fitted to response variables that do not satisfy the assumptions required by ANOVA and multiple linear regression models.

4.6 Summary

- Multiple linear regression is used to assess the relationship between a set of explanatory variables and a continuous-response variable.
- The response variable is assumed to be normally distributed with a mean that is a linear function of the explanatory variables, and a variance that is independent of them.
- The explanatory variables are strictly assumed to be fixed. In practice, where this is almost never the case, the results of multiple regression are to be interpreted conditional on the observed values of these variables.
- It may be possible to find a more parsimonious model for the data, that is, one with fewer explanatory variables using all subsets of regression or one of the "stepping" methods. Care is required when using the latter.
- An extremely important aspect of a regression analysis is the inspection of a number of regression diagnostics in a bid to identify any departures from assumptions, outliers, etc.
- The models used in ANOVA are equivalent to those used in multiple linear regression.
- By using dummy variables to appropriately code the levels of the factors in an ANOVA design, the model for the design can be put in the form of a multiple linear regression model.

- ANOVA software essentially transforms the required analysis into a regression format and then gives results in the form of an ANOVA table. Consequently, such software remains useful for undertaking the required analysis.

4.7 Exercises

4.1 The data were collected to investigate the determinants of pollution. For 41 cities in the United States, seven variables were recorded:

SO_2:	SO_2 content of air in micrograms per cubic meter
Temp:	Average annual temperature in degrees Fahrenheit
Manuf:	Number of manufacturing enterprises employing 20 or more workers
Pop:	Population size (according to 1970 census) in thousands
Wind:	Average annual wind speed in miles per hour
Precip:	Average annual precipitation in inches
Days:	Average number of days with precipitation per year

Construct a scatterplot matrix of the data and use it to guide the fitting of a multiple linear regression model with SO_2 as the response variable and the remaining variables as explanatory. Find the variance inflation factor (VIF) for each explanatory variable and use the factors to decide if there are any problems in using all six of the explanatory variables. Use the procedure involving the AIC described in the text to search for a more parsimonious model for the data. For the final model chosen, use some regression diagnostics to investigate the assumptions made in fitting the model.

4.2 The data arise from a study of the quality of statements elicited from young children reported by Hutcheson et al. (1995). The variables are statement quality, child's gender, age and maturity, how coherently the children gave their evidence, the delay between witnessing an incident and recounting it, the location of the interview (the child's home, school, a formal interviewing room, or an interview room specially constructed for children), and whether or not the case proceeded to prosecution. Carry out a complete regression analysis on these data to see how statement quality depends on the other variables, including selection of the best subset of explanatory variables and examining residuals and other regression diagnostics. Pay careful attention to how the categorical explanatory variables with more than two categories are coded.

4.3 Four sets of bivariate data from Anscombe (1973) are given. Fit a simple linear regression to each data set. What do you find? Now construct regression graphics and describe what you conclude from these plots.

4.4 The age, percentage fat, and gender of 20 normal adults are given. Investigate multiple linear regression models with the percentage of fat as the response variable, and age and gender as explanatory variables. Illustrate the models you fit with informative graphics.

4.5 The data arise from a survey of systolic blood pressure in individuals classified according to smoking status and family history of circulation and heart problems. Analyze the data using multiple linear regression and find the residuals from the fitted model, and use them to check the assumptions made by the model you have fitted.

4.6 The data were collected in a clinical trial of the use of estrogen patches in the treatment of postnatal depression. Using posttreatment depression score as the response, formulate a suitable model for examining the effects of the baseline measurements and treatment group on the response. Construct a suitable 95% confidence interval for the treatment effect and state your conclusions based on the analyses you have carried out.

5

Generalized Linear Models

5.1 Introduction

In the previous chapter we showed that analysis of variance (ANOVA) models and the multiple linear regression model are essentially completely equivalent. Both involve a linear combination of a set of explanatory variables (dummy variables coding factor levels in the case of ANOVA) as a model for the observed response variable. And both include residual terms assumed to have a normal distribution. Also analysis of covariance (ANCOVA) is the same model, with a mixture of continuous and categorical explanatory variables.

But situations often arise when a response variable of interest cannot be assumed to have the conditional normal distribution required by the models described in Chapter 4. For example, a response such as improved/not improved as a rating of a patient's condition is clearly not normally distributed—it is not even continuous. Such a variable may be coded, say, one for improvement and zero for not improved and is generally known as a *binary variable.* A further type of non-normal response is when it is a count and so takes only positive (or non-negative) integer values, for example, the number of correct scores in a testing situation. The question then arises as to how the effects of explanatory variables on such non-normal responses can be modelled? This question was largely answered in a landmark paper by Nelder and Wedderburn (1972) who introduced what is known as the *generalized linear model* (GLM), which enables a wide range of seemingly disparate problems of statistical modeling and inference to be set in an elegant unifying framework of great power and flexibility. A comprehensive account of GLMs is given in McCullagh and Nelder (1989) and a more concise description in Dobson and Barnett (2018). Here a brief description is given in Technical Section 5.1.

Technical Section 5.1: Generalized Linear Models

Essentially GLMs consist of three main features:

1. An *error distribution* giving the distribution of the response around the mean. For ANOVA and multiple linear regression this will be the normal distribution. For binary responses (see Section 5.2) it is the binomial distribution, and for responses

that are counts (see Section 5.3) it is the Poisson distribution. Each of these (and others used in other situations) come from the same *exponential family* of probability distributions, and it is this family that is used in generalized linear modeling (see Everitt and Pickles, 2004).

2. A *link function*, g, that shows how the linear function of the explanatory variables is related to the expected value of the response:

$$g(\mu) = \beta_0 + \beta_1 x_1 + \beta_2 x_2 + \cdots + \beta_q x_q$$

For ANOVA and multiple linear regression the link function is simply the identity function, that is, the explanatory variables are then directly related to the expected value of the response. A suitable link function for binary variables is described in the next section and one for count variables is introduced in Section 5.3.

3. The *variance function* $V(\mu)$ that captures how the variance of the response variable depends on the mean. We will return to this aspect of GLMs later in this chapter.

Estimation of the parameters in a GLM is usually achieved through a maximum likelihood approach—see McCullagh and Nelder (1989) for details. Having estimated a GLM for a data set the question of the quality of its fit arises. Clearly the investigator needs to be satisfied that the chosen model describes the data adequately, before drawing conclusions about the parameter estimates themselves. In practice, most interest will lie in comparing the fit of competing models, particularly in the context of selecting subsets of explanatory variables that describe the data in a parsimonious manner. In GLMs a measure of fit is provided by a quantity known as the *deviance* which measures how closely the model-based fitted values of the response approximate the observed value. Comparing the deviance values for two models gives a likelihood ratio test of the two models that can be compared by using a statistic having a chi-squared distribution with degrees of freedom equal to the difference in the number of parameters estimated under each model. More details are given in Cook (2005).

In the next two sections, we will go through two important special cases of the GLM. The first one, introduced in Section 5.2, is widely used for modeling binary response variables and is known as *logistic regression.* The second example of a GLM is applicable to count data and is known as *Poisson regression*; this is described in Section 5.3.

5.2 Binary Response Variables

The case of a binary response variable is essential in a huge number of applications in behavioral sciences, but also in medicine, social sciences, and many other areas.

In any regression problem, the key quantity is the population mean (or expected value) of the response variable given the values of the explanatory variables. As we have learned in Chapter 4, in multiple linear regression, the mean of the response variable is modeled directly as a linear function of the explanatory variables, that is, using the E operator to denote expected value

$$E(y|x_1, x_2, \ldots, x_q) = \beta_0 + \beta_1 x_1 + \beta_2 x_2 + \cdots + \beta_q x_q$$

where $E(y|x_1, x_2, \ldots, x_q)$ represents the expected value (population mean) of the response variable given the values of the explanatory variables. However, we now want to consider how to model appropriately a binary response variable with its two categories labeled 0 and 1. The first question we need to ask is: what is the expected value of such a variable? It is easy to show that the mean (expected value) in this case is simply the probability that the response variable takes the value 1. (We shall denote this probability by π.) So, to investigate the effects of a set of explanatory variables on a binary response, why not simply continue to use the multiple linear regression approach and consider a model of the form

$$\pi = \beta_0 + \beta_1 x_1 + \beta_2 x_2 + \cdots + \beta_q x_q$$

There are two problems with this model:

- The predicted value of the probability π must satisfy $0 \leq \pi < 1$, whereas a linear predictor can yield any values from minus infinity to plus infinity.
- The observed values of the response variable y, conditional on the values of the explanatory variables, will not now follow a normal distribution with mean π but rather what is known as a Bernoulli distribution.

Details of how these problems are overcome by a type of GLM known as logistic regression are described in Technical Section 5.2.

Technical Section 5.2: Logistic Regression

We have a binary response variable y, coded 0 or 1, and a set of explanatory variables $x_1, x_2, \ldots, x_q$. The mean of the response variable given the values of the explanatory variables is the probability that it takes the value 1; we represent this probability as π. Because of the problems identified earlier, π cannot be modeled directly as a linear function of the explanatory variables; instead, some suitable transformation of π must be modeled. In the GLM context, this transformation is called the *link function.* For logistic regression, the link function most often used is the *logit function* of the probability, which is simply the logarithm of the odds, that is, $\log(\frac{\pi}{1-\pi})$, and this leads to the logistic regression model having the form

$$\text{logit}\,(\pi) = \log\frac{\pi}{1-\pi} = \beta_0 + \beta_1 x_1 + \beta_2 x_2 + \cdots + \beta_q x_q$$

The logit link is chosen because, from a mathematical point of view, it is extremely flexible and, from a practical point of view, it leads to meaningful and convenient interpretation, as we will discuss in Chapter 6 when looking at examples of the application of logistic regression. The logit transformation of π can take any values ranging from minus infinity to plus infinity and thus overcome the first problem associated with modeling π directly.

In a logistic regression model, the parameter β_i associated with the explanatory variable x_i represents the expected change in the log odds when x_i is increased by one unit, conditional on the other explanatory variables remaining the same. Interpretation is simpler using $\exp(\beta_i)$, which corresponds to an *odds ratio*, as we will see later when discussing some examples in Chapter 6.

The preceding logistic regression model can be rearranged to give the following model for π:

$$\pi = \frac{\exp\,(\beta_0 + \beta_1 x_1 + \beta_2 x_2 + \cdots + \beta_q x_q)}{1 + \exp\,(\beta_0 + \beta_1 x_1 + \beta_2 x_2 + \cdots + \beta_q x_q)}$$

This equation can be used to predict the probability that the response variable takes the value 1 for any values of the explanatory variables, but it would, of course, only make sense for values in the observed range of the explanatory variables in the data set being modeled.

In linear regression, the observed value of the outcome variable is expressed as its expected value, given the explanatory variables plus an error term. The error terms are assumed to have a normal distribution with mean 0 and a variance that is constant across levels of the explanatory variables. With a binary response, we can express an observed value y in the same way as

$$y = \pi(x_1, x_2, \ldots, x_q) + \varepsilon$$

but here ε can only assume one of two possible values (note that here we have introduced a slight change in the nomenclature to remind us that the expected value of the response is dependent on the explanatory variables). If $y = 1$ then $\varepsilon = 1 - \pi(x_1, x_2, \ldots, x_q)$ with probability $\pi(x_1, x_2, \ldots, x_q)$, and if $y = 0$, then $\varepsilon = \pi(x_1, x_2, \ldots, x_q)$ with probability $1 - \pi(x_1, x_2, \ldots, x_q)$. Consequently, ε has a distribution with mean

0 and variance equal to $\pi(x_1, x_2, \ldots, x_q)(1 - \pi(x_1, x_2, \ldots, x_q))$. So, the distribution of the response variable conditional on the explanatory variables follows what is known as a Bernoulli distribution (which is simply a binomial distribution for a single trial) with probability $\pi(x_1, x_2, \ldots, x_q)$.

The importance of the logistic regression model, and its connection with the concept of the *odds ratio*, merits discussion of its application in a separate chapter—see Chapter 6.

5.3 Response Variables That Are Counts

Giardiello et al. (1993) and Piantadosi (1997) describe the results of a placebo-controlled trial of a non-steroidal anti-inflammatory drug in the treatment of familial andenomatous polyposis (FAP). The trial was halted after a planned interim analysis had suggested compelling evidence in favor of the treatment. The data shown in Table 5.1 give the number of colonic polyps after a 12-month treatment period. The question of interest is whether the number of polyps is related to treatment and/or age of patients. The data are plotted in Figure 5.1; this graph suggests quite strongly that people given the placebo tend to have more polyps.

In this example the response variable, namely the number of colonic polyps, is a count which can take only positive (or non-negative) integer values, and which is unlikely to have a normal distribution. For such a response variable a GLM with a log link function will ensure that fitted values of the response are positive (non-negative).

TABLE 5.1
The Colonic Polyps Data Giving the Number of Polyps for Two Treatments

Number	Treatment	Age	Number	Treatment	Age
63	placebo	20	3	drug	23
2	drug	16	28	placebo	22
28	placebo	18	10	placebo	30
17	drug	22	40	placebo	27
61	placebo	13	33	drug	23
1	drug	23	46	placebo	22
7	placebo	34	50	placebo	34
15	placebo	50	3	drug	23
44	placebo	19	1	drug	22
25	drug	17	4	drug	42

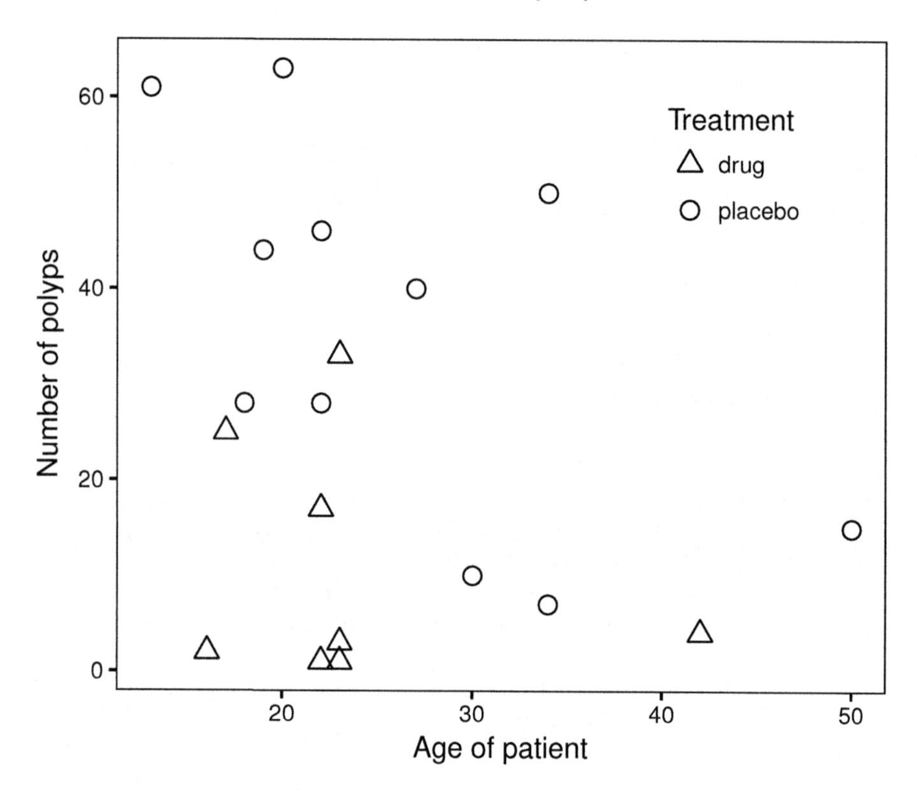

FIGURE 5.1
Scatterplot of number of polyps and age of the patient in the data shown in Table 5.1.

The usual error distribution assumed for a count variable is the Poisson distribution given by

$$P(y) = \frac{\lambda^y}{y!}e^{-\lambda}$$

where λ represents the expected value of the distribution. This type of GLM model is often known as *Poisson regression.* The results of fitting this GLM are shown in Table 5.2. The regression coefficients for both Age and Treatment are highly significant. However, there is a problem with the model, but before we can deal with it we need a short digression to describe in more detail the third component of GLMs mentioned in Technical Section 5.1, namely their variance functions, $V(\mu)$.

TABLE 5.2
Results from Fitting the Poisson Regression Model to the Polyps Data

	Estimate	Standard Error	z-Value	Pr(>|z|)
Intercept	4.5290	0.1469	30.84	<2e-16
Treatment	−1.3591	0.1176	−11.55	<2e-16
Age	−0.0388	0.0060	−6.52	7-02e-11

Note: Dispersion parameter taken to be 1.
Null deviance: 378.66 on 19 DF; Residual deviance: 179.54 on 17 DF.

5.3.1 Overdispersion and Quasi-Likelihood

The variance function of a GLM captures how the variance of a response variable depends upon its mean. The general form of the relationship is

$$\mathrm{Var}(\mathrm{response}) = \phi V(\mu)$$

where ϕ is constant and $V(\mu)$ specifies how the variance depends on the mean. For the error distributions considered previously this general form becomes:

- Normal: $V(\mu) = 1,\ \phi = \sigma^2$; here, the variance does not depend on the mean, and so can be freely estimated.
- Binomial: $V(\mu) = \mu(1 - \mu),\ \phi = 1$.
- Poisson: $V(\mu) = \mu,\ \phi = 1$.

In the case of a Poisson variable, we see that the mean and variance are equal, and in the case of a binomial variable, where the mean is the probability of, say, the event of interest π, the variance is $\pi(1 - \pi)$.

Both the Poisson and binomial distributions have variance functions that are completely determined by the mean. There is no free parameter for the variance since, in applications of GLM with binomial or Poisson error distributions the dispersion parameter, ϕ, is defined to be one (see previous results for Poisson regression in Table 5.2). But in some applications this becomes too restrictive to fully account for the empirical variance in the data; in such cases it is common to describe the phenomenon as *overdispersion*. For example, if the response variable is the proportion of family members who have been ill in the past year, observed in a large number of families, then the individual binary observations that make up the observed proportions are likely to be correlated rather than independent. The non-independence can lead to a variance that is greater (less) than on the assumption of binomial variability. And observed counts often exhibit larger variances than would be expected from the Poisson assumption, a fact noted almost 100 years ago by Greenwood and Yule (1920).

When fitting GLMs with binomial or Poisson error distributions, overdispersion can often be spotted by comparing the residual deviance with its degrees of freedom. For a well-fitting model the two quantities will be approximately equal. If the deviance is far greater than its degrees of freedom overdispersion may be indicated. This is indeed the case for the results in Table 5.2, as the residual deviance 179.5 is about tenfold its 17 degrees of freedom. So what can we do?

We can deal with overdispersion by using a procedure known as *quasi-likelihood*, which allows the estimation of model parameters without fully knowing the error distribution of the response variable. McCullagh and Nelder (1989) give full details of the quasi-likelihood approach. In many respects, it simply allows for the estimation of ϕ from the data rather than defining it to be unity for the binomial and Poisson distributions.

Applying the quasi-likelihood estimation to the colonic polyps data, we obtain the results shown in Table 5.3. The regression coefficients remain the same, but their estimated standard errors are now about 3.3 times greater than the values given in Table 5.2, as a result of being multiplied by the square root of the estimated dispersion or scale parameter ϕ that is now about 10.7 instead of one. A possible reason for overdispersion in these data is that polyps do not occur independently of one another, but instead may 'cluster' together.

TABLE 5.3
Results from Fitting the Poisson Regression Model to the Polyps Data Using Quasi-Likelihood Estimation

	Estimate	**Standard Error**	**t-Value**	**Pr(>|t|)**
Intercept	4.5290	0.4811	9.42	3.72e-08
Treatment	−1.3591	0.3853	−3.53	0.00259
Age	−0.0388	0.0195	−1.99	0.06284

Note: Dispersion parameter taken to be 10.728.
Null deviance: 378.66 on 19 DF; Residual deviance: 179.54 on 17 DF.

5.4 Summary

- The generalized linear model (GLM) allows a suitable transform of the mean of the response variable to be modeled as a linear function of the explanatory variables, and to have an error distribution appropriate for the type of response involved.

- Two important cases of the GLM are logistic regression for the binary response variables and Poisson regression for responses that are counts.

- Logistic regression is a GLM with a binomial error distribution, and typically used with the logit link function.

- Poisson regression is a GLM with a Poisson error distribution, and typically used with the log link function.
- Overdispersion is common with both the above models, as their variance functions are completely determined by the mean. The problem can usually be dealt with the quasi-likelihood procedure, which allows the dispersion or scale parameter ϕ to be estimated from the data.
- Applications of the logistic regression model are considered in Chapter 6.

5.5 Exercises

5.1 The data shown in Table 5.4 are from Seeber (2005). Here, 31 patients treated for superficial bladder cancer have recorded both the number of recurrent tumors during a particular time-period after removal of the primary and the size of the primary tumor (whether smaller or larger than 3 cm). Fit a suitable GLM model to assess the effect of the size of the primary tumor on the number of recurrent tumors.

TABLE 5.4
Bladder Cancer Data (Taken with Permission of the Publishers, John Wiley & Sons Ltd, from Seeber, 2005)

Time	X	n	Time	X	n
11	1	1	7	0	2
2	0	1	13	0	2
3	0	1	15	0	2
6	0	1	18	0	2
8	0	1	23	0	2
9	0	1	20	0	3
10	0	1	24	0	4
11	0	1	1	1	1
13	0	1	5	1	1
14	0	1	17	1	1
16	0	1	18	1	1
21	0	1	25	1	1
22	0	1	18	1	2
24	0	1	25	1	2
26	0	1	4	1	3
27	0	1	19	1	4

Note: X: 0 = size of tumor < 3 cm

TABLE 5.5
Data on Incidence of CHD and Associated Risk Factors

Person-years	Smoking	Blood pressure	Behavior	n of CHD cases
5268.2	0	0	0	20
2542.0	10	0	0	16
1140.7	20	0	0	13
614.6	30	0	0	3
4451.1	0	0	1	41
2243.5	10	0	1	24
1153.6	20	0	1	27
925.0	30	0	1	17
1366.8	0	1	0	8
497.0	10	1	0	9
238.1	20	1	0	3
146.3	30	1	0	7
1251.9	0	1	1	29
640.0	10	1	1	21
374.5	20	1	1	7
338.2	30	1	1	12

Note: Smoking: 0 = non-smoker, 10 = 1–10, 20 = 11–20, 30 = 20+ cigarettes a day
Blood pressure: 0 = < 140, 1 = ≥ 140
Behavior: 0 = Type B, 1 = Type A personality

* 5.2 The data shown in Table 5.5 arise from a prospective study of potential risk factors for coronary heart disease (CHD) (Rosenman et al., 1975). The study looked at 3154 men aged 40–50 for an average of 8 years and recorded the incidence of cases of CHD. The potential risk factors included smoking, blood pressure, and personality/behavior type. The type A is characterized by impatience, competitiveness, aggressiveness, a sense of time urgency and tenseness, while type B is characterized by easygoing, relaxed about time, not competitive and not easily angered or agitated.

1. Fit an appropriate GLM model using only the smoking risk factor.
2. Now extend the model to include all risk factors. What are your conclusions?

6

Applying Logistic Regression

6.1 Introduction

In the previous chapter, the basic theory behind a regression model for dealing with a binary response variable, logistic regression, was covered. In this chapter, a number of examples of the application of this type of regression will be described. But before this we need to digress for a short account of *odds* and *odds ratios* both of which are central to interpreting the results of fitting logistic regression models.

6.2 Odds and Odds Ratios

Table 6.1 shows part of a data set collected in a study of a psychiatric screening questionnaire known as the GHQ (General Health Questionnaire; see Goldberg, 1972 for details). Each person in the study was given a score on the GHQ and also categorized as being a psychiatric case or not. Here, the question of interest to the researcher is how being judged to be a "case" is related to gender and GHQ score. In Table 6.1, the binary responses (case/not case) of individuals with the same values of the two explanatory variables, GHQ score and gender, have been grouped together.

If we collapse these data over the GHQ score, we get the following 2×2 contingency table of caseness against gender:

	Case	**Noncase**
Male	25	79
Female	43	131

Such a table would usually be analyzed to assess the independence or otherwise of gender and caseness using a chi-squared test. However, here we will use the table to explain the meanings of the terms odds and odds ratios.

First odds, which is defined for an event with probability p as $p/(1-p)$. For women in the 2×2 table in the preceding text, the probability of being judged

TABLE 6.1
Psychiatric Caseness Data

GHQ Score	Gender	Number of Cases	Number of Noncases
0	F	4	80
1	F	4	29
2	F	8	15
...			
...			
10	F	1	0
0	M	1	36
1	M	2	25
2	M	2	8
...			
...			
10	M	2	0

a case is estimated to be $43/174 = 0.247$, and so, for women, the odds of being judged a case versus being judged a noncase is $0.247/(1-0.247) = 0.328$. The same calculations for men show that the probability of being judged a case is 0.240, and the corresponding value for the odds is 0.316. It is easy to see that the odds for women can be calculated directly from the frequencies in the 2×2 table as $43/131$; similarly, for men, the odds are $25/79$.

Further, having found the odds for caseness versus noncaseness for women and for men, the odds ratio is simply what it says—the ratio of the two odds, that is, $0.316/0.328 = 0.963$. When the two variables forming the contingency table are independent, the odds ratio in the population will be 1. So, is it possible to use the estimated odds ratio to test the hypothesis that the population value is 1 and, more importantly, is it possible to construct a confidence interval (CI) for the odds ratio? Technical Section 6.1 shows how to do the latter.

Technical Section 6.1: CI for the Odds Ratio

Consider the general 2×2 table given by

	Variable 1	
Variable 2	**Category 1**	**Category 2**
Category 1	a	b
Category 2	c	d

The odds ratio in the population will be denoted by ψ, and it can be estimated from the observed frequencies in the table as

$$\hat{\psi} = \frac{a/b}{c/d} = \frac{ad}{bc}$$

A CI for ψ can be constructed relatively simply by using the following estimator of the variance of $\log(\psi)$:

$$\widehat{\text{var}}(\log \psi) = 1/a + 1/b + 1/c + 1/d$$

So, an approximate 95% confidence interval for $\log(\psi)$ is given by

$$\log(\hat{\psi}) \pm 1.96 \times \sqrt{\widehat{\text{var}}(\log \hat{\psi})}$$

If the limits of the CI for $\log(\psi)$ obtained in this way are ψ_L, ψ_U, then the corresponding confidence interval for ψ is $[\exp(\psi_L), \exp(\psi_U)]$.

We can illustrate the construction of the CI for the odds ratio using the data from the 2×2 table of gender and caseness given earlier in the chapter. First, the odds ratio is estimated to be

$$\hat{\psi} = \frac{25 \times 131}{43 \times 79} = 0.964$$

and so $\log(\hat{\psi}) = -0.037$ and the estimated variance of $\log(\psi)$ is

$$1/25 + 1/43 + 1/79 + 1/131 = 0.084,$$

leading to a 95% CI for $\log(\psi)$ of

$$[-0.037 - 1.96 \times 0.290, -0.037 + 1.96 \times 0.290], \quad \text{that is,} \quad [-0.604, 0.531]$$

Finally, the CI for ψ is found as $[\exp(-0.604), \exp(0.531)]$, that is, [0.546, 1.701]. As this interval contains the value 1, we can conclude that there is no evidence of an association between caseness and gender.

6.3 Applying Logistic Regression to the GHQ Data

To begin, we will fit both a logistic regression and a linear regression to the data using the GHQ score as the single explanatory variable. So, for the linear model, the probability of being a case is modeled as a linear function of the GHQ score, whereas in the logistic model, the logit transformation of the probability of being a case is modeled as a linear function of the GHQ score. The results from fitting both models are shown in Table 6.2.

TABLE 6.2
Results of Fitting a Linear and a Logistic Regression Model to the GHQ Data with Only a Single Explanatory Variable, the GHQ Score

	Logistic Model[a]			
	Estimate	**Standard Error**	**z-Value**	**Pr(>|z|)**
Intercept	−2.71073	0.27245	−9.950	<2e-16
GHQ	0.73604	0.09457	7.783	7.1e-15
	Linear Model[b]			
	Estimate	**Standard Error**	**z-Value**	**Pr(>|z|)**
Intercept	0.11434	0.05923	1.931	0.0678
GHQ	0.10024	0.01001	10.012	3.1e-09

[a] Null deviance: 130.306 on 21 DF; residual deviance: 16.237 on 20 DF; AIC: 56.211; number of Fisher scoring iterations: 5.

[b] Residual standard error: 0.1485 on 20 DF; multiple R-squared: 0.8337; F-statistic: 100.2 on 1 and 20 DF; *p*-value: 3.099e-09

The results from both models show that the GHQ score is a highly significant predictor of the probability of being judged a case. In the linear model, the estimated regression coefficient is 0.10; the estimated increase in the probability of being a case is 0.10 for each increase of 1 in the GHQ score. For an individual with a GHQ score of, say, 10, the linear model would predict that the probability of the individual being judged a case is $0.114 + 0.100 \times 10 = 1.114$; with this model, fitted values of the probabilities are not constrained to lie in the interval (0,1). Now consider the fitted logistic regression model, that is,

$$\log \frac{\Pr(\text{case})}{\Pr(\text{not case})} = -2.71 + 0.74 \times \text{GHQ score}$$

This equation can be rearranged to give the predicted probabilities for the fitted logistic regression model as

$$\Pr(\text{case}) = \frac{\exp(-2.71 + 0.84 \times \text{GHQ score})}{1 + \exp(-2.71 + 0.84 \times \text{GHQ score})}$$

For the individual with a GHQ score of 10, this model predicts the probability of the individual being judged a case as 0.99.

Now, let us consider the fitted logistic regression model for individuals with GHQ scores of, say, S and S + 1; the corresponding models for the logits can be written as

$$\log \frac{\Pr(\text{case})}{[\Pr(\text{not case}) \mid \text{GHQ} = \text{S}]} = -2.71 + 0.74 \times \text{S}$$

$$\log \frac{\Pr(\text{case})}{[\Pr(\text{not case}) \mid \text{GHQ} = \text{S} + 1]} = -2.71 + 0.74 \times (\text{S} + 1)$$

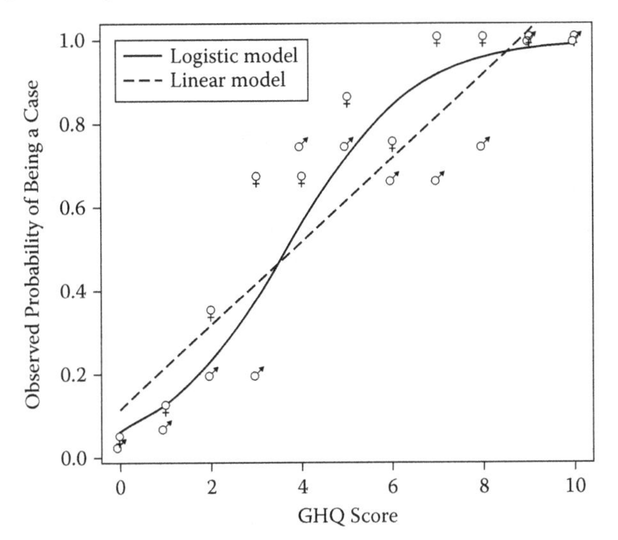

FIGURE 6.1
Plot of predicted probabilities of caseness from both linear and logistic regression models with GHQ score as the single explanatory variable and observed probabilities labeled by gender.

So, subtracting the first equation from the second, we get

$$\log \frac{\Pr(\text{case})}{[\Pr(\text{not case}) \mid \text{GHQ} = \text{S} + 1]} - \log \frac{\Pr(\text{case})}{[\Pr(\text{not case}) \mid \text{GHQ} = \text{S}]} = 0.74$$

This can be rewritten as

$$\log \left\{ \frac{\Pr(\text{case})}{[\Pr(\text{not case}) \mid \text{GHQ} = \text{S} + 1]} \Big/ \frac{\Pr(\text{case})}{[\Pr(\text{not case}) \mid \text{GHQ} = \text{S}]} \right\} = 0.74$$

and thus, by exponentiating, we get

$$\frac{\Pr(\text{case})}{[\Pr(\text{not case}) \mid \text{GHQ} = \text{S} + 1]} \Big/ \frac{\Pr(\text{case})}{[\Pr(\text{not case}) \mid \text{GHQ} = \text{S}]} = \exp(0.74)$$

The left-hand side is simply the odds ratio for being rated a case for a subject with a GHQ score 1 higher than another subject, and this is estimated to be $\exp(0.74) = 2.10$. The corresponding approximate 95% CI is calculated as $[\exp(0.74 - 1.96 \times 0.09), \exp(0.74 + 1.96 \times 0.09)]$ to give [1.76, 2.50]. The increase in the odds of being judged a case associated with a 1 unit increase in the GHQ is estimated to be between 76% and 150%.

The null deviance given in Table 6.2 is for a model with no explanatory variables, and the residual deviance is that for a model with GHQ as the single explanatory variable; the reduction in deviance is considerable, and could be tested as a chi-squared with 1 DF. The value of Akaike's fit criterion AIC (see Chapter 4) is also given, and this might be useful when comparing competing models as we shall see later.

In Figure 6.1, we plot the predicted probabilities of being a case from each model against GHQ, along with the observed probabilities labeled by gender. (The observed probabilities are found from the data by dividing number of cases by number of cases plus number of noncases.) The plot again demonstrates that the linear regression model leads to predicted probabilities greater than 1 for some values of GHQ.

Next, we will fit a logistic regression model that includes only gender as an explanatory variable. The results are shown in Table 6.3. The estimated regression coefficient of −0.037 is the odds ratio calculated earlier in the chapter for the gender × case/not case cross-classification, and the estimated standard error of the estimated regression coefficient is also the same as we calculated earlier.

The next logistic model we will consider for the GHQ data is one where the gender and GHQ scores are both included as explanatory variables. The results are shown in Table 6.4. In this model, both GHQ and gender are significant. The regression coefficient for gender shows that, conditional on GHQ score, the log odds of caseness for men is −0.94 lower than for women, which gives an estimated odds ratio of $\exp(-0.94) = 0.39$ with 95% CI of [0.167, 0.918]. For a given GHQ score, the odds of a man being diagnosed as a case is between about 0.17 and 0.92 of the corresponding odds for a woman, although we know from previous analyses that the overall odds ratio, ignoring the GHQ score, does not differ significantly from 1. We might ask: why the difference? The reason is that the overall odds ratio is dominated by the large number of cases for the lower GHQ scores.

The estimated regression coefficient for GHQ conditional on gender is 0.779. This is very similar to the value for the model having only the GHQ score, and so the interpretation of the conditional coefficient is very similar to that given previously.

TABLE 6.3
Results from Fitting a Logistic Regression Model to the GHQ Data with Only a Single Explanatory Variable, Gender

	Estimate	Standard Error	z-Value	Pr(>|z|)
Intercept	−1.11400	0.17575	−6.338	2.32e-10
Gender	−0.03657	0.28905	−0.127	0.9

Note: Null deviance: 130.31 on 21 DF; residual deviance: 130.29 on 20 DF; AIC: 170.26

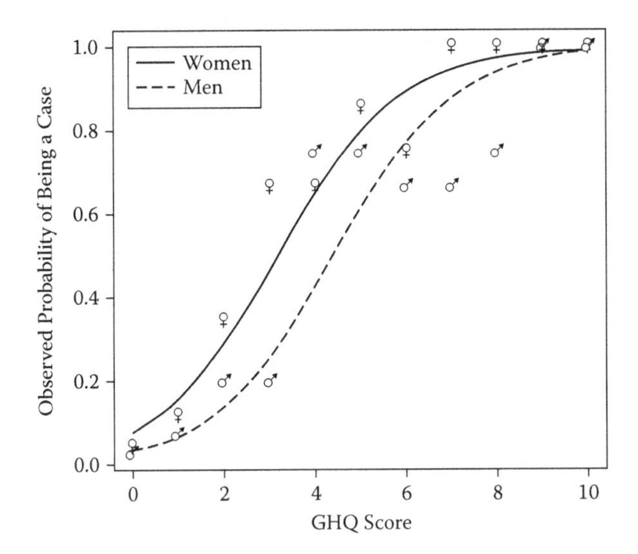

FIGURE 6.2
Plot of predicted probabilities of caseness from logistic regression model with gender and GHQ score as explanatory variables and observed probabilities labeled by gender.

The AIC value of 53.1 is lower than that for the models with only the GHQ score or gender, so the model with both is considered a better model. The fitted model is displayed graphically in Figure 6.2.

Finally, for the GHQ data we will fit a model with gender, GHQ score, and gender × GHQ score interaction. The result is shown in Table 6.5. The model is illustrated graphically in Figure 6.3. Although the AIC value is a little lower than for the previously fitted model with no interaction term, the regression coefficient for the interaction term in Table 6.5 is not significant, and a reasonable conclusion is that the model with only gender and GHQ score provides the best description of these data.

TABLE 6.4
Results from Fitting a Logistic Regression Model to the GHQ Data with Gender and GHQ Scores as Explanatory Variables

	Estimate	Standard Error	z-Value	Pr(>|z|)
Intercept	−2.49351	0.28164	−8.854	<2e-16
Gender	−0.93609	0.43435	−2.155	0.0311
GHQ	0.77910	0.09903	7.867	3.63e-15

Note: Null deviance: 130.306 on 21 DF; residual deviance: 11.113 on 19 DF; AIC: 53.087.

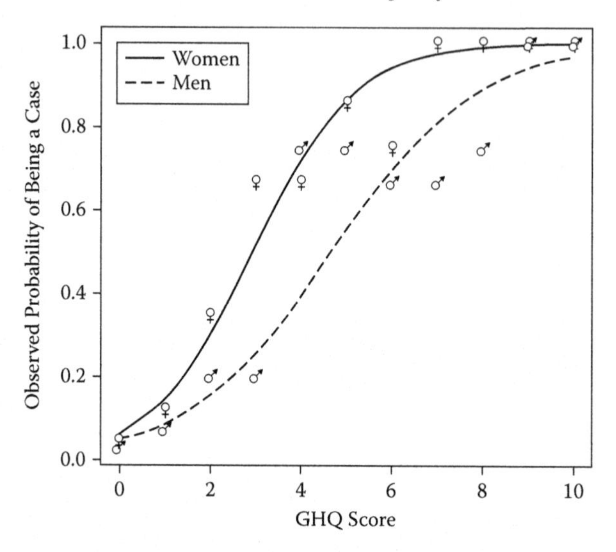

FIGURE 6.3
Plot of predicted probabilities of caseness from logistic regression model with gender, GHQ score, gender × GHQ as explanatory variables, and observed probabilities labeled by gender.

TABLE 6.5
Results from Fitting a Logistic Regression Model to the GHQ Data with Gender, GHQ Score, and the Interaction of Gender and GHQ Score as Explanatory Variables

	Estimate	Standard Error	z-Value	Pr(>|z|)
Intercept	−2.7732	0.3586	−7.732	1.06e-14
Gender	−0.2253	0.6093	−0.370	0.712
GHQ	0.9412	0.1569	6.000	1.97e-09
Gender×GHQ	−0.3020	0.1990	−1.517	0.129

Note: Null deviance: 130.306 on 21 DF; residual deviance: 8.767 on 18 DF; AIC: 52.741.

6.4 Selecting the Most Parsimonious Logistic Regression Model

As with the fitting of multiple linear regression models, the aim when fitting logistic regression models is to select the most parsimonious model that gives an adequate description of the data. We can use exactly the same approach as given in Chapter 4, as we shall illustrate on data collected from a survey of employed men aged between 18 and 67 years who were asked whether in the preceding year they had carried out work on their home for which they would

TABLE 6.6
Do-It-Yourself Data

			Accommodation					
			Apartment Age			House Age		
Work	**Tenure**	**Response**	**<30**	**31–45**	**46+**	**<30**	**31–45**	**46+**
Skilled	Rent	Yes	18	15	6	34	10	2
		No	15	13	9	28	4	6
	Own	Yes	5	3	1	56	56	35
		No	1	1	1	12	21	8

have employed a craftsman previously. The response variable is their answer, yes/no, to this question. In addition to age, the respondents' accommodation type, apartment or house; whether this accommodation was rented or owned; and their type of work, skilled, unskilled, or office were recorded. Part of the data is shown in Table 6.6.

The first point to consider about these data is how to deal with the categorical variables, work and age, that have more than two categories. We could simply label the categories for work one, two, three, and likewise the age categories, and use these numerical values in the model-fitting process. But this would be a mistake, particularly for the work variable; such coding would imply that changing, say, from work category one to work category two, and from work category two to work category three, has an equal effect on the probability of responding yes in the survey, which is not necessarily the case. The appropriate method is to use dummy variable coding for both work and age. So, for the work variables, we define two dummy variables D1 and D2 to label the three categories such that

Work	**D1**	**D2**
Skilled	0	0
Unskilled	1	0
Office	0	1

So, in the fitted logistic regression model, the estimated regression coefficient for D1 will quantify the difference between unskilled and skilled workers, and D2 will quantify the difference between office and skilled workers. A similar coding will be used for the age variable. We will now fit a logistic model for the probability of responding yes in the survey with work, tenure, accommodation type, and age as explanatory variables. The results are shown in Table 6.7. We will wait to interpret the estimated regression coefficients until we have explored whether a simpler model might be adequate for these data. This we will do by using a backward search procedure using the AIC

TABLE 6.7
Results from Fitting a Logistic Regression Model to the Do-It-Yourself Data with Explanatory Variables Work, Tenure, Type, and Age

	Estimate	Standard Error	z-Value	Pr(>|z|)
Intercept	0.30606	0.15428	1.984	0.0473
Work(D1)	−0.76267	0.15197	−5.018	5.21e-07
Work(D2)	−0.30535	0.14088	−2.167	0.0302
Type	−0.00249	0.14717	−0.017	0.9865
Tenure	1.01570	0.13787	7.367	1.74e-13
Age(D1)	−0.11304	0.13697	−0.825	0.4092
Age(D2)	−0.43661	0.14059	−3.106	0.0019

Note: Null deviance: 158.884 on 35 DF; residual deviance: 29.671 on 29 DF; AIC: 167.87.

criterion to guide the search, that is, the same method was used in Chapter 4 for multiple linear regression models.

The results of the backward search are as follows:

Start: AIC = 167.87

Explanatory variables in the model are Work, Tenure, Type, and Age.

Step 1: Removing one explanatory variable at a time and leaving the other three in the model

Remove Type: AIC = 165.87

Remove Age: AIC = 174.76

Remove Work: AIC = 191.72

Remove Tenure: AIC = 221.80

The variable Type can be removed because the model without Type but with the other three explanatory variables has a lower AIC value than the four-variable model.

Current: AIC = 165.87

Explanatory variables in the model are now Work, Tenure, and Age.

Step 2: Removing one explanatory variable at a time and leaving the other two in the model

Remove Age: AIC = 172.81

Remove Work: AIC = 189.19

Remove Tenure: AIC = 244.98

TABLE 6.8
Results from Fitting the Model with Work, Age, and Tenure as Explanatory Variables

	Estimate	Standard Error	z-Value	Pr(>|z|)
Intercept	0.3048	0.1347	2.262	0.02370
Work(D1)	−0.7627	0.1520	−5.019	5.21e-07
Work(D2)	−0.3053	0.1408	−2.168	0.03012
Tenure	1.0144	0.1144	8.866	<2e-16
Age(D1)	−0.1129	0.1367	−0.826	0.40877
Age(D2)	−0.4364	0.1401	−3.116	0.00183

Note: Null deviance: 158.884 on 35 DF; residual deviance: 29.671 on 30 DF; AIC: 165.87.

TABLE 6.9
Estimated Odds Ratios and CIs of Each Explanatory Variable in the Final Model Chosen for the Do-It-Yourself Data

Variable	Estimated Odds Ratio	95% CI
Work(D1)	0.466	[0.346, 0.628]
Work(D2)	0.737	[0.559, 0.972]
Tenure	2.757	[2.205, 3.447]
Age(D1)	0.893	[0.683, 1.168]
Age(D2)	0.647	[0.491, 0.851]

Removing any of the three explanatory variables currently in the model leads to an AIC value greater than that for the current model and so we accept the current model that has Work, Age, and Tenure as explanatory variables. Fitting this model gives the results shown in Table 6.8. In Table 6.9 we list the estimated odds ratios and 95% CIs for each variable. The results show the following:

- The odds of unskilled workers responding yes to the questions asked are between about 35% and 63% of the odds of skilled workers.
- The odds of office workers responding yes to the question asked are between about 56% and 97% of the odds of skilled workers.
- The odds of home owners responding yes to the question asked are between about twice to three-and-a-half times the odds of non-home owners.
- The odds of respondents in the age range 31–45 responding yes to the question asked does not differ from the odds of those respondents less than 30.

- The odds of respondents aged over 46 responding yes to the question asked are between about 50% and 85% of the odds of respondents less than 30.

6.5 Driving and Back Pain: A Matched Case–Control Study

A frequently used design in psychology and other areas of the behavioral sciences is the *matched case–control study* in which each patient suffering from a particular condition of interest included in the study is matched to one or more people without the condition. The most commonly used matching variables are age, ethnic group, mental status, etc. A design with m controls per case is known as $1 : m$ (one case, m controls) matched study. In many cases, m will be one, and it is the $1 : 1$ (one case, one control) matched study that we shall concentrate on here where we analyze the data on low back pain.

The data in Table 6.10 arise from a study reported in Kelsey and Hardy (1975) which was designed to investigate whether driving a car is a risk factor for low back pain resulting from acute herniated lumbar intervertebral discs (AHLID). A case–control study was used with cases selected from people who had recently had X-rays taken of the lower back and had been diagnosed as having AHLID. The controls were taken from patients admitted to the same hospital as a case with a condition unrelated to the spine. Further matching was made on age and gender and a total of 217 matched pairs were recruited, consisting of 89 female pairs and 128 male pairs. As a further potential risk

TABLE 6.10
Low Back Pain Data: Number of Non-drivers and Drivers, City and Suburban (Sub) Inhabitants either Suffering from a Herniated Disc (Cases) or Not (Controls)

			Controls				
			Non-drivers		Drivers		
			City	Sub	City	Sub	Total
Cases	Non-drivers	City	9	0	10	7	26
		Sub	2	2	1	1	6
	Drivers	City	14	1	20	29	64
		Sub	22	4	32	63	121
	Total		47	7	63	100	217

factor, the variable Suburban indicates whether each member of the pair lives in the suburbs or in the city.

To begin, we shall describe the form of the logistic regression model appropriate for case–control studies in the simplest case where there is only one binary explanatory variable. With matched pairs data the form of the logistic model involves the probability, φ, that in matched pair number i, for a given values of the explanatory variable the member of the pair is a case. Specifically, the model is

$$\text{logit}(\varphi_i) = \alpha_i + \beta x$$

The odds that a subject with $x = 1$ is a case equals $\exp(\beta)$ times the odds that a subject with $x = 0$ is a case.

The model generalizes to the situation where there are q explanatory variables as

$$\text{logit}(\varphi_i) = \alpha_i + \beta_1 x_1 + \beta_2 x_2 + \ldots + \beta_q x_q$$

Typically, one x is an explanatory variable of real interest, such as past exposure to a risk factor, with the others being used as a form of statistical control in addition to the variables already controlled by virtue of using them to form matched pairs. This is the case in our back pain example where it is the effect of car driving on lower back pain that is of most interest.

The problem with the model above is that the number of parameters increases at the same rate as the sample size with the consequence that maximum likelihood estimation is no longer viable. We can overcome this problem if we regard the parameters α_i as of little interest and so are willing to forgo their estimation. If we do, we can then create a *conditional likelihood function* that will yield maximum likelihood estimators of the coefficients $\beta_1, \ldots, \beta_q$, that are consistent and asymptotically normally distributed. The details behind this are described in Collett (2003).

The results of fitting the model are shown in Table 6.11. The estimate of the odds ratio of a herniated disc occuring in a driver relative to a nondriver is $\exp(0.658) = 1.93$ with a 95% confidence interval of [1.09, 3.44]. Conditional on the residence we can say that the risk of a herniated disc occurring in a driver is about twice that of a nondriver. There is no evidence that where a person lives affects the risk of lower back pain.

TABLE 6.11
Results from Fitting a Conditional Logistic Regression Model to the Low Back Pain Data with Driver and Suburban as Explanatory Variables

	Estimate	**Standard Error**	**z-Value**	**Pr(>|z|)**
Driver	0.658	0.294	2.24	0.025
Suburban	0.255	0.226	1.13	0.258

Note: Likelihood ratio test: 9.55 on 2 DF; p=0.008.

6.6 Summary

- The logistic regression model is a GLM that can be used to assess the effects of a set of explanatory variables on a binary response variable (see details in Chapter 5).
- The estimated parameters in the logistic regression model can be interpreted in terms of odds and odds ratios.
- Parsimonious models can be selected by the same approach as for the multiple linear regression model.
- There are a number of diagnostics available for logistic regression that can be used to assess various aspects of the model. Details are available in Collett (2003); however, the binary nature of the response variable often makes the use of these diagnostics somewhat difficult to interpret.
- Logistic regression is also very useful in the analysis of matched case–control studies.

6.7 Exercises

6.1 The data arise from a survey carried out in 1974/1975 in which each respondent was asked if he or she agreed or disagreed with the statement "Women should take care of running their homes and leave running the country up to men." The years of education of each respondent was also recorded. Use logistic regression to assess the effects of gender and years of education on the probability of agreeing with the statement, and construct suitable graphics for illustrating the models you fit. What are your conclusions?

6.2 Return to the data on do-it-yourself used in the text and use a backward search approach to assess models that allow interactions between each pair of explanatory variables. What conclusion do you reach, and how do these differ from those given in the text?

6.3 The data were obtained from a study of the relationship between car size and car accident injuries. Accidents were classified according to their type, severity, and whether or not the driver was ejected. Using severity as the response variable, derive and interpret a suitable logistic model for these accidents.

6.4 The data (taken from Johnson and Albert, 2013) are for 30 students in a statistics class. The response variable y indicates whether or not the student passed ($y = 1$) or failed ($y = 0$) the statistics examination at the end of the course. Also given are the student's scores on a previous mathematics test and their grade for a prerequisite probability course. Fit a logistic model to the data with mathematics test and probability course grade as explanatory variables. Cross-tabulate the predicted passes and failures with the observed passes and failures.

6.5 The data relate to a sample of girls in Warsaw, the response variable indicating whether or not the girl has begun menstruation and the exploratory variable age in years (measured to the month). Plot the estimated probability of menstruation as a function of age, and show the linear and logistic regression fits to the data on the plot.

* 6.6 Donati et al. (2013) studied the gambling behavior of high school students in a suburban area in Tuscany, Italy. The data set includes the gender of the 994 students and their answers to various gambling-related questions. The response variable indicates the gambling problem severity of the respondent (classified as a non-problem gambler or at-risk/problem gambler). The explanatory variables measure probabilistic reasoning, knowledge of basic mathematical principles, sensation seeking, and superstitious thinking of the participants, their perception of the economic profitability of gambling and their parental and peer gambling behavior. Fit a logistic model to the data starting from a model where gender is the only explanatory variable, and proceed by adding more explanatory variables one at a time. What would be your final model?

7

Survival Analysis

7.1 Introduction

In many studies, particularly in medicine, the main outcome variable is the time from a well-defined time origin to the occurrence of a particular event or end point. In medical studies, the end point is frequently the death of a patient, and the resulting data are, quite literally, survival times. Behavioral research studies are, fortunately, rarely life threatening, but end points other than death may be of importance in such studies—for example, the time to relief of symptoms, to the recurrence of a specific condition, or simply, to the completion of an experimental task. Such observations should perhaps properly be referred to as *time-to-event data*, although the generic term *survival data* is commonly used even when literal survival is not the issue.

Where time-to-event data are collected, the interest generally lies in assessing the effects of some explanatory variables on survival times, and it might be thought that multiple linear regression would fit the bill; but it would not because survival data require special techniques for their analysis for two main reasons:

- The distribution of survival times is very often positively skewed, and so, assuming normality for an analysis (as is done in multiple linear regression, for example) is almost always not reasonable.
- More critical than the probable normality problem, however, is the likely presence of *censored observations*, which occur when, at the completion of the study or experiment, the end point of interest has not been reached. When censoring occurs, all that is known about the individual's survival time is that it is larger than the time elapsed up to censoring.

An example of survival data from a behavioral study is provided by an investigation of the age at which women experience their first sexual intercourse (Schmidt et al., 1995). Data were collected from 84 women in two diagnostic groups: restricted anorexia nervosa (RAN) and normal controls (NC). Part of the data is shown in Table 7.1. Some women at the time the study took place had not had intercourse; consequently, these observations are censored (to be precise, the observations are *right* censored).

TABLE 7.1
Age at First Sexual Intercourse for Women in Two Diagnostic Groups

	Diagnosis	Age at First Sex	Age	Status
1	RAN	—*	30	0
2	RAN	—	24	0
3	RAN	12	18	1
4	RAN	21	42	1
5	RAN	—	19	0
...	...	...	...	...
71	NC	19	22	1
72	NC	19	22	1
73	NC	18	21	1
74	NC	17	19	1
75	NC	19	21	1
...	...	...	...	...

Note: *Status 0 = censored; first sex has not yet taken place at time of interview; 1 = age at first sex, occurred at age given.

When the response variable of interest is a survival time, the censoring problem requires special regression techniques for modeling the relationship of the survival time to the explanatory variables of interest. A number of procedures are available, but the most widely used by some margin is the one known as *Cox's proportional hazards model*, or *Cox regression* for short. Introduced by Sir David Cox in 1972 (Cox, 1972), the method has become one of the most commonly used in medical statistics, and the original paper one of the most heavily cited. But, before discussing Cox regression, we must give an account of two approaches most often used to characterize and describe survival times, namely, the *survival function* and the *hazard function*.

7.2 The Survival Function

As with all data, an initial step in the analysis of a set of survival data is to construct an informative graphic for the data. However, the presence of censored observations makes the use of conventional descriptive graphics such as boxplots rather problematic, and survival data containing censored observations are best displayed graphically using an estimate of what is known as the data's survival function. This function and how it is estimated from sample data is described in Technical Section 7.1.

Technical Section 7.1: The Survival Function

The survival function $S(t)$ is defined as the probability that the survival time T is greater than or equal to t, that is,

$$S(t) = \Pr(T \geq t)$$

A plot of an estimate of $S(t)$ against t is often a useful way of describing the survival experience of a group of individuals. When there are no censored observations in the sample of survival times, a nonparametric survival function can be estimated simply as

$$\hat{S}(t) = \frac{\text{Number of individuals with survival times } \geq t}{\text{Number of individuals in the data set}}$$

As every subject is "alive" at the beginning of the study and no one is observed to "survive" longer than the largest of the observed survival times t_{max}, then

$$\hat{S}(0) = 1 \quad \text{and} \quad \hat{S}(t) = 0 \quad \text{for} \quad t > t_{\text{max}}$$

The estimated survival function is assumed to be constant between two adjacent times of death so that a plot of $\hat{S}(t)$ against t is a step function that decreases immediately after each "death." Because the estimator above is simply a proportion, confidence intervals (CIs) can be obtained for each time t by using the usual estimate of the variance of a proportion to give

$$\text{var}[\hat{S}(t)] = \frac{\hat{S}(t)[1 - \hat{S}(t)]}{n}$$

However, the simple proportion estimator cannot be used to estimate the survival function when the data contains censored observations. In the presence of censoring, the survival function is generally estimated using the Kaplan–Meier estimator (Kaplan and Meier, 1958) which is based on the calculation and use of conditional probabilities. Assume again we have a sample on n observations from the population for which we wish to estimate the survival function. Some of the survival times in the sample are *right censored*—for those individuals we know only that their true survival times exceed the censoring time. We denote by $t_1 < t_2 < \ldots$ the times when deaths are observed and let d_j be the number of individuals who "die" at time t_j. The Kaplan–Meier estimator for the survival function then takes the form

$$\hat{S}(t) = \prod_{j:t_j \leq t} \left(1 - \frac{d_j}{r_j}\right)$$

where r_j is the number of individuals at risk, that is, alive and not censored, just prior to time t_j. If there are no censored observations, the estimator reduces to the simple proportion given earlier. The essence of the Kaplan–Meier estimator is the use of the continued product of a series of conditional probabilities. For example, to find the probability of surviving, say, 2 years, we use Pr(surviving 2 years) = Pr(surviving 1 year)×Pr(surviving 2 years | having survived 1 year), and the probability of surviving 3 years is found from Pr(3) = Pr(3|2) × Pr(2) = Pr(3|2) × Pr(2|1) × Pr(1). In this way, censored observations can be accommodated correctly.

The variance of the Kaplan–Meier estimator is given by

$$\text{var}[\hat{S}(t)] = [\hat{S}(t)]^2 \sum_{j:t_j \leq t} \frac{d_j}{r_j\,(r_j - d_j)}$$

When there is no censoring, this reduces to the standard variance of a proportion used earlier for the simple proportion estimator of the survival function.

7.2.1 Age at First Sexual Intercourse for Women

Let us begin by constructing and plotting the Kaplan–Meier estimated survival function of all the ages at first sexual intercourse (see Table 7.1), ignoring for the moment the two diagnostic groups. Table 7.2 shows the numerical results, and Figure 7.1 is a plot of the estimated survival function against age at first sex. Also shown in Figure 7.1 are the upper and lower 95% CI bounds for the survival function. Table 7.2 shows the number at risk at the time of each "event" (age at first sex in this example), the estimated survival function at this time, and the 95% CI. The plot in Figure 7.1 summarizes the numerical information in Table 7.2. From this plot, we can read off the median event time (the median is the preferred measure of location for survival time data because of their likely skewness) as 18 with 95% CI of about [17, 19].

We can now estimate and plot the survival curves for age at first sexual intercourse in each diagnostic group for the data, again using the Kaplan–Meier estimator. The resulting plot is shown in Figure 7.2. This plot suggests very strongly that the time to first sexual intercourse in women diagnosed as RAN is later than in NC. For those not convinced by the plot, a formal test is available to test the hypothesis that the time to first intercourse is the same in each diagnostic group; the required test is known as a *log-rank test*, which assesses the null hypothesis that the survival functions of the two groups are the same. Essentially, this test compares the observed number of "deaths" occurring at each time point with the number to be expected if the survival experience of the groups being compared is the same. Details of the test are given in Collett (2015); here, we shall simply give the value of the chi-squared test statistic that arises from the log-rank test, namely, 46.4 with one degree of

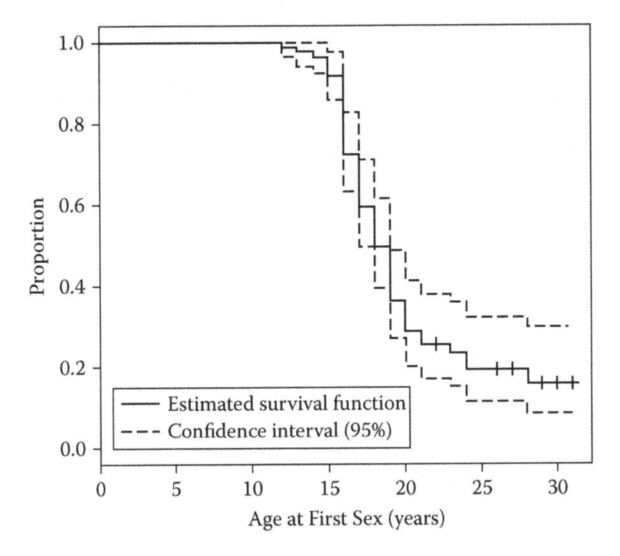

FIGURE 7.1
Plot of estimated survival function for age at first sex data, along with the 95% confidence bands for the survival function.

TABLE 7.2
Kaplan–Meier Estimate of the Survival Function of the Age at First Sex Data

Time	N.risk	N.event	Survival	Standard Error	Lower 95% CI	Upper 95% CI
12	84	1	0.988	0.0118	0.9652	1.000
13	83	1	0.976	0.0166	0.9441	1.000
14	82	1	0.964	0.0202	0.9254	1.000
15	81	4	0.917	0.0302	0.8594	0.978
16	77	16	0.726	0.0487	0.6368	0.828
17	61	11	0.595	0.0536	0.4990	0.710
18	47	8	0.494	0.0551	0.3969	0.615
19	34	9	0.363	0.0551	0.2697	0.489
20	24	5	0.288	0.0530	0.2003	0.413
21	18	2	0.256	0.0517	0.1719	0.380
23	12	1	0.234	0.0516	0.1521	0.361
24	11	2	0.192	0.0503	0.1147	0.320
28	6	1	0.160	0.0510	0.0854	0.299

Note: N. risk = Number of subjects at risk; N. event = Number of events; Survival = Proportion “survived”

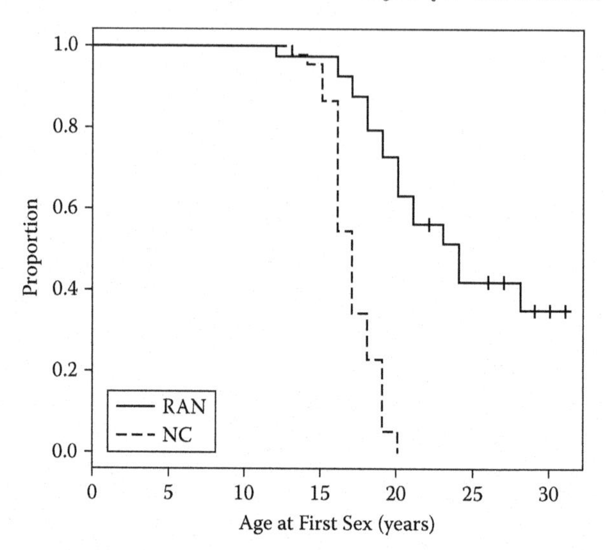

FIGURE 7.2
Estimated survival functions for age at first sex for two diagnostic groups.

freedom and associated p-value <0.00001. There is very convincing evidence that the age at first sex differs in the two diagnostic groups.

7.3 The Hazard Function

In the analysis of survival data, it is often of interest to assess which periods have high or low chances of death (or whatever the event of interest may be) among those still active at the time. A suitable approach to characterizing such risks is the *hazard function* $h(t)$, which is described in Technical Section 7.2. (The hazard function is also known as the intensity function, instantaneous failure rate, and age-specific failure rate.)

Technical Section 7.2: Hazard Function

The hazard function $h(t)$ is defined as the probability that an individual experiences the event of interest in a small time interval s, given that the individual has survived up to the beginning of the interval, when the size of the interval approaches zero. Mathematically, this is written as

$$h(t) = \lim_{s \to 0} \Pr\left(t \leq T \leq t + s | T \geq t\right)$$

where T is the individual's survival time. The conditioning feature of this

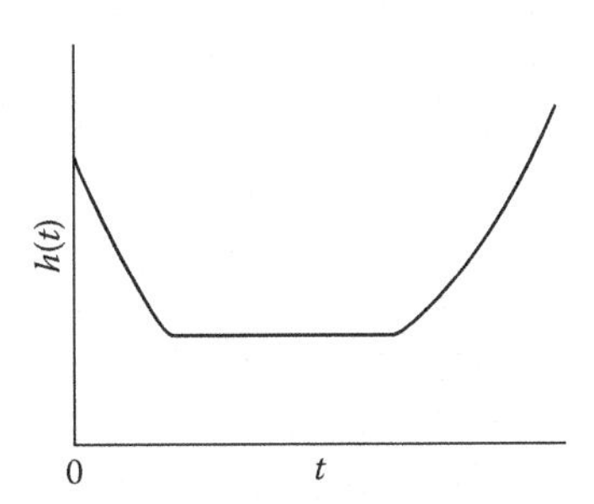

FIGURE 7.3
Bath tub hazard function.

definition is very important. For example, the probability of dying at age 100 is very small because most people die before that age; in contrast, the probability of a person dying at age 100 for one who has reached that age is much greater (for an example of this fact, see Chapter 17). In practice, the hazard function may increase, decrease, remain constant, or have a more complex shape. The hazard function for death in human beings, for example, has the "bathtub" shape shown in Figure 7.3. It is relatively high immediately after birth, declines rapidly in the early years, and then remains approximately constant before beginning to rise again during late middle age.

The hazard function can also be defined in terms of the cumulative distribution function $F(t)$ and the probability density function $f(t)$ of survival times as

$$h(t) = \frac{f(t)}{1 - F(t)} = \frac{f(t)}{S(t)}$$

It then follows that

$$h(t) = -\frac{d}{dt}\left[\log S(t)\right]$$

and so

$$S(t) = \exp\left[-H(t)\right]$$

where $H(t)$ is the integrated or *cumulative hazard function* and is given by

$$H(t) = \int_0^t h(u)du$$

The cumulative hazard function gives the accumulated risk until a specific time.

The hazard function can be estimated as the proportion of individuals experiencing the event of interest in an interval per unit time, given that they have survived to the beginning of the interval. That is,

$$\hat{h}(t) = \frac{d_j}{n_j\left(t_{j+1} - t_j\right)}$$

where d_j is the number of individuals experiencing an event in the interval beginning at time t, and n_j is the number of patients surviving at time t. The sampling variation in the estimate of the hazard function within each interval is usually considerable, and so, it is rarely plotted directly. Instead, plots of the cumulative hazard function are used because they are typically smoother and easier to interpret. Everitt and Rabe-Hesketh (2001) show that the cumulative hazard function can be estimated as

$$\hat{H}(t) = \sum_j \frac{d_j}{n_j}$$

The estimated survival function is more helpful than the estimated hazard function for an initial assessment of survival data, but the hazard function assumes great importance when we consider how to investigate the effect of explanatory variables on survival times, as we shall see in the next section.

7.4 Cox's Proportional Hazards Model

One of the main aims of many behavioral studies is the investigation of how a number of explanatory variables of interest relate to the chosen response variable. The same is true for studies involving survival time data, but because of the special features of such data, the regression methods considered in earlier chapters are no longer appropriate. A number of models that can be applied to survival data have been developed, of which perhaps the most successful is that described by Cox (1972). In these models, it is the hazard function that serves as the response variable because this is a simpler vehicle for modeling the joint effects of explanatory variables as it does not involve the cumulative history of events. Here, we shall concentrate on the model first suggested by Cox.

(For interest, Sir David Cox was the inaugural recipient of the International Prize in Statistics considered the highest honour in its field. The prize was awarded to Sir David largely for his development of the proportional hazards model which bears his name, a method which has had tremendous impact on the analysis of survival data in a wide variety of settings.)

To introduce the basic concepts of the Cox model, suppose first that there are two explanatory variables of interest, x_1 and x_2. We might begin by considering a model in which the hazard function is a linear function of the two explanatory variables. But, it is easy to see that such a model is unsuitable because the hazard function is restricted to being positive, whereas the postulated linear function need not necessarily be likewise constrained (compare the discussion of generalized linear models in Chapter 5). A more sensible model is one that models the log of the hazard function as a linear function of the explanatory variables, and it is a description of this model that begins Technical Section 7.3.

Technical Section 7.3: Cox Regression

A possible model for the log of the hazard function when there are q explanatory variables is

$$\log[h(t)] = \beta_0 + \beta_1 x_1 + \beta_2 x_2 + \cdots + \beta_q x_q$$

In this model, the hazard function does not depend on time, which can be shown to imply that the survival times have an *exponential distribution*. Such a model is very restrictive because hazard functions that increase or decrease with time are much more likely; a suitable model is

$$\log[h(t)] = \beta_0 + \beta_1 x_1 + \beta_2 x_2 + \cdots + \beta_q x_q + \alpha t$$

Depending on the sign of this model, it can represent both a hazard function that increases with time and one that decreases with time. But, what if the hazard function has some more complicated form, for example, the bathtub shape illustrated in Figure 7.3? In most practical situations, it will be very difficult to determine the correct function of time to include in the model, but this problem is overcome in Cox regression where the dependence of the hazard function on time does not have to be specified explicitly. In Cox regression, the model is

$$\log h(t) = \log h_0(t) + \beta_1 x_1 + \beta_2 x_2 + \cdots + \beta_q x_q$$

where $h_0(t)$ is known as the baseline hazard function, being the hazard function for individuals with all explanatory variables equal to zero (this makes more sense if you suppose that each explanatory variable is centered at its sample mean). The model can be rewritten as

$$h(t) = h_0(t) \exp(\beta_1 x_1 + \beta_2 x_2 + \cdots + \beta_q x_q)$$

Written in this way, it is possible to show that the Cox model forces the hazard ratio between two individuals with vectors of covariate values $\mathbf{x}_1$ and $\mathbf{x}_2$ to be constant over time:

$$\frac{h(t|\mathbf{x}_1)}{h(t|\mathbf{x}_2)} = \frac{h_0(t)\exp(\boldsymbol{\beta}'\mathbf{x}_1)}{h_0(t)\exp(\boldsymbol{\beta}'\mathbf{x}_2)} = \frac{\exp(\boldsymbol{\beta}'\mathbf{x}_1)}{\exp(\boldsymbol{\beta}'\mathbf{x}_2)}$$

where $\boldsymbol{\beta}$ is the vector of regression coefficients. So, we see that the Cox model implies that if, at some early time point, say, an individual has a risk of "death" that is twice as high as another individual, then at all later times, the risk of death remains twice as high. Hence, Cox regression is often labeled *Cox's proportional hazards model.* While the proportionality assumption may sound complicated, it is actually very simple and implies that the single regression coefficient associated with each explanatory variable represents the effect of this variable throughout the time period. If the risk of outcome associated with a particular variable is higher at one point in time and lower at another, a single coefficient cannot represent that relationship.

In the Cox model, the baseline hazard describes the common shape of the survival time distribution for all individuals, while the relative risk function $\exp(\boldsymbol{\beta}'\mathbf{x})$ gives the level of each individual's hazard. The interpretation of a particular element of the vector $\boldsymbol{\beta}$, say the parameter β_j, is that $\exp(\beta_j)$ gives the relative risk change associated with an increase of one unit in x_j, all other explanatory variables remaining constant. An alternative aid to interpretation is to calculate $100(\exp(\beta_j)-1)$ to get the percentage change in the hazard function with each unit change in the appropriate explanatory variable, conditional on the other explanatory variables remaining constant.

The parameters in a Cox model can be estimated by maximizing what is known as a *partial likelihood.* Details are given in Kalbfleisch and Prentice (2002). The partial likelihood is derived by assuming continuous survival times. In reality, however, survival times are measured in discrete units, and there are often ties. There are three common methods for dealing with ties, which are described briefly in Everitt and Rabe-Hesketh (2001).

The Cox model can be extended to allow the baseline hazard function to vary with the levels of a stratification variable. Such a *stratified proportional hazards model* is useful in situations where the stratifier is thought to affect the hazard function but is not of primary interest. A common example of a stratifier variable is gender.

As a first example of applying Cox regression, we will apply it to the data on age at first sex (see Table 7.1). The estimated regression coefficient for the single explanatory variables diagnosis coded 0 for RAN and 1 for NC is 2.14, with an estimated standard error of 0.346; dividing 2.14 by 0.346 gives a z-statistic with a value of 6.18 and an associated very, very small p-value. The log hazard for first sex is estimated to be 2.14 greater in the NC than in the RAN patients. The value 2.14 can be exponentiated to give a value of 8.47 for the relative risk; a 95% CI for the relative risk is found

TABLE 7.3
Data for Heroin Addicts Being Treated with Methadone

	Clinic	Status	Time (Days)	Prison	Dose (mg/Day)
1	Clinic 1	1	428	No prison record	50
2	Clinic 1	1	275	Prison record	55
3	Clinic 1	1	262	No prison record	55
4	Clinic 1	1	183	No prison record	30
5	Clinic 1	1	259	Prison record	65
98	Clinic 2	1	708	Prison record	60
99	Clinic 2	0	713	No prison record	50
100	Clinic 2	0	146	No prison record	50
101	Clinic 2	1	450	No prison record	55
102	Clinic 2	0	555	No prison record	80

Note: Status = 1 implies methadone cessation; otherwise, status = 0.

as $[\exp(2.14 - 1.96 \times 0.346), \exp(2.14 + 1.96 \times 0.346)]$, giving [4.3, 16.7]. The hazard function for first sex for normal controls is about 4 to 17 times that for the women diagnosed as restricted anorexia nervosa.

7.4.1 Retention of Heroin Addicts in Methadone Treatment

Our second example of the use of Cox regression will involve data collected on the retention of heroin addicts in maintenance treatment using methadone. The end point in this study is methadone cessation (either by the choice of the patient or the treating doctor) and departure from the treating clinic. The time in days until methadone cessation was recorded for 238 heroin addicts. Censored observations occurred when patients departed for reasons other than methadone cessation (status = 1 if methadone cessation, and 0 otherwise). So, in this example, methadone cessation is equivalent to "death" in a study in which the end point is actual death. In addition, the maximum methadone dose prescribed (milligrams per day) and whether or not the addict had a prison record were recorded (0 = no prison record, 1 = prison record). The patients were treated in one of two clinics (coded as clinic 1 = 0, clinic 2 = 1). The data for five patients from each of the two clinics are shown in Table 7.3.

To begin, let us take a look at the estimated survival functions for the two clinics (see Figure 7.4). The plot demonstrates that the second clinic appears to be more successful in keeping heroin addicts on their methadone treatment. From the graph, the median methadone treatment in clinic 1 can be estimated as 428 days; the estimate is not available for clinic 2 because more than 50% of the patients continued treatment throughout the study period.

But our real interest for these data lies in assessing the effects of all the explanatory variables by using Cox regression. The results of fitting the model for the heroin data are shown in Table 7.4. Looking at the exponentiated results given in the lower part of Table 7.4, we see that

TABLE 7.4
Results from Fitting the Cox Regression Model to the Data on Heroin Addicts

	Coef	Exp(coef)	Se(coef)	z	p
Prison record	0.3266	1.386	0.16722	1.95	0.051
Dose	−0.0354	0.965	0.00638	−5.54	2.9e-08
Clinic	−1.0099	0.364	0.21489	−4.70	2.6e-06

	Exp(coef)	Exp(−coef)	Lower 95% CI	Upper 95% CI
Prison record	1.386	0.721	0.999	1.924
Dose	0.965	1.036	0.953	0.977
Clinic	0.364	2.745	0.239	0.555

Note: CI = Confidence interval.

- Heroin addicts with a criminal record are estimated to have a hazard of immediate methadone cessation of between about 0.999 and 1.924 times the corresponding hazard for addicts without a criminal record, conditional on

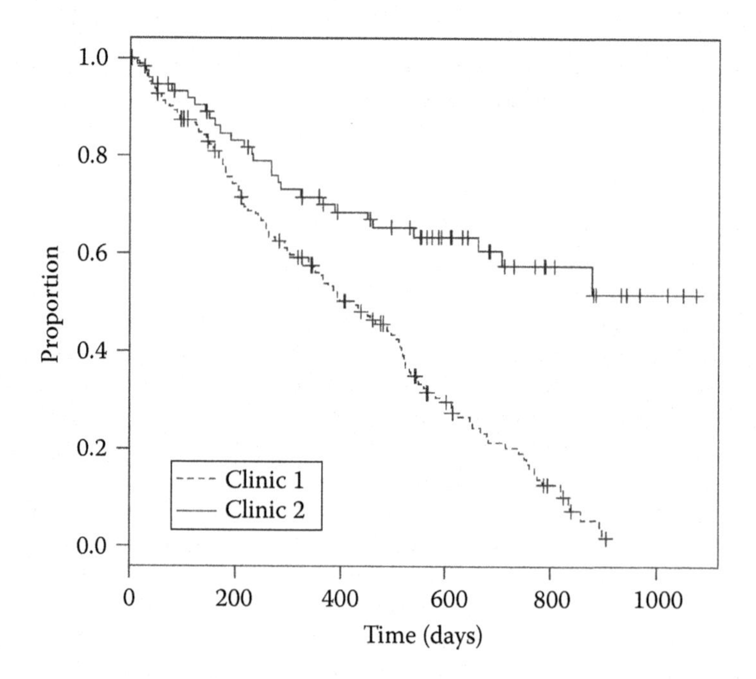

FIGURE 7.4
Estimated survival functions for the two clinics in the heroin addicts data.

clinic and dose. As the CI contains the value 1, there is no strong evidence that a prison record has any real effect on the hazard function, although there may be a tendency for addicts with a criminal record to have an increased risk of immediate methadone cessation.

- Each milligram per day increase in maximum methadone dose prescribed leads to a decrease in the hazard function of immediate methadone cessation of between about 2% and 5%, conditional on prison record and clinic. Higher doses of methadone tend to keep addicts on treatment.
- In clinic 2, the hazard function of immediate methadone cessation is about 0.3 to 0.6 times the corresponding hazard function in clinic 1, conditional on dose and prison record. Clinic 2 is more successful in keeping addicts on treatment; the risk of immediate methadone cessation is higher in clinic 1.

As in multiple linear regression and logistic regression, we might often wish to find a more parsimonious Cox regression model than the one containing all the explanatory variables under investigation. We can use the same backward elimination approach as we used for both multiple linear and logistic regressions, and if we do, we get the following results:

Start: AIC = 1352.52

Step 1: Model with all three explanatory variables and one variable at a time considered for removal.

Drop prison: AIC = 1354.3

Drop clinic: AIC = 1376.9

Drop dose: AIC = 1381.3

We see in this example that dropping any of the three explanatory variables results in a model with larger AIC value than the model that includes all three variables. Consequently, it is this model that is accepted despite the regression coefficient for the prison variable not being significant at the 5% level.

The Cox model is based on the assumption of proportional hazards, but this assumption will not hold for all sets of survival data. For example, if two individuals with a heart condition receive different treatments, one medical and one surgical, then the individual treated surgically may be at higher risk initially because of the possibility of operative mortality. At a later stage, however, the risk may become the same or even less than that for a medically treated individual. In this case, if one of the explanatory variables is a dummy variable indicating which treatment a person receives, then the proportional hazards assumption will not hold, and the regression coefficient for treatment effect will change over time. Examining whether or not any of the regression estimates in a Cox regression change over time is one way of checking the proportional hazards assumption. Therneau and Grambsch (2000) describe a test of the hypothesis of constant regression coefficients over time that can be applied individually to each coefficient and globally to the set of coefficients. The test statistic is a chi-squared; however, we will not give details of the test here but simply look at the results, which are as follows:

	chisq	**p**
Prison	0.22	0.64
Dose	0.70	0.40
Clinic	11.19	<0.01
Global	12.62	<0.01

It appears that the proportional hazards function does not hold for the clinics in this example. Consequently, a better model for the data may be one in which the clinic is used as a stratifier. Consideration of such a model is left as an exercise for the reader (see Exercise 7.1).

7.5 Summary

- Survival data need specialized techniques for their analysis because of the presence of censored observations. Such data arise frequently in medical studies but less often in behavioral research. However, behavioral researchers need to recognize such data and be aware of the correct methods of analysis.

- The first step in the analysis of survival data is, generally, to plot the estimated survival functions for particular groups of observations.

- The hypothesis that two survival functions are the same can be tested formally using the log-rank test.

- Cox regression is the model most often used to assess the effects of explanatory variables on survival times. The assumption of proportional hazards on which the model is based needs to be checked.

- Diagnostics for survival analysis, in particular for checking the assumptions made by Cox regression, are described in Collett (2015).

7.6 Exercises

7.1 Fit a Cox regression model to the data on heroin addicts using the clinic as a stratifier. How do the results compare with those derived in the text?

7.2 The data are the survival times (in months) after mastectomy of women with breast cancer. The cancers are classified as having metastasized or not based on a histochemical marker. Censoring is indicated by the event variable that takes the value TRUE in the case of death and FALSE otherwise. Plot the Kaplan–Meier estimated survival functions for each type of cancer on the same graph and comment on the differences. Use a log-rank test to formally compare the survival experience of each class.

7.3 Grana et al. (2002) report the results from a nonrandomized clinical trial investigating a novel radioimmunotherapy in malignant glioma patients. The overall survival, that is, the time from the beginning of therapy to the disease-caused death of the patient, is compared for two groups of patients—one, the standard therapy plus the novel therapy; and the other, the standard therapy alone. In addition to the treatment groups a number of other explanatory variables are recorded, namely age, gender, and histology. Fit a Cox model to the data and find a CI for the treatment effect conditional on age, gender, and histology.

* 7.4 Palotie et al. (2017) investigated the "survival" or longevity of teeth restorations in Finland. The data that were extracted from the electronic patient files of the Helsinki City Public Dental Service (PDS) and prepared for analyses consist of 5542 restorations conducted on 3051 patients in 2002. Information about the size, type, and material for the restorations of each particular tooth as well as the dates of the placement, possible re-intervention, and the patient's most recent visit were recorded. The fate of the restorations was followed as long as the patient continued visiting the Helsinki City PDS, but no longer than the end of 2015. The researchers focused on the premolars and molars (excluding the wisdom teeth), one of their hypotheses being that the longevity of restorations in the premolars exceeds that in the molars. Dig into the data, plot the Kaplan–Meier curves for each type of teeth on the same graph, and comment on the differences. Use a log-rank test to compare the survival functions.

8

Analysis of Longitudinal Data I: Graphical Displays and Summary Measure Approach

8.1 Introduction

Previous chapters have looked at ways to model and analyse data where there is a single random response variable and a number of explanatory variables considered to be fixed rather than random; it is sometimes helpful to refer to such data as *multivariable* rather than *multivariate* keeping the latter for data sets containing a number of variables but with no division into response and explanatory and where all variables are considered random. Analysing such data will be the subject of Chapters 13–18. In this chapter and the following two chapters we consider data that comes somewhere between multivariable and multivariate. As with the former there is a division into explanatory and response variables, but now the response variable (and possibly some or all of the explanatory variables) is observed more than once on each individual in the study and so could be labeled multivariate.

In behavioral research, data with a multivariate response are common; for example, a response variable may be measured under a number of different experimental conditions, leading to what are generally called *repeated measures data*, or a response variable may be recorded on several different occasions over some period of time, in which case we have *longitudinal data*. (Although we think distinguishing two different types of data here is useful, it has to be said that the repeated measures label is often used for both types of data.) The variation among the repeated measures of the response is *within-subject variation*. But often one of the covariates will be a factor such as gender or treatment that will give rise to *between-subject variation*. Longitudinal data occur most frequently in the behavioral sciences in the clinical trials frequently undertaken by psychologists, psychiatrists and others to assess the effectiveness or otherwise of different treatments; it is largely this type of data which will be the subject of this chapter and the two following chapters.

As several observations of the response variable are made on the same individual, it is likely that the repeated measurements of the response will be correlated rather than independent even after conditioning on the explanatory variables. Consequently, for the analysis of repeated measures data and for longitudinal data, models are needed that can both assess the effects of

TABLE 8.1
BPRS Measurements from 40 Subjects

		Week								
Treatment	**Subject**	**0**	**1**	**2**	**3**	**4**	**5**	**6**	**7**	**8**
1	1	42	36	36	43	41	40	38	47	51
1	2	58	68	61	55	43	34	28	28	28
1	3	54	55	41	38	43	28	29	25	24
1	4	55	77	49	54	56	50	47	42	46
1	5	72	75	72	65	50	39	32	38	32
1	6	48	43	41	38	36	29	33	27	25
1	7	71	61	47	30	27	40	30	31	31
1	8	30	36	38	38	31	26	26	25	24
1	9	41	43	39	35	28	22	20	23	21
1	10	57	51	51	55	53	43	43	39	32
1	11	30	34	34	41	36	36	38	36	36
1	12	55	52	49	54	48	43	37	36	31
1	13	36	32	36	31	25	25	21	19	22
1	14	38	35	36	34	25	27	25	26	26
1	15	66	68	65	49	36	32	27	30	37
1	16	41	35	45	42	31	31	29	26	30
1	17	45	38	46	38	40	33	27	31	27
1	18	39	35	27	25	29	28	21	25	20
1	19	24	28	31	28	29	21	22	23	22
1	20	38	34	27	25	25	27	21	19	21
2	1	52	73	42	41	39	38	43	62	50
2	2	30	23	32	24	20	20	19	18	20
2	3	65	31	33	28	22	25	24	31	32
2	4	37	31	27	31	31	26	24	26	23
2	5	59	67	58	61	49	38	37	36	35
2	6	30	33	37	33	28	26	27	23	21
2	7	69	52	41	33	34	37	37	38	35
2	8	62	54	49	39	55	51	55	59	66
2	9	38	40	38	27	31	24	22	21	21
2	10	65	44	31	34	39	34	41	42	39
2	11	78	95	75	76	66	64	64	60	75
2	12	38	41	36	27	29	27	21	22	23
2	13	63	65	60	53	52	32	37	52	28
2	14	40	37	31	38	35	30	33	30	27
2	15	40	36	55	55	42	30	26	30	37
2	16	54	45	35	27	25	22	22	22	22
2	17	33	41	30	32	46	43	43	43	43
2	18	28	30	29	33	30	26	36	33	30
2	19	52	43	26	27	24	32	21	21	21
2	20	47	36	32	29	25	23	23	23	23

explanatory variables on the multiple measures of the response variable and account for the likely correlations between these multiple measures. Suitable models for longitudinal data will be described in Chapters 9 and 10. But before consideration of such models we will in this chapter look at how to display such data graphically and a possible "quick and dirty" approach to their analysis.

8.2 Graphical Displays of Longitudinal Data

Graphical displays of data are almost always useful for exposing patterns in the data, particularly when these are unexpected; this might be of great help in suggesting which class of models might be most sensibly applied in the later more formal analysis. According to Diggle et al. (2002), there is no single prescription for making effective graphical displays of longitudinal data, although they do offer the following simple guidelines:

- Show as much of the relevant raw data as possible rather than only data summaries;
- Highlight aggregate patterns of potential scientific interest;
- Identify both cross-sectional and longitudinal patterns;
- Try to make the identification of unusual individuals or unusual observations simple.

A number of graphical displays which can be useful in the preliminary assessment of longitudinal data from clinical trials will now be illustrated using the data shown in Table 8.1 taken from Davis (2002). Here 40 male subjects were randomly assigned to one of two treatment groups and each subject was rated on the brief psychiatric rating scale (BPRS) measured before treatment began (week 0) and then at weekly intervals for eight weeks. The BPRS assesses the level of 18 symptom constructs such as hostility, suspiciousness, hallucinations and grandiosity; each of these is rated from one (not present) to seven (extremely severe). The scale is used to evaluate patients suspected of having schizophrenia.

To begin we shall plot the BPRS values for all 40 men, differentiating between the treatment groups into which the men have been randomized. The resulting diagram is shown in Figure 8.1. This simple graph makes a number of features of the data readily apparent. First, the BPRS score of almost all the men is decreasing over the eight weeks of the study. Second, the men who have higher BPRS values at the beginning tend to have higher values throughout the study. This phenomenon is generally referred to as *tracking*.

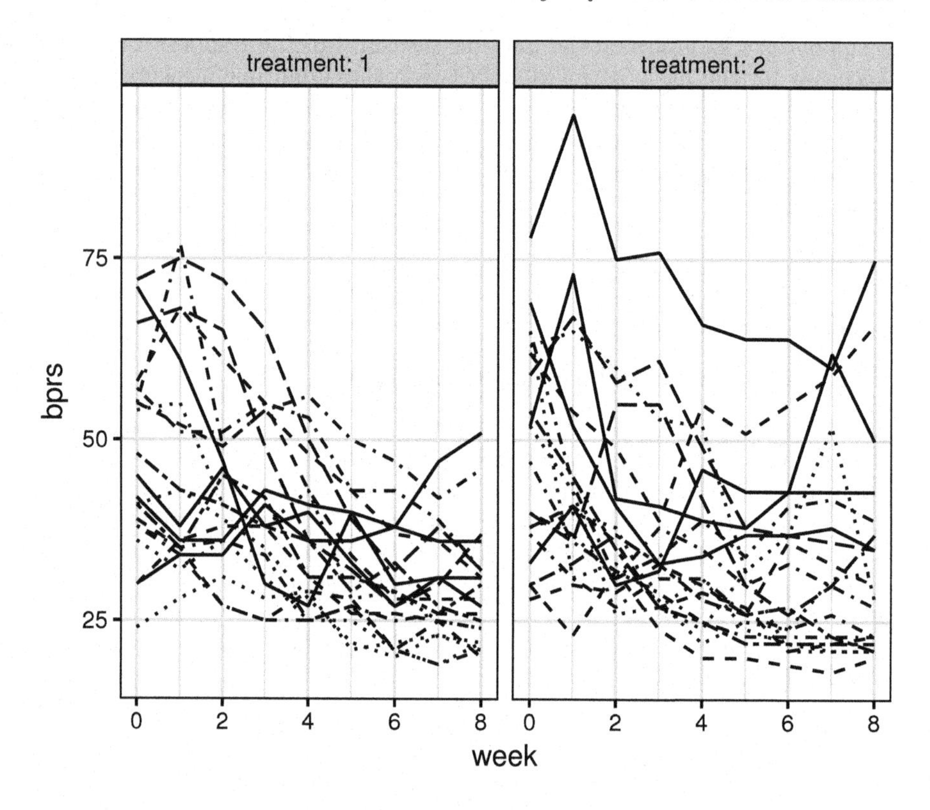

FIGURE 8.1
Individual response profiles by treatment group for the BPRS data.

Third, there are substantial individual differences and variability appears to decrease with time.

The tracking phenomenon can be seen more clearly in a plot of the standardized values of each observation, i.e., the values obtained by subtracting the relevant occasion mean from the original observation and then dividing by the corresponding visit standard deviation. This plot is shown in Figure 8.2.

With large numbers of observations, graphical displays of individual response profiles are of little use and investigators then commonly produce graphs showing average profiles for each treatment group along with some indication of the variation of the observations at each time point. The result is Figure 8.3. There is considerable overlap in the mean profiles of the two treatment groups suggesting perhaps that there is little difference between the two groups with respect to the mean BPRS values.

A possible alternative to plotting the mean profiles as in Figure 8.3 is to graph side-by-side box plots of the observations at each time point. The resulting plot is shown in Figure 8.4. The plot suggests the presence of some possible "outliers" at a number of time points and indicates again the general

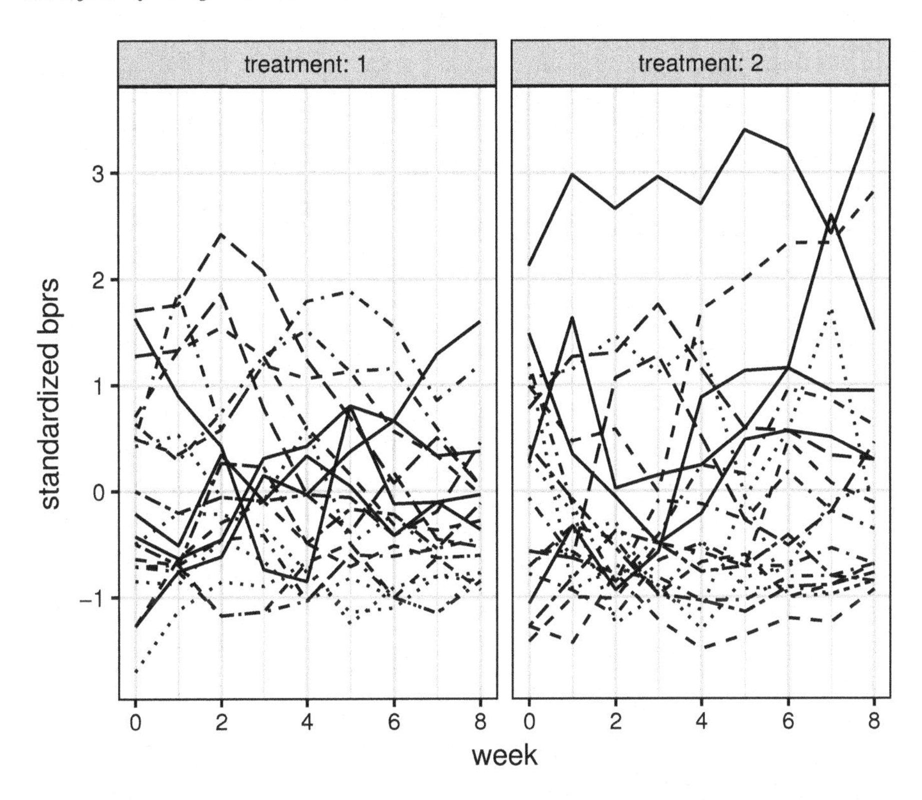

FIGURE 8.2
Individual response profiles for BPRS data after standardization.

decline in BPRS values over the eight weeks of the study in both treatment groups.

Another graphic for longitudinal data that is often helpful in making informed decisions about models that might be appropriate for the data is the scatterplot matrix. But we shall leave consideration of this type of plot until the next chapter when we begin to discuss possible models for longitudinal data.

8.3 Summary Measure Analysis of Longitudinal Data

According to Matthews (2005) "the use of summary measures is one of the most important and straightforward methods for the analysis of longitudinal data". The approach is certainly straightforward but as to "most important" we think not. The models to be described in the next two chapters are of far

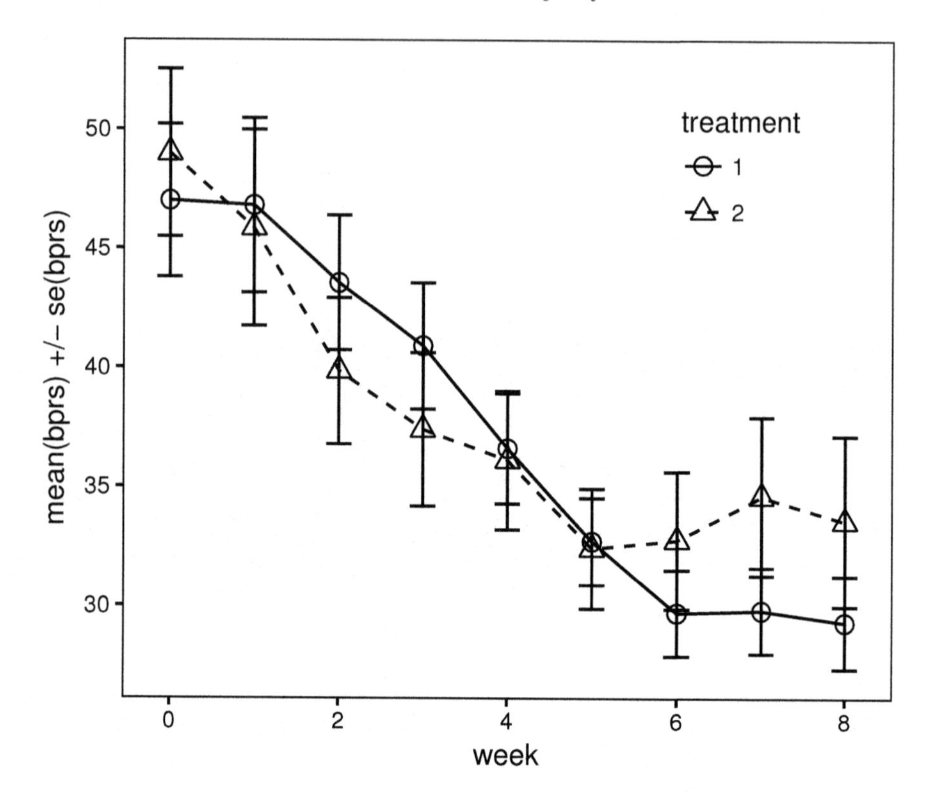

FIGURE 8.3
Mean response profiles for the two treatment groups in the BPRS data.

greater importance for dealing appropriately with longitudinal data. Nevertheless we will describe the summary measure method (often also called the response feature method) here because it may be helpful in some cases for an initial assessment of longitudinal data.

The summary measure method operates by transforming the repeated measurements made on each individual in the study into a single value that captures some essential feature of the individual's response over time. Analysis then proceeds by applying standard univariate methods to the summary measures from the sample of subjects (see later examples). The approach has been in use for many years, and is described in Oldham (1962), Yates (1982) and Matthews et al. (1990).

8.3.1 Choosing Summary Measures

The key step to a successful summary measure analysis of longitudinal data is the choice of a relevant summary measure. The chosen measure needs to be

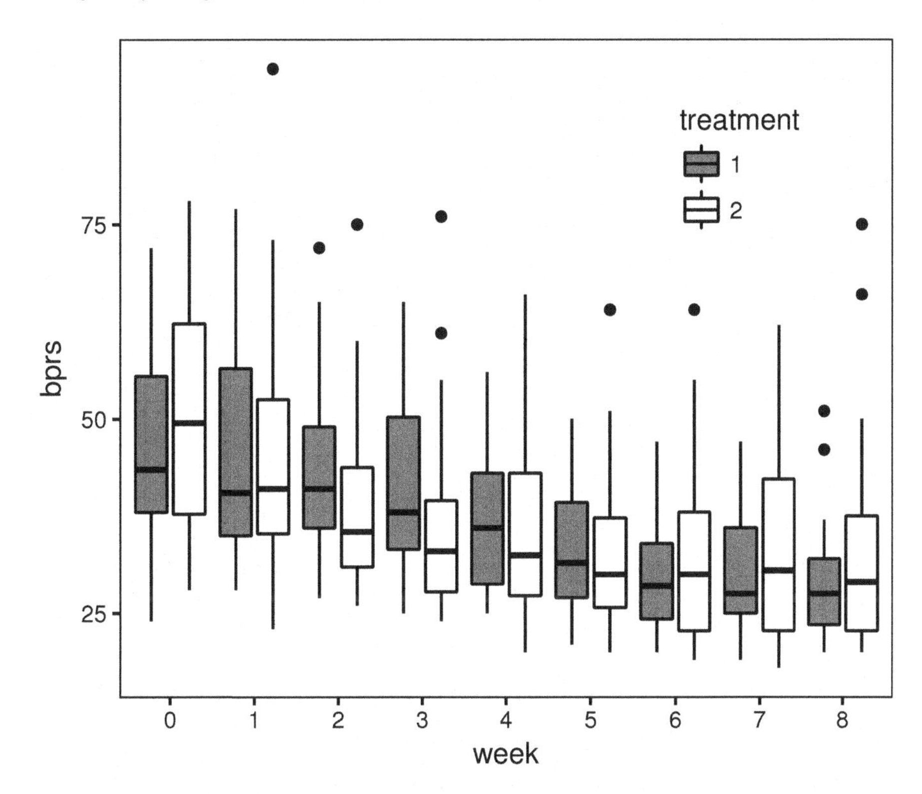

FIGURE 8.4
Boxplots for the BPRS data.

relevant to the particular questions of interest in the study and in the broader scientific context in which the study takes place. In some longitudinal studies, more than a single summary measure might be deemed relevant or necessary, in which case the problem of combined inference may need to be addressed. More often in practice, however, it is likely that the different measures will deal with substantially different questions so that each will have a notional interpretation in its own right. In most investigations, the decision over what summary measure to use needs to be made before the data are collected.

A wide range of possible summary measures have been proposed. Those given in Table 8.2, for example, were suggested by Matthews et al. (1990). Frison and Pocock (1992) argue that the average response to treatment over time is often likely to be the most relevant summary statistic in treatment trials. In some cases the response on a particular visit may be chosen as the summary statistic of most interest, but this must be distinguished from the generally flawed approach which separately analyses the observations at each and every time point.

TABLE 8.2
Possible Summary Measures

Type of Data	Question of Interest	Summary Measure
Peaked	Is overall value of outcome variable the same in different groups?	Overall mean (equal time intervals) or area under curve (unequal intervals)
Peaked	Is maximum (minimum) response different between groups?	Maximum (minimum) value
Peaked	Is time to maximum (minimum) response different between groups?	Time to maximum (minimum) response
Growth	Is rate of change of outcome different between groups?	Regression coefficient
Growth	Is eventual value of outcome different between groups?	Final value of outcome or difference between last and first values or percentage change between first and last values
Growth	Is response in one group delayed relative to the other?	Time to reach a particular value (e.g., a fixed percentage of baseline)

8.3.2 Applying the Summary Measure Approach

As our first example of the summary measure approach it will be applied to the post treatment values of the BPRS in Table 8.1. The mean of weeks 1 to 8 will be the chosen summary measure. We first calculate this measure and then look at boxplots of the measure for each treatment group. The resulting plot is shown in Figure 8.5. The diagram indicates that the mean summary measure is more variable in the second treatment group and its distribution in this group is somewhat skew. The boxplot of the second group reveals an outlier, a subject whose mean BPRS score of the eight weeks is over 70. It might bias the conclusions from further comparisons of the groups, so we decide to remove that subject from the data. The new version of the boxplots is shown in Figure 8.6. Without the outlier, the eight-week mean of the second treatment group is lower than of the first group, but there is still little evidence of a difference in location of the summary measure distributions in each group.

Although the informal graphical material presented up to now has all indicated a lack of difference in the two treatment groups, most investigators would still require a formal test for a difference. Consequently we shall now

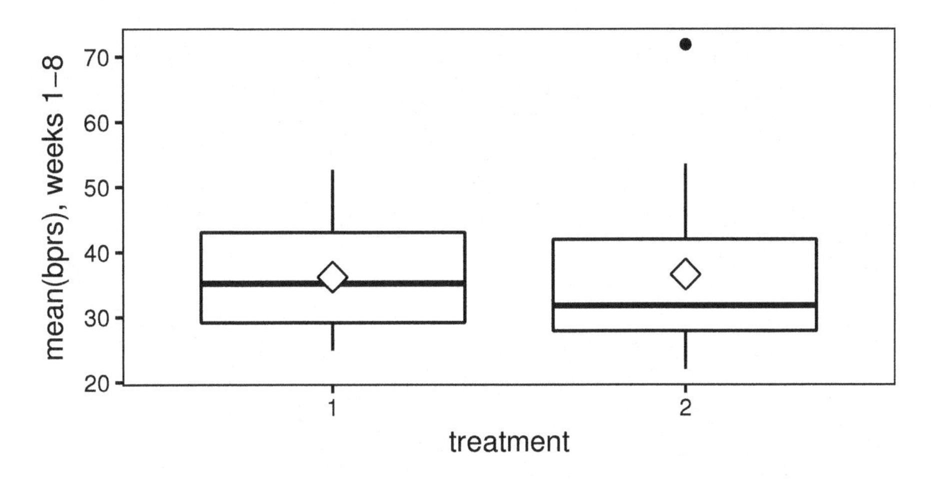

FIGURE 8.5
Boxplots of mean summary measures for the two treatment groups in the BPRS data.

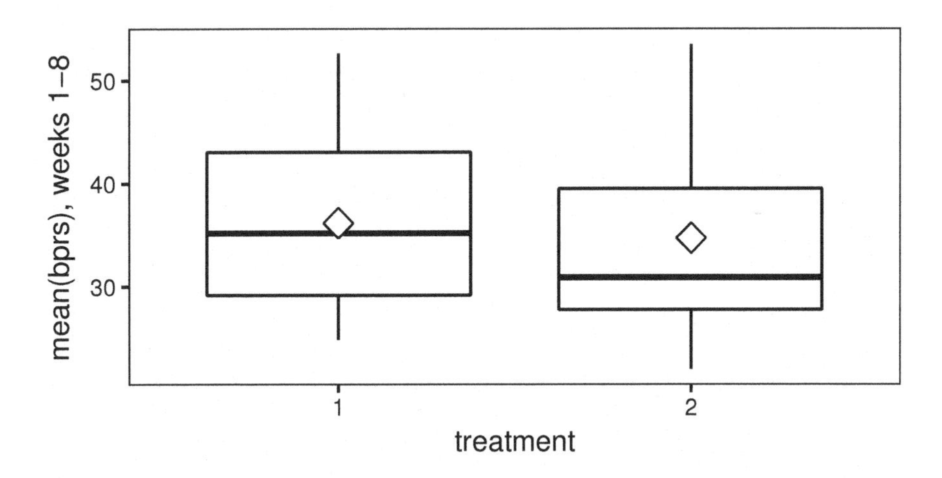

FIGURE 8.6
Boxplots of mean summary measures for the two treatment groups in the BPRS data, without the outlier shown in Figure 8.5.

apply a t-test to assess any difference between the treatment groups, and also calculate a confidence interval for this difference. We use the data without the outlier that was revealed in Figure 8.5. The results are shown in Table 8.3. The t-test confirms the lack of any evidence for a group difference. Also the 95% confidence interval is wide and includes the zero, allowing for similar conclusions to be made.

TABLE 8.3
Results from an Independent Samples t-test on the Mean Summary Measure for the BPRS Data, without the Outlier Shown in Figure 8.5

	Treatment 1	Treatment 2	Difference	t-Value	Pr(>|t|)
Mean	36.16875	34.70395	1.4648	0.52095	0.6055
95% confidence interval for the mean difference: [−4.23, 7.16]					

Note: Student's two sample t-test (two-sided) on 37 DF.

8.3.3 Incorporating Pre-Treatment Outcome Values into the Summary Measure Approach

Baseline measurements of the outcome variable in a longitudinal study are often correlated with the chosen summary measure and using such measures in the analysis can often lead to substantial gains in precision when used appropriately as a covariate in an analysis of covariance (see Everitt and Pickles, 2004). We can illustrate the analysis on the data in Table 8.1 using the BPRS value corresponding to time zero taken prior to the start of treatment as the baseline covariate. The results are shown in Table 8.4. We see that the baseline BPRS is strongly related to the BPRS values taken after treatment has begun, but there is still no evidence of a treatment difference even after conditioning on the baseline value.

TABLE 8.4
Results from an Analysis of Covariance of the BPRS Data with Baseline BPRS and Treatment Group as Covariates

Source	Sum of Squares	Df	Mean Square	F	p-Value
Baseline	1868.07	1	1868.07	30.1437	3.077e-06
Treatment	3.45	1	3.45	0.0557	0.8148
Residuals	2292.97	37	61.97	–	–

8.3.4 Dealing with Missing Values When Using the Summary Measure Approach

One of the problems that often occurs in the collection of longitudinal data is that a subject may not have values of the outcome measure recorded on all the occasions intended. This problem will be considered in detail in later chapters, but as an example of where it has arisen we can examine the data shown in Table 8.5 (taken from Davis, 2002). The data come from a clinical trial comparing two treatments for maternal pain relief during labor. In this study 83 women in labor were randomized to receive an experimental pain

TABLE 8.5
Pain Scores from 83 Women in Labor: 43 Subjects in Group 1 (Medication) and 40 Subjects in Group 2 (Placebo); First 20 Subjects in Each Group

		Self-reported Pain Scores at 30-minute intervals						
Group	**Subject**	**0**	**30**	**60**	**90**	**120**	**150**	**180**
1	1	0.0	0.0	0.0	0.0			
1	2	0.0	0.0	0.0	0.0	2.5	2.3	14.0
1	3	38.0	5.0	1.0	1.0	0.0	5.0	
1	4	6.0	48.0	85.0	0.0	0.0		
1	5	19.0	5.0					
1	6	7.0	0.0	0.0	0.0			
1	7	44.0	42.0	42.0	45.0			
1	8	1.0	0.0	0.0	0.0	0.0	6.0	24.0
1	9	24.5	35.0	13.0				
1	10	1.0	30.5	81.5	67.5	98.5	97.0	
1	11	35.5	44.5	55.0	69.0	72.5	39.5	26.0
1	12	0.0	0.0	0.0	0.0	0.0	0.0	0.0
1	13	8.0	30.5	26.0	24.0	29.0	45.0	91.0
1	14	7.0	6.5	7.0	4.0	10.0		
1	15	6.0	8.5	19.5	16.5	42.5	45.5	48.5
1	16	32.5	9.5	7.5	5.5	4.5	0.0	7.0
1	17	10.5	10.0	18.0	32.5	0.0	0.0	0.0
1	18	11.5	20.5	32.5	37.0	39.0		
1	19	72.0	91.5	4.5	32.0	10.5	10.5	10.5
1	20	0.0	0.0	0.0	0.0	13.54	7.0	
2	1	4.0	9.0	30.0	75.0	49.0	97.0	
2	2	0.0	0.0	1.0	27.5	95.0	100.0	
2	3	9.0	6.0	25.0				
2	4	52.5	18.0	12.5				
2	5	90.5	99.0	100.0	100.0	100.0	100.0	100.0
2	6	74.0	70.0	81.5	94.5	97.0		
2	7	0.0	0.0	0.0	1.5	0.0	18.0	71.0
2	8	0.0	51.5	56.0				
2	9	6.5	7.0	7.0	9.0	25.0	36.0	20.0
2	10	19.0	31.0	41.0	58.0			
2	11	6.0	23.0	45.0	67.0	90.5		
2	12	42.0	64.0	6.0				
2	13	86.5	53.0	88.0	100.0	100.0		
2	14	50.0	100.0	100.0	100.0	100.0		
2	15	27.5	36.5	74.0	97.0	100.0	100.0	95.0
2	16	0.0	0.0	6.0	6.0			
2	17	62.0	79.0	80.5	85.0	90.0	97.5	97.0
2	18	17.5	27.5	21.0	60.0	80.0	97.0	
2	19	6.5	5.5	18.5	20.0	36.5	63.5	81.5
2	20	8.0	9.0	35.5	39.0	70.0	92.0	98.0

medication (43 subjects) or placebo (40 subjects). Treatment was initiated when the cervical dilation was 8 cm. At 30-minute intervals, the amount of pain was self-reported by placing a mark on a 100 mm line (0 = no pain, 100 = very much pain). Table 8.5 gives the data for the first 20 subjects in each group.

If we use the mean as a summary measure for these data, the missing values can be dealt with by (a) simply leaving out the subjects with any missing values or (b) calculating, for each subject, the mean of their available values. So, for example, for subject one in group one this would be the mean of four values, and for subject five in the same group, the mean of only two values. Both approaches are very straightforward but there are a number of possible problems which for the moment we shall conveniently ignore. The difficulties of missing values in data in general, and longitudinal data in particular will be considered in Chapter 11. Here we shall carry out both possibilities mentioned to see how the results compare. These results are shown in Table 8.6.

In both cases, the mean difference is statistically significant, but in alternative (a) about 60% of the values are discarded, when the means are computed. The alternative (b) uses all available values, and therefore the p-value is even smaller and the corresponding confidence interval for the mean difference (that now does not include the zero) is narrower.

TABLE 8.6
Results from an Independent Samples t-test for the Mean Summary Measure Used on the Data Partially Shown in Table 8.5. (a) Leaving out Subjects with Any Missing Value, (b) Mean of Available Values for Each Subject

	Treatment 1	**Treatment 2**	**Difference**	**t-Value**	**Pr(>|t|)**
Mean (a)	16.74286	46.80000	−30.05714	−3.8465	0.00054
Mean (b)	18.34181	41.03771	−22.69590	−4.1455	0.00008

(a) 95% confidence interval for the mean difference: [−45.97, −14.14]
(b) 95% confidence interval for the mean difference: [−33.59, −11.80]

Note: Student's two sample t-test (two-sided) on (a) 32 DF, (b) 81 DF.

8.4 Summary

- The methods described in this chapter are most (only) suitable for an initial exploration and initial analysis of longitudinal data.
- The graphical methods can provide insights into both potentially interesting patterns of response over time and the structure of any treatment differences. In addition, they can indicate possible outlying observations that may need special attention.

- The response feature approach to analysis has the distinct advantage that it is straightforward, can be tailored to consider aspects of the data thought to be particularly relevant, and produces results which are relatively simple to understand.
- Depending on the chosen summary measure, the approach can often accommodate data containing missing values without difficulty, although it might be misleading if the observations are anything other than missing completely at random (see Chapter 11).
- But, although simple to apply, the summary measure approach has a number of distinct drawbacks; one such is that it forces the investigator to focus on only a single aspect of the repeated measurements over time.
- It seems intuitively clear that when several repeated measures are replaced by a single number summary, there must necessarily be some loss of information. And it is possible for individuals with quite different response profiles to have the same or similar values for the chosen summary measure.
- Finally, the simplicity of the summary measure method is lost when there are missing data or the repeated measures are irregularly spaced as is the methods efficiency; the methods to be described in the next two chapters are more efficient than a summary measure analysis and can also handle missing data with minimal difficulty.

8.5 Exercises

8.1 Investigate the use of the other summary measures listed in Table 8.2 for the BPRS data and summarize your conclusions.

8.2 For the pain score data replace each missing value for a subject by the mean of the values the subject actually has for the 30–180 minute observations and then apply the summary measure method to the data using the mean of the six observations that each subject now has. What do you consider to be the problems of this approach?

8.3 Five different types of electrodes were applied to the arms of 16 subjects and resistance measured in kilohms (the first eight subjects are women and the remaining eight men). The resulting data are shown in Table 8.7. The experiment was performed to see whether all electrode types performed similarly and whether men and women differed in skin resistance. Use a summary measure approach to analyse the data after you have used suitable graphics to display the data and considered whether any observations should be removed before analysis.

TABLE 8.7
Skin Resistance in Kilohms and Electrode Type: 16 Subjects (the First 8 Women, the Remaining 8 Men)

	Electrode type				
Subject	**1**	**2**	**3**	**4**	**5**
1	500	400	98	200	250
2	660	600	600	75	310
3	250	370	220	250	220
4	135	300	450	430	70
5	27	84	135	190	180
6	100	50	82	73	78
7	105	180	32	58	32
8	90	180	220	34	64
9	200	290	320	280	135
10	15	45	75	88	80
11	160	200	300	300	220
12	250	400	50	50	92
13	170	310	230	20	150
14	66	1000	1050	280	220
15	107	48	26	45	51
16	100	80	230	280	150

9

Analysis of Longitudinal Data II: Linear Mixed Effects Models for Normal Response Variables

9.1 Introduction

The summary measure approach to the analysis of longitudinal data described in the previous chapter sometimes provides a useful first step in making inferences about the data, but it is really only ever a first step, and a more complete and a more appropriate analysis will involve fitting a suitable model to the data and estimating parameters that link the explanatory variables of interest to the repeated measures of the response variable.

The main objective in the analysis of data from a longitudinal study is to characterize change in the repeated values of the response variable and to determine the explanatory variables most associated with any change. Because several observations of the response variable are made on the same individual, it is likely that the measurements will be correlated rather than independent, even after conditioning on the explanatory variables. Consequently models for longitudinal data need to include parameters analogous to the regression coefficients in the usual multiple regression model (see Chapter 4) that relate the explanatory variables to the repeated measurements, and, in addition, parameters that account adequately for the correlational structure of the repeated measurements of the response variable.

It is the regression coefficients that are generally of most interest with the correlational structure parameters often being regarded as *nuisance parameters*. However, providing an adequate model for the correlational structure of the repeated measures is necessary to avoid misleading inferences about those parameters that are of primary relevance to the researcher, as is made clear in Fitzmaurice et al. (2011); these authors emphasize that although the estimation of the correlational structure of the repeated measurements is usually regarded as a secondary aspect of any analysis (relative to the mean response over time), the estimated correlational structure must describe the actual correlational structure present in the data relatively accurately to avoid making misleading inferences on the substantive parameters.

Over the last decade or two, methodology for the analysis of repeated measures data has been the subject of much research and development, and there are now a variety of powerful techniques available. A comprehensive account of these methods is given in Diggle et al. (2002) and Davis (2002). In this chapter we will concentrate on a single class of methods, *linear mixed effects models*, suitable for responses that can be assumed to be approximately normally distributed after conditioning on the explanatory variables. Non-normal responses will be the subject of Chapter 10.

9.2 Linear Mixed Effects Models for Repeated Measures Data

Linear mixed effects models for repeated measures data formalize the idea that an individual's pattern of responses is likely to depend on many characteristics of that individual, including some that are unobserved. These unobserved variables are included in the model as random variables, that is, *random effects.* The essential feature of the model is that the (usually positive) correlation among the repeated measurements on the same individual arises from shared, unobserved variables. Fitzmaurice et al. (2011) suggest several possible sources of correlation in longitudinal data including the following:

- Between-individual heterogeneity reflecting natural variation in individuals' propensity to respond—some consistently respond higher than the average and others consistently lower—the result being positive correlation between the repeated measurements of the response.

- Within-individual biological variation—the notion here is that some underlying biological process (or processes) that change through time in a relatively smooth and continuous fashion leads to random deviations from an individual's underlying response trajectory being more similar when measurements are obtained very close together in time—with the consequence that measurements taken closely together will typically be more highly correlated than measurements that are further separated in time.

Conditional on the values of the random effects, the repeated measurements are assumed to be independent, the so-called *local independence assumption.*

Two examples of linear mixed effects models, namely the *random intercept model* and the *random intercept and slope model* are introduced and described in Technical Section 9.1.

Technical Section 9.1: Introducing Linear Mixed Effects Models

Consider a simple set of longitudinal data in which a number of individuals each has values of a response variable recorded at times $t_1, t_2, \ldots, t_r$. (We assume the same set of time points for each individual to make the description of linear mixed effects models simpler, but longitudinal data in which each individual is observed at a different set of time points present no problems for such models.)

Let y_{ij} represent the value of the response for individual i at time t_j with $j = 1, 2, \ldots, r$ and $i = 1, 2, \ldots, n$. If the repeated measurements of the response variable are independent of one another, then the fact that sets of r observations come from the same individual could be ignored, and the data might be described by a simple linear regression model of the form

$$y_{ij} = \beta_0 + \beta_1 t_j + \varepsilon_{ij}$$

but for repeated measures data, independence is very unlikely, so this model is not appropriate. A possible model for the y_{ij} that does not assume independence is

$$y_{ij} = \beta_0 + \beta_1 t_j + u_i + \varepsilon_{ij}$$

which is often helpful to write as

$$y_{ij} = (\beta_0 + u_i) + \beta_1 t_j + \varepsilon_{ij}$$

for reasons that will (hopefully) become clear later.

In the model above, the total residual that would be present in the usual linear regression model has been partitioned into a subject-specific random component u_i, which is constant over time, plus a residual ε_{ij}, which varies randomly over time. The u_i is assumed to be normally distributed with zero mean and variance σ_u^2. Similarly, the ε_{ij} is, as always, assumed normally distributed with zero mean and variance σ^2. The u_i and ε_{ij} are assumed to be independent of each other and of the time t_j. This model is known as a *random intercept model*, the u_i terms producing the random intercepts. The repeated measurements made over time for an individual vary about that individual's own regression line, which can differ in intercept but not in slope from the regression lines of other individuals. The random effects u_i model possible heterogeneity in the intercepts of the individuals' regression lines.

Let us now look at how the presence of the random effects introduces covariance between the repeated measurements over time. The random intercept model implies that the total variance of each repeated measurement is

$$\text{Var}(y_{ij}) = \text{Var}(u_i + \varepsilon_{ij}) = \sigma_u^2 + \sigma^2$$

Due to this decomposition of the total residual variance into a between-subject component, σ_u^2, and a within-subject component, σ^2, the model is sometimes referred to as a *variance component model.* The covariance between the total residuals at two time points t_j and t_k, in the same individual i is given by

$$\text{Cov}(u_i + \varepsilon_{ij}, u_i + \varepsilon_{ik}) = \sigma_u^2$$

Note that these covariances are induced by the shared random intercept; for individuals with $u_i > 0$, the total residuals will tend to be greater than the mean; for individuals with $u_i < 0$, they will tend to be less than the mean. It follows from the two relations above that the residual correlations are given by

$$\text{Cor}(u_i + \varepsilon_{ij}, u_i + \varepsilon_{ik}) = \frac{\sigma_u^2}{\sigma_u^2 + \sigma^2}$$

This is an *intraclass correlation* interpreted as the proportion of the total residual variance that is due to residual variability between subjects.

The formulae given earlier for the variance of each repeated measurement and for the covariance of each pair of repeated measurements do not involve time, and demonstrate that the random intercept model constrains the variance of each repeated measurement to be the same and the correlation of each pair of measurements to be equal. This particular correlational structure is known as *compound symmetry.* Fitting a random intercept model to longitudinal data implies that the compound symmetry structure is considered appropriate for the data. However, this is very often not the case; for example, it is more common for measures taken closer to each other in time to be more highly correlated than those taken further apart. In addition, the variances of the measurements taken later in time are often greater than those measurements taken earlier. Consequently, for many longitudinal data sets, the random intercept model will not do justice to the observed pattern of variances and correlations between the repeated measurements, and will, therefore, not be the most appropriate model for the data.

A model that allows a more realistic structure for covariances is one that allows heterogeneity in both intercepts and slopes, namely, the *random slope and intercept model.* In this model, there are two types of random effects: the first modeling heterogeneity in intercepts, u_i, and the second modeling heterogeneity in slopes, v_i. Explicitly, the model is

$$y_{ij} = \beta_0 + \beta_1 t_j + u_i + v_i t_j + \varepsilon_{ij}$$

which may also be written as

$$y_{ij} = (\beta_0 + u_i) + (\beta_1 + v_i)t_j + \varepsilon_{ij}$$

so as to show more clearly how the two subject random effects alter the intercept and the slope in the model. The two random effects are assumed to have a bivariate normal distribution with zero means for both variables and variances σ_u^2 and σ_v^2 with covariance σ_{uv}. With this model, the total residual is $u_i + v_i t_j + \varepsilon_{ij}$ with variance

$$\text{Var}(u_i + v_i t_j + \varepsilon_{ij}) = \sigma_u^2 + 2\sigma_{uv} t_j + \sigma_v^2 t_j^2 + \sigma^2$$

which is now no longer constant for different values of t_j. Similarly, the covariance between two total residuals of the same individual is

$$\text{Cov}(u_i + v_i t_j + \varepsilon_{ij}, u_i + v_i t_k + \varepsilon_{ik}) = \sigma_u^2 + \sigma_{uv}(t_j + t_k) + \sigma_v^2 t_j t_k$$

and this is now not constrained to be the same for all pairs t_j and t_k. The random intercept and slope model allows for both variances of the repeated measurements that change with time, and covariances of pairs of repeated measurements that are not all the same.

In the model we have been considering, time has a fixed effect measured by the parameter β_1. It is this parameter that is likely to be of more interest to the investigator than the other parameters in the model, namely, the variance of the error terms and the variance (and possibly covariance) of the random effects. However, if the estimate of β_1 and its estimated standard error are derived from the simple regression model assuming independence, the standard error will be larger than it should be because of ignoring the likely within-subject dependences that will reduce the error variance in the model. Consequently, use of the simple regression model may give a misleading inference for β_1. As we shall see later, the situation for between-subject fixed effects is the reverse of that for within-subject fixed effects, with the estimated standard error of the effect being smaller in the (usually) inappropriate independence model than in a linear mixed effects model.

Linear mixed effects models can be estimated by maximum likelihood (ML). However, this method tends to underestimate the variance components. A modified version of ML, known as restricted maximum likelihood (REML), is therefore often recommended. Details are given in Diggle et al. (2002) and Longford (1993). Competing linear mixed effects models for a data set, for example, a random intercept model and a random intercept and slope model, can be compared using a likelihood ratio test (although see later comments about this test). The distinction between ML and REML is relevant when using the likelihood ratio test to compare two nested models because, unlike ML, which places no restrictions on likelihood ratio tests involving fixed and random effects, when REML is used, such tests are only appropriate when both models have the same set of fixed effects (see Longford, 1993).

9.3 How Do Rats Grow?

To begin to see how linear mixed effects models are applied in practice, we shall use some data from a nutrition study conducted in three groups of rats (Crowder and Hand, 1990). The three groups were put on different diets, and each animal's body weight (grams) was recorded repeatedly (approximately weekly, except in week seven when two recordings were taken) over a 9-week period. The question of most interest is whether the growth profiles of the three groups differ. The data are shown in Table 9.1.

9.3.1 Fitting the Independence Model to the Rat Data

To begin, we shall ignore the repeated-measures structure of the data and assume that all the observations are independent of one another. It is easier to imagine this if we write the data in what is known as the long form (in Table 9.1, the data are in their wide form); the data for the first two rats in group 1 are shown in their long form in Table 9.2. Now if we simply ignore that the sets of 11 weights come from the same rat, we have a data set consisting of 176 weights, times, and group memberships that we see can easily be analyzed using multiple linear regression. To begin, we will plot the data, identifying the observations in each group but ignoring the longitudinal nature of the

TABLE 9.1
Body Weights of Rats Recorded over a 9-Week Period

		Day										
ID	**Group**	**1**	**8**	**15**	**22**	**29**	**36**	**43**	**44**	**50**	**57**	**64**
1	1	240	250	255	260	262	258	266	266	265	272	278
2	1	225	230	230	232	240	240	243	244	238	247	245
3	1	245	250	250	255	262	265	267	267	264	268	269
4	1	260	255	255	265	265	268	270	272	274	273	275
5	1	255	260	255	270	270	273	274	273	276	278	280
6	1	260	265	270	275	275	277	278	278	284	279	281
7	1	275	275	260	270	273	274	276	271	282	281	284
8	1	245	255	260	268	270	265	265	267	273	274	278
9	2	410	415	425	428	438	443	442	446	456	468	478
10	2	405	420	430	440	448	460	458	464	475	484	496
11	2	445	445	450	452	455	455	451	450	462	466	472
12	2	555	560	565	580	590	597	595	595	612	618	628
13	3	470	465	475	485	487	493	493	504	507	518	525
14	3	535	525	530	533	535	540	525	530	543	544	559
15	3	520	525	530	540	543	546	538	544	553	555	548
16	3	510	510	520	515	530	538	535	542	550	553	569

data to give Figure 9.1. Clearly, there is a difference between the weights of the group 1 rats and those in the other two groups. Continuing to ignore the repeated-measures structure of the data, we might fit a multiple linear regression model with weight as response and time and group (coded as two dummy variables D_1 and D_2, with both D_1 and D_2 taking the value 0 for rats in group 1, D_1 being 1 and D_2 being 0 for rats in group 2, and D_1 being 0 and D_2 being 1 for rats in group 3) as explanatory variables. Fitting the model gives the results shown in Table 9.3. As we might have anticipated from Figure 9.1, both group 2 and group 3 differ significantly from group 1 conditional on time; the regression on time is also highly significant. We might go on to fit a model with a group × time interaction, but we will not do this because we know from the structure of the data that the model considered here is wrong. The model assumes independence of the repeated measures of weight, and this assumption is highly unlikely. So, now we will move on to consider both some more appropriate graphics and appropriate models.

TABLE 9.2
Long Form of the Data for the First Two Rats in Group 1 in Table 9.1

	ID	Group	Time	Weight
1	1	1	1	240
2	1	1	8	250
3	1	1	15	255
4	1	1	22	260
5	1	1	29	262
6	1	1	36	258
7	1	1	43	266
8	1	1	44	266
9	1	1	50	265
10	1	1	57	272
11	1	1	64	278
12	2	1	1	225
13	2	1	8	230
14	2	1	15	230
15	2	1	22	232
16	2	1	29	240
17	2	1	36	240
18	2	1	43	243
19	2	1	44	244
20	2	1	50	238
21	2	1	57	247
22	2	1	64	245

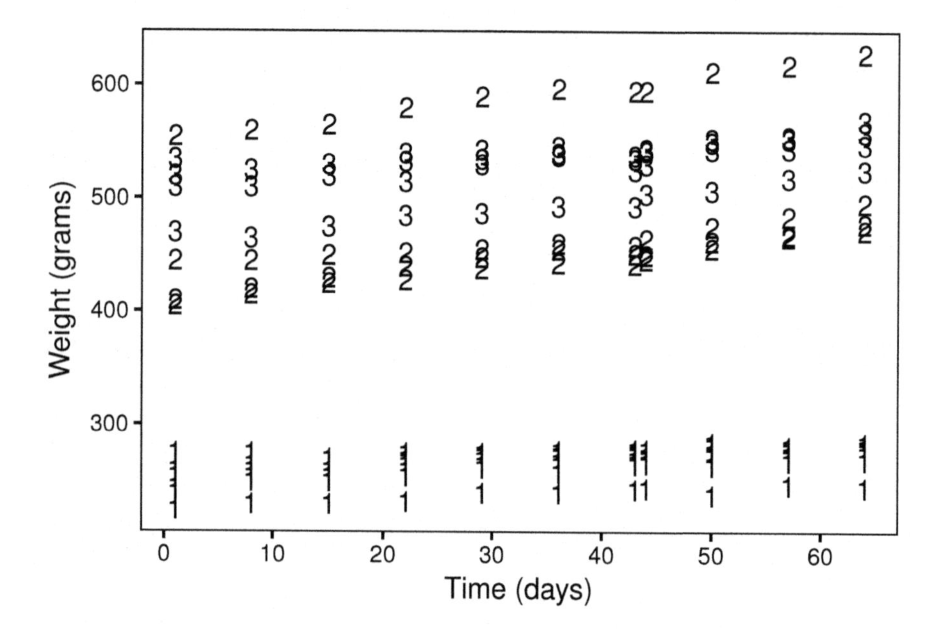

FIGURE 9.1
Plot of weight against time for rat data, ignoring the repeated-measures structure of the data but identifying the group to which each observation belongs.

9.3.2 Fitting Linear Mixed Models to the Rat Data

We begin with a graphical display of the rat growth data that takes into account the longitudinal structure of the data by joining together the points belonging to each rat to show the weight growth profiles of individual rats; the plot appears in Figure 9.2. In Figure 9.3, a scatterplot matrix of the repeated measures of weight, although not a terribly helpful graphic, does demonstrate that the repeated measures are certainly not independent of one another.

TABLE 9.3
Results from Fitting a Linear Regression Model to Rat Data with Weight as Response Variable, and Group and Time as Explanatory Variables, and Ignoring the Repeated-Measures Structure of the Data

	Estimate	Standard Error	t-Value	Pr(>|t|)
Intercept	244.0689	5.7725	42.281	<2e-16
Time	0.5857	0.1331	4.402	1.88e-05
D_1	220.9886	6.3402	34.855	<2e-16
D_2	262.0795	6.3402	41.336	<2e-16

Note: Multiple R-squared: 0.9283; F-statistic: 742.6 on 3 and 172 DF; p-value: <2.2e-16.

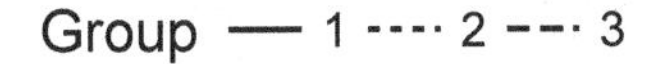

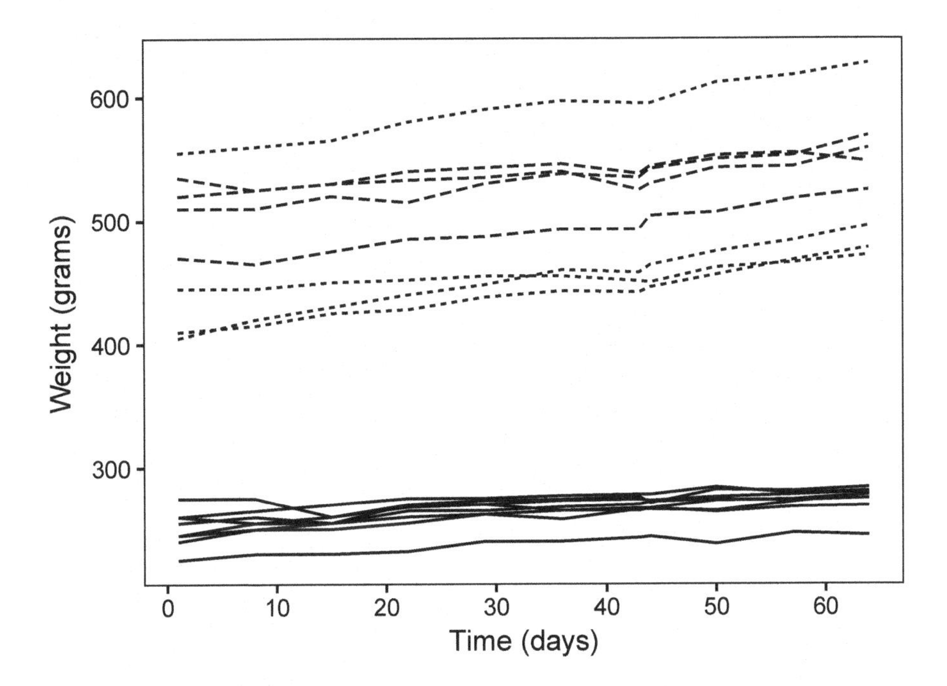

FIGURE 9.2
Plot of individual rat growth profiles.

To begin the more formal analysis of the rat growth data, we will first fit the random intercept model and include the two explanatory variables: time and group (coded as two dummy variables). If we represent the weight of the ith rat at time t_j by y_{ij}, the model can be written as

$$y_{ij} = (\beta_0 + u_i) + \beta_1 t_j + \beta_2 D_{i1} + \beta_3 D_{i2} + \varepsilon_{ij}$$

where u_i is the random effect specific to the ith subject, with these random effects having a normal distribution with zero mean and variance σ_u^2; the ε_{ij} are the usual "error" terms with a normal distribution with zero mean and variance σ^2, and D_{i1} and D_{i2} are the same two dummy variables used to code the group membership of the ith rat as in the independence model fitted earlier. This model allows the linear regression fit for each rat to differ in intercept from other rats. Fitting this model gives the results shown in Table 9.4. The estimated variance of the rat random effects is quite large, indicating the considerable variation in the intercepts of the regression fits of the individual rat growth profiles. The estimated regression parameters for time and the two dummy variables are very similar to those from fitting the independence model shown in Table 9.3, and all are highly significant again

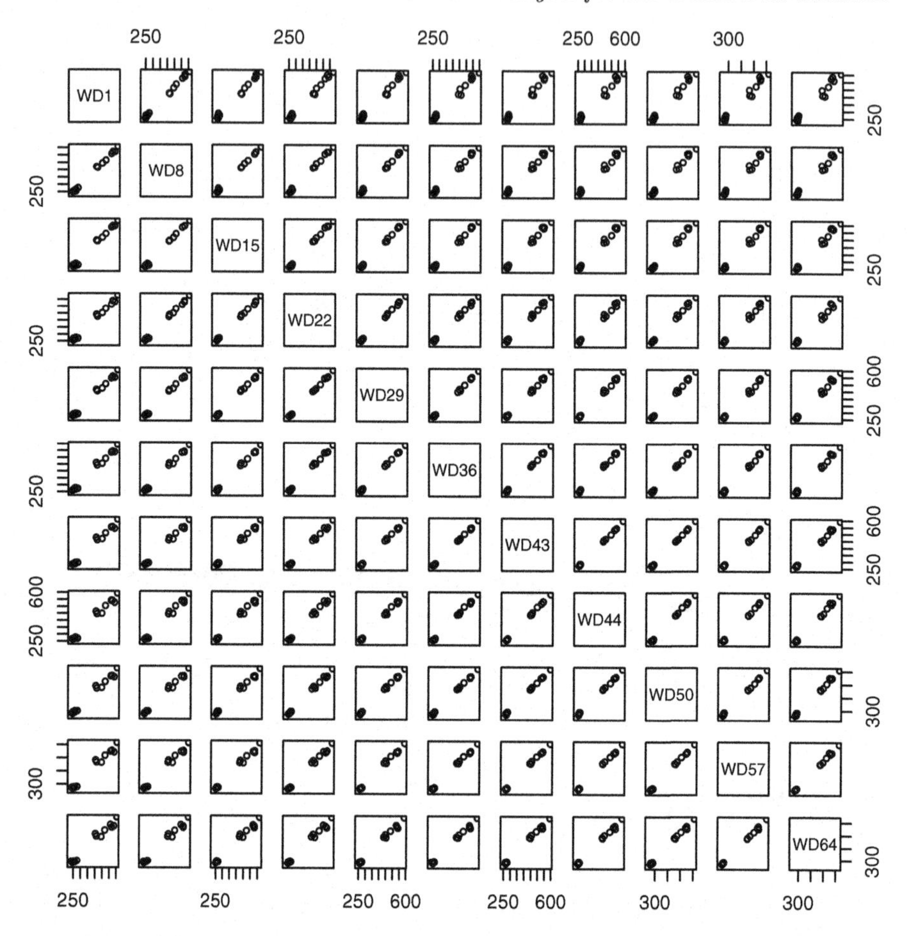

FIGURE 9.3
Scatterplot matrix of repeated measures in rat growth data.

as they were in Table 9.3. However, the estimated standard error of time is much smaller in Table 9.4 than it is in Table 9.3, reflecting the point made in Technical Section 9.1 that assuming independence will lead to the standard error of a within-subject covariate such as time being larger than it should be because of ignoring the likely within-subject dependences, which will reduce the error variance in the model. In contrast, the standard errors of each dummy variable in Table 9.4 are about three times the size of those in Table 9.3. The dummy variables are between-subject effects, and the reason for the smaller standard errors with the independence model is that the effective sample size for estimating these effects is less than the actual sample size because of the correlated nature of the data, and so the estimates for the independence model are unrealistically precise. In this example, the conclusions from the independence model and the random intercept model are the same, but in other examples this will not necessarily be so, as we shall see later.

Now we can move on to fit the random intercept and random slope model to the rat growth data; explicitly, the model is

$$y_{ij} = (\beta_0 + u_i) + (\beta_1 + v_i)t_j + \beta_2 D_{i1} + \beta_3 D_{i2} + \varepsilon_{ij}$$

where the extra term from the random intercept model is the random effect v_i that allows the linear regression fits for each individual to differ in slope; these random effects are assumed to have a normal distribution with zero mean and variance σ_v^2, and are allowed to be correlated with the u_i random intercept effects (see Technical Section 9.1). The results from fitting the random intercept and slope model to the rat growth data are shown in Table 9.5. The results for the fixed effects are very similar to those in Table 9.4, but the likelihood ratio test for the random intercept model versus the random intercept and slope model gives a chi-squared statistic of 142.94 with 2 degrees of freedom (DF) (the two additional parameters in the latter model are the variance of the v random effects and the covariance of the u and v random effects), and the associated p-value is very small. The random intercept and slope model provides a better fit for these data. (There are some technical problems with this likelihood ratio test, which are discussed in detail in Rabe-Hesketh and Skrondal, 2012; fortunately, the correct p-value for testing which of the two models is to be preferred can be found simply by dividing the p-value from the flawed likelihood ratio test by 2.)

Finally, we can fit a random intercept and slope model that allows for a group $\times$ time interaction. Explicitly, this model can be written as

$$y_{ij} = (\beta_0 + u_i) + (\beta_1 + v_i)t_j + \beta_2 D_{i1} + \beta_3 D_{i2} + \beta_4(D_{i1} \times t_j) + \beta_5(D_{i2} \times t_j) + \varepsilon_{ij}$$

TABLE 9.4
Results from Fitting Random Intercept Model, with Time and Group as Explanatory Variables, to Rat Growth Data

Random Effects			
$\hat{\sigma}_u^2 = 1085.92$, $\hat{\sigma}^2 = 66.44$			
Fixed Effects			
	Estimate	**Standard Error**	**t-Value**
Intercept	244.06890	11.73107	20.80
Time	0.58568	0.03158	18.54
D_1	220.98864	20.23577	10.92
D_2	262.07955	20.23577	12.95

TABLE 9.5
Results from Fitting the Random Intercept and Slope Model, with Time and Group as Explanatory Variables, to Rat Growth Data

Random Effects[a]			
$\hat{\sigma}_u^2 = 1140.54$, $\hat{\sigma}_v^2 = 0.11$, $\hat{\sigma}^2 = 19.75$			
Fixed Effects			
	Estimate	**Standard Error**	**t-Value**
Intercept	246.45727	11.81526	20.859
Time	0.58568	0.08548	6.852
D_1	214.58735	20.17983	10.634
D_2	258.92732	20.17983	12.831

[a] Estimated correlation between the u and v random effects is −0.22.

Fitting this model gives the results in Table 9.6. The likelihood ratio test of the interaction random intercept and slope model against the corresponding model without an interaction is 12.36 with 2 DF; the associated p-value is very small, and we can conclude that the interaction model provides a better fit for the rat growth data. The estimated regression parameters for the interaction in Table 9.6 indicate that the growth rate slopes are considerably higher for rats in group 2 than for rats in group 1 (on average 0.61 higher with an approximate 95% confidence interval [CI] of [0.33, 0.89]) but less so when comparing group 3 rats with those in group 1 (on average 0.30 higher, CI [0.02, 0.58]).

We can find the fitted values from the interaction model and plot the fitted growth rates for each rat; these are shown in Figure 9.4 alongside the

TABLE 9.6
Results from Fitting the Random Intercept and Slope Model That Allows for a Group × Time Interaction to Rat Growth Data

Random Effects[a]			
$\hat{\sigma}_u^2 = 1107.0$, $\hat{\sigma}_v^2 = 0.05$, $\hat{\sigma}^2 = 19.75$			
Fixed Effects			
	Estimate	**Standard Error**	**t-Value**
Intercept	251.65165	11.80279	21.321
Time	0.35964	0.08215	4.378
D_1	200.66549	20.44303	9.816
D_2	252.07168	20.44303	12.330
$D_1 \times$ Time	0.60584	0.14229	4.258
$D_2 \times$ Time	0.29834	0.14229	2.097

[a] Estimated correlation between the u and v random effects is −0.15.

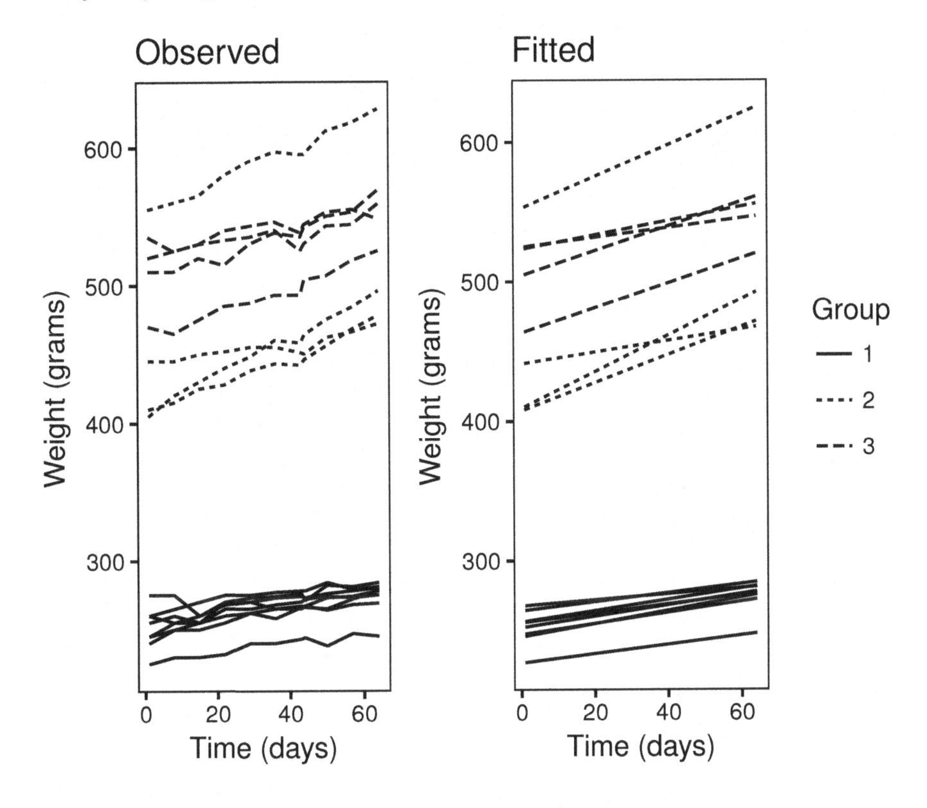

FIGURE 9.4
Fitted growth rate profiles from the interaction model and observed growth rate profiles.

observed values. This graphic underlines how well the interaction model fits the observed data. (The fitted values for each rat include "predicted" values of the u and v random effects for the rat; details of how these predicted values are calculated are given in Rabe-Hesketh and Skrondal, 2012.)

9.4 Computerized Delivery of Cognitive Behavioral Therapy—Beat the Blues

Depression is a major public health problem across the world. Antidepressants are the frontline treatment, but many patients either do not respond to them or do not like taking them. The main alternative is psychotherapy, and the modern "talking treatments" such as cognitive behavioral therapy (CBT) have been shown to be as effective as drugs, and probably more so when it comes

TABLE 9.7
First Five Patients in Each Treatment Group of the "Beat the Blues" (BtB) Clinical Trial of CBT for Depression

Drug	Length	Treatment	BDIpre	BDI2m	BDI4m	BDI6m	BDI8m
No	>6m	TAU	29	2	2	NA	NA
Yes	>6m	BtheB	32	16	24	17	20
Yes	<6m	TAU	25	20	NA	NA	NA
No	>6m	BtheB	21	17	16	10	9
Yes	>6m	BtheB	26	23	NA	NA	NA
Yes	<6m	BtheB	7	0	0	0	0
Yes	<6m	TAU	17	7	7	3	7
No	>6m	TAU	20	20	21	19	13
Yes	<6m	BtheB	18	13	14	20	11
No	>6m	TAU	30	32	24	12	2

Note: BDIpre, pretreatment Beck Depression Inventory.

to relapse (Watkins and Williams, 1998). But there is a problem, namely, availability—there are simply nothing like enough skilled therapists to meet the demand, and little prospect at all of this situation changing.

A number of alternative modes of delivery of CBT have been explored, including interactive systems making use of new computer technologies. The principles of CBT lend themselves reasonably well to computerization, and, perhaps surprisingly, patients adapt well to this procedure, and do not seem to miss the physical presence of the therapist as much as one might expect. Workers at the Institute of Psychiatry in the United Kingdom have developed one particular program, known as Beating the Blues (BtB). Full details are given by Proudfoot et al. (2004), but in essence, BtB is an interactive program using multimedia techniques, in particular, video vignettes. The computer-based intervention consists of nine sessions, followed by eight therapy sessions, each lasting about 50 minutes. Nurses are used to explain how the program works, but are instructed to spend no more than 5 minutes with each patient at the start of each session, and are there simply to assist with the technology. In a randomized controlled trial of the program, patients with depression recruited in primary care were randomized to either the BtB program or to "Treatment as Usual" (TAU). Patients randomized to BtB also received pharmacology and/or general practitioner (GP) support and practical/social help, offered as part of TAU, with the exception of any face-to-face counseling or psychological intervention. Patients allocated to TAU received whatever treatment their GP prescribed. The latter included, besides any medication, discussion of problems with GP, provision of practical/social help, referral to a counselor, referral to a practice nurse, referral to mental health professionals (psychologist, psychiatrist, community psychiatric nurse, counselor, etc.), or further physical examination.

A number of outcome measures were used in the trial, but here we concentrate on the Beck Depression Inventory II (BDI; Beck et al., 1996). Measurements on this variable were made on the following five occasions:

- Prior to treatment,
- 2, 4, 6, and 8 months after treatment began.

Data from 100 patients will be analyzed in this section; these data are a subset of the original and are used with the kind permission of the organizers of the study, in particular, Dr. Judy Proudfoot. Data for the first five patients from each treatment group are shown in Table 9.7. Two additional explanatory variables are also available for each patient: the first, drug, is whether the patient was taking antidepressant drugs (yes or no), and the second, length, is the length of the current episode of depression categorized into less than six months (<6m) or more than six months (>6m). The NAs (not available) in Table 9.7 indicate where a protocol-specified measurement of the BDI was not made; here, all the NAs are due to patients dropping out of the study. How dropouts might affect the results obtained from the analysis of the data will be discussed later in the chapter. The main question of interest here is to estimate the treatment effect of the BtB program.

In the distant past, the analysis of the BtB data would have involved only those patients with a complete set of five BDI values. At best, a "completers only" analysis would have been inefficient because the subset of BDI values for patients who dropped out are not used, thus lowering the sample size on which the analysis is based. But using only the completers in an analysis could have more dire consequences such as giving rise to biased parameter estimates and thus incorrect inferences. By considering the data in the long form, however, we see that analyses that use all the available data, including the BDI values that are recorded for patients who eventually drop out of the study, are straightforward. However, before considering models for the data, we should, as with any data set, try to discover some features of the data from some graphical material. So, we will begin by looking at boxplots of the BDI scores at each occasion of recording for each treatment group; the plot is shown in Figure 9.5. We see that the BDI scores decrease over time in each treatment group, but perhaps a little more in the BtB group, and the variance of the observations in the TAU group appears to be greater than those in the BtB group on each posttreatment time of recording. As a second graphic for these data, Figure 9.6 shows the scatterplot matrix of the five BDI scores; clearly, the repeated BDI values are not independent of one another.

We now move on to considering models for the BtB data. Again, we begin with an unrealistic multiple linear regression model that assumes that the repeated measures of the BDI are independent and contains the explanatory variables drug (coded 0 for no and 1 for yes), length (coded 0 for <6 months and 1 for >6 months), treatment (coded 0 for TAU and 1 for BtB), time, and BDI prevalue. The results are shown in Table 9.8. As we might have expected, the regression parameters for time and pretreatment BDI (BDIpre) are highly

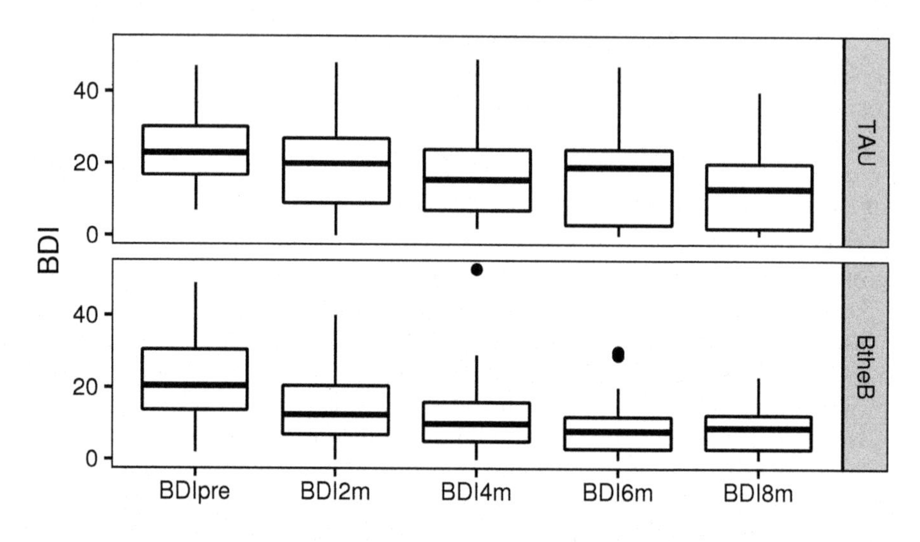

FIGURE 9.5
Box plots of BDI scores by occasion of recording and treatment group.

significant. The negative value for the time coefficient tells us what we have already surmised from the boxplots in Figure 9.5, namely, that the BDI scores decrease over time. The positive regression coefficient for BDIpre simply indicates that patients with, say, a higher-than-average BDI score before treatment begins, will tend to have higher-than-average values posttreatment. The regression coefficient for drug is also very significant, and its negative value tells us that patients taking antidepressant drugs will tend to have lower BDI scores than those not taking such medication. The regression coefficient for length is not significant at the 5% level; there is no evidence that the length of the current episode of depression affects the BDI score. Finally, we see that the regression coefficient for treatment is also highly significant; treatment with BtB rather than TAU is estimated to lower depression scores on average by 3.36 BDI units conditional on the other explanatory variables with an approximate 95% CI of $[-3.36-2\times1.10, -3.36+2\times1.10]$, that is, $[-5.56, -1.16]$ (this estimated treatment difference applies to all posttreatment occasions because, in the model fitted, there is no allowance for a treatment $\times$ time interaction). However, we know from Figure 9.6 that the independence assumption for the repeated BDI scores is almost certainly incorrect, and so we need to consider some linear mixed models for these data that do allow for departures from independence.

So, we start with a random intercept model including the same explanatory variables as the independence model. The results are shown in Table 9.9. The regression parameters for both time and BDIpre remain highly significant, but those for treatment and drug are now not significant primarily because

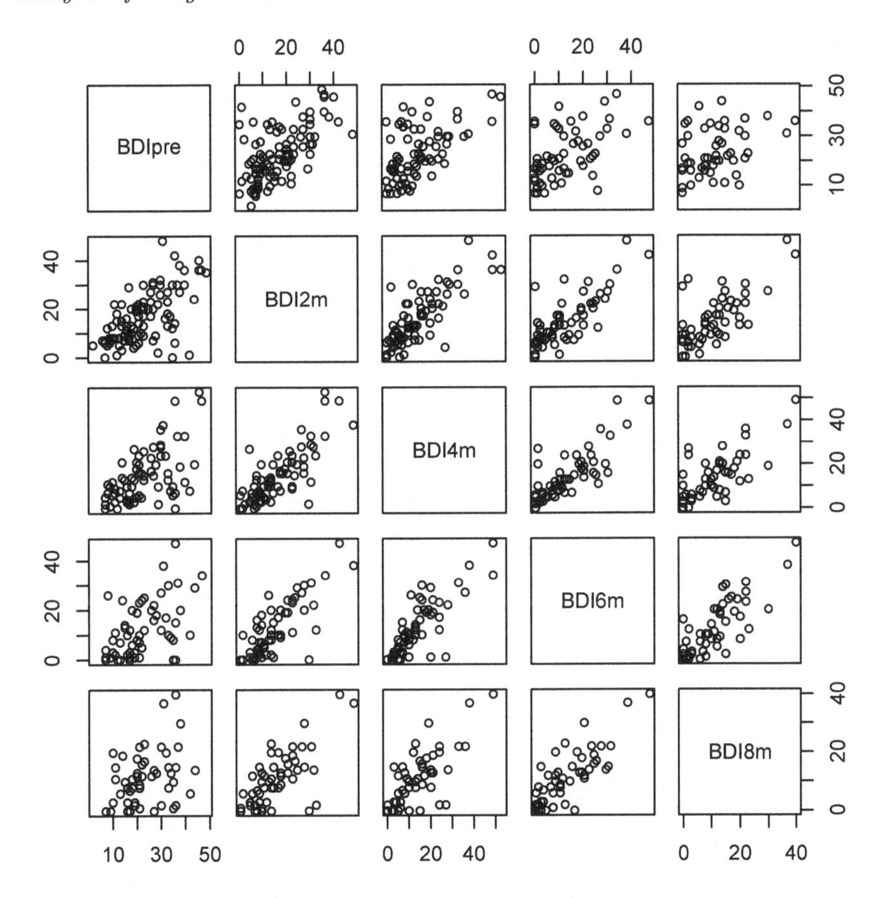

FIGURE 9.6
Scatterplot matrix of BDI scores.

TABLE 9.8
Results from Fitting a Multiple Linear Regression Model to BtB Data Assuming the Repeated Measurements of BDI Are Independent

	Estimate	Standard Error	t-Value	Pr(>|t|)
Intercept	7.88307	1.78049	4.427	1.38e-05
BDIpre	0.57237	0.05486	10.433	<2e-16
Time	−0.96081	0.23263	−4.130	4.82e-05
Treatment	−3.35397	1.09832	−3.054	0.00248
Drug	−3.54601	1.14469	−3.098	0.00215
Length	1.75308	1.10850	1.581	0.11492

Note: Multiple R-squared: 0.3978; F-statistic: 36.2 on 5 and 274 DF; *p*-value: <2.2e-16; *BDIpre:* pretreatment Beck Depression Inventory.

the associated estimated standard errors of these parameters have increased considerably. However, before using the estimates in Table 9.9 for interpretation, we should perhaps consider whether a random intercept and slope model gives a better fit. If we fit such a model, the likelihood ratio test comparing the random intercept model with the random intercept and slope model has a value of 0.82 with 2 DF; it appears that the more complicated model is not needed for these data. So, returning to Table 9.9 and, in particular, the estimated regression parameter for treatment, we find that treatment with BtB is estimated to lower the average BDI score by 2.36 units conditional on the other covariates with 95% CI of $[-5.78, 1.06]$; there is no compelling evidence that treatment with BtB is effective, a different conclusion from that produced by using the independence model.

TABLE 9.9
Results from Fitting a Random Intercept Model to BtB Data

Random Effects			
$\hat{\sigma}_u^2 = 51.44$, $\hat{\sigma}^2 = 25.27$			
Fixed Effects			
	Estimate	**Standard Error**	**t-Value**
Intercept	5.92148	2.30586	2.568
BDIpre	0.63888	0.07961	8.025
Time	−0.71353	0.14664	−4.866
Treatment	−2.35900	1.70841	−1.381
Drug	−2.78885	1.76594	−1.579
Length	0.23810	1.67537	0.142

Note: BDIpre, pretreatment Beck Depression Inventory.

9.5 Summary

- Repeated measurements of a response under different experimental conditions or over a period of time occur often in behavioral research.
- The analysis of such data requires special techniques because of the likely nonindependence of the repeated measurements.
- Linear mixed models allow for correlations between the repeated measurements by introducing random effects for subjects.
- The essential feature of such models is that there is natural heterogeneity across individuals in their responses over time and that this heterogeneity

can be represented by an appropriate probability distribution. Correlation between observations from the same individual arises from unobserved or unmeasured characteristics of the individual that remain the same over time, for example, an increased propensity to the condition under investigation, or perhaps a predisposition to exaggerate symptoms.

- Conditional on the values of the random effects, the repeated measurements of the response variable are assumed independent—the local independence assumption.

- Linear mixed effects models can be fitted by maximum likelihood, and competing models can be assessed by a likelihood ratio test.

- Longitudinal data often suffer from missing values of the response variable. In particular, patients often drop out of a study for a variety of reasons. Every effort needs to be made to avoid the problem as far as possible because if there are a large number of dropouts the conclusions drawn from any analysis of the data may be compromised. Missing values are considered in detail in Chapter 11, but it is worth pointing out here that the models described in this chapter can be shown to give valid results under the relatively weak assumption that the missing data mechanism is MAR as defined in Chapter 11 (see Carpenter et al., 2002).

- In this chapter, only responses that can be assumed to have a normal distribution conditional on the explanatory variables have been considered. Non-normal responses will be the subject of the next chapter.

9.6 Exercises

9.1 For the BtB data, construct a plot that shows the mean profiles over time for each treatment group and has appropriate error bars at each time point.

9.2 Investigate whether there is any evidence of a treatment × time interaction in the BtB data.

* 9.3 The data arise from a trial of estrogen patches in the treatment of postnatal depression. Women who had suffered an episode of postnatal depression were randomly allocated to two groups: the members of one group received an estrogen patch, and the members of the other group received a "dummy" patch—the placebo. The dependent variable was a composite measure of depression, which was recorded on two occasions prior to randomization and for each of 6 months posttreatment. A number of observations are missing (indicated by −9) because some women dropped out of the study. Begin by

fitting an independence model to the data using multiple linear regression with the depression score as dependent variable, and treatment time and the two baseline measurements as explanatory variables. Next, fit both a random intercept and random intercept and slope model, and test which model is best. Use the estimated treatment effect from the model you choose to find a confidence interval for the treatment effect. What do you conclude from this confidence interval?

* 9.4 The data give the plasma inorganic phosphate levels for 33 subjects, 20 of whom are controls and 13 of whom have been classified as obese (Davis, 2002). Produce separate plots of the profiles of the individuals in each group and, guided by these plots, fit what you think might be a sensible linear mixed model to the data.

10

Analysis of Longitudinal Data III: Non-Normal Responses

10.1 Introduction

In many longitudinal studies it will be clear that the assumption of normality for the response variable is simply not justified. Two examples are shown in Tables 10.1 and 10.2. The first, in Table 10.1, results from a clinical trial comparing two treatments for a respiratory illness (Davis, 1991). In each of two centres, eligible patients were randomly assigned to active treatment or placebo. During treatment, the respiratory status (categorized as 0 = poor, 1 = good) was determined at each of four monthly visits. A total of 111 patients were entered into the trial, 54 into the active group and 57 into the placebo group. The sex and age of each participant was also recorded along with their baseline respiratory status. Here the response variable is binary making the models described in the previous chapter inappropriate for these data. The observations for the first five patients in the data set are shown in Table 10.1.

The data in Table 10.2 also arise from a clinical trial reported in Thall and Vail (1990). Here 59 patients with epilepsy were randomized to receive either the anti-epileptic drug Progabide or a placebo in addition to standard chemotherapy. The number of seizures was counted over four two-week periods. In addition, a baseline seizure rate was recorded for each patient, based on the eight-week pre-randomization seizure count. Finally the age of each patient was recorded. Data for the first five patients only are given in Table 10.2.

TABLE 10.1
Respiratory Disorder Data (First Five Patients)

Patient	Centre	Treatment	Sex	Age	BL	V1	V2	V3	V4
1	1	1	1	46	0	0	0	0	0
2	1	1	1	28	0	0	0	0	0
3	1	2	1	23	1	1	1	1	1
4	1	1	1	44	1	1	1	1	0
5	1	1	2	13	1	1	1	1	1

Note: Treatment: 1 = placebo, 2 = active; Sex: 1 = male, 2 = female.

TABLE 10.2
Data for Five Patients from a Clinical Trial of Patients Suffering from Epilepsy

ID	Time1	Time2	Time3	Time4	Treatment	Baseline	Age
1	5	3	3	3	0	11	31
2	3	5	3	3	0	11	30
3	2	4	0	5	0	6	25
4	4	4	1	4	0	8	36
5	7	18	9	21	0	66	22

In this example the observations are counts which can take only positive values and so again make the normality assumption needed for the linear mixed effects models of the previous chapter difficult to justify.

In the models for normal responses described in Chapter 9, estimation of the regression parameters linking explanatory variables to the response variable and their standard errors needed to take account of the correlational structure of the data, but their interpretation could be undertaken *independently* of this structure. When modeling non-normal responses this independence of estimation and interpretation no longer holds; different assumptions about the source of the within-subject correlation can lead to regression coefficients with quite different interpretations.

The essential difference is between *marginal models* (also known as *population-average models*) and *conditional models* (also known as *subject-specific-models*). There is no automatic way of choosing between these two types of model for the analysis of non-normal longitudinal data. Instead, the choice has to be made on subject-matter considerations knowing that the different model types have different inferential targets and address often subtly different scientific questions, as we shall attempt to make clear in the following section. (A detailed account of these subtle differences in the two types of models is given in Fitzmaurice et al., 2011).

10.2 Marginal Models and Conditional Models

10.2.1 Marginal Models

Marginal models are essentially an extension of generalized linear models (GLM, Chapter 5) to longitudinal data. The "marginal" term is used in this context to describe that the mean response depends only on the covariates of interest and not on any random effects, in contrast to the linear mixed effects models of the previous chapter where the mean response depends not only on the covariates but also on a number of random effects. In essence longitudinal data can be considered as a series of cross-sections, and marginal models for such data use the GLM to fit each cross section. In such models the relationship of the marginal mean response to the covariates is modelled separately from the within-subject correlation among the repeated responses and the goal when fitting these models is to make inferences about population means; the within-subject correlation is regarded as a "nuisance" characteristic of the data that nevertheless has to be properly accounted for to make correct inferences about changes in the population mean response. The marginal regression coefficients have the same interpretation as coefficients from a cross-sectional analysis, and marginal models are natural analogues for correlated data of GLMs for independent data. Details of the marginal models are given in Technical Section 10.1.

Technical Section 10.1: Marginal Models

A marginal model for longitudinal data can be specified in terms of the following three components:

1. The expectation or mean of each response conditional on the covariates which is assumed to depend on these covariates through a known link function.
2. As in conventional GLMs covered in Chapter 5, the variances of the responses given the covariates are assumed to be of the form $V(y) = \phi V(\mu)$ where the variance function is determined by the choice of distribution family. The dispersion or scale parameter ϕ may be known or may have to be estimated. Because overdispersion is common in longitudinal data, estimation of ϕ is often needed even if the distribution requires $\phi = 1$.
3. The conditional within-subject correlation among the repeated responses, given the covariates is assumed to be a function of an additional set of association parameters.

The third component is needed to take care of the characteristic lack of independence of the repeated measurements of the response variable in longitudinal data. It should be noted that for a binary response the correlation is not the most useful measure of departure from independence because for such responses its values are restricted to ranges determined by the means of the response, i.e., the probability of a "success". The odds ratio (or log odds ratio) is a much more preferable measure of association among pairs of binary responses. (For more details, see Fitzmaurice et al., 2011.)

The problem with applying a direct analogue of the GLM to longitudinal data with non-normal responses is that there is usually no suitable likelihood function with the required combination of the appropriate link function, error distribution, and correlation structure to allow maximum likelihood to be used. To overcome this problem, Liang and Zeger (1986) introduced a general method for incorporating within-subject association in GLMs, which is essentially an extension of the quasi-likelihood approach mentioned in Chapter 5. The feature of this general method that differs from the usual GLM is that different responses on the same individual are, given the covariates, allowed to depart from independence with a relatively small number of parameters defining a relatively simple pairwise correlation structure for the repeated measurements. The covariance matrix of the repeated measurements implied by the assumed correlation structure is known as a "working" covariance matrix for the repeated measures with the implication that it may not accurately represent the variances and within-subject associations of the repeated measures. The estimated regression coefficients are "robust" in the sense that any misspecification of the model for the covariance has very little impact on the estimates of the regression coefficients, in particular they remain both unbiased and consistent assuming that the mean structure is correctly specified.

However, misspecification of the covariance leads to incorrect values for the estimated standard errors of the estimates of the regression coefficients and can lead to misleading inferences about the regression coefficients with confidence intervals which are too narrow (or in some cases too wide) and p-values that are too small (or sometimes too large). Where there is some doubt about the model used for the covariance structure of the repeated measurements valid estimates of the standard errors can be obtained using the so-called *sandwich estimator*; standard error estimates obtained in this way are robust to misspecification of the covariance model. Details of the sandwich estimator are given in, for example, Fitzmaurice et al. (2011).

Given that the sandwich estimator of the standard errors is available, an obvious question arises: "Why not use the estimator in all cases and so avoid the effort to model the within subject association?" For example, why not simply assume the repeated measurements of the response are independent and then use the sandwich estimators of the standard errors of the estimated regression coefficients? Fitzmaurice et al. (2011) give two main reasons for modeling the covariance structure:

1. In general, the closer the "working" covariance matrix approximates the true underlying covariance matrix, the greater the efficiency or precision with which the regression coefficients can be estimated.

2. The robustness property of the sandwich estimator is a large sample (or asymptotic) property and so the use of the estimator is best suited to balanced longitudinal data where the number of subjects is relatively large and the number of re-

peated measures relatively small. Reliance on the sandwich estimator is not to be recommended when the number of subjects is modest or the design is unbalanced.

So, in general, modeling the correlation structure of the repeated measurements will be worthwhile and the following possibilities are commonly used:

1. An identity matrix leading to the independence working model, obtained by assuming that the repeated measurements are independent.
2. An *exchangeable* correlation matrix with a single parameter. Here, the correlation between each pair of repeated measurements is assumed to be the same, i.e., $\text{Corr}(Y_{ij}, Y_{ik}) = \alpha$, where Y_{ij} is the jth repeated measurement for the ith individual.
3. An AR-1 autoregressive correlation matrix, also with a single parameter, but in which $\text{Corr}(Y_{ij}, Y_{ik}) = \alpha^{|k-j|}, j \neq k$. This can allow the correlations of measurements taken further apart to be less than those taken closer to one another.
4. An unstructured correlation matrix in which $\text{Corr}(Y_{ij}, Y_{ik}) = \alpha_{jk}$.

For given values of the regression parameters $\beta_1, \ldots, \beta_q$, the α-parameters of the working correlation matrix can be estimated along with the dispersion parameter ϕ (see Zeger and Liang, 1986, for details). These estimates can then be used in the so-called *generalized estimating equations* (GEE) to obtain estimates of the regression parameters. The GEE algorithm proceeds by iterating between (1) estimation of the regression parameters using the correlation and dispersion parameters from the previous iteration and (2) estimation of the correlation and dispersion parameters using the regression parameters from the previous iteration. (For more details see Fitzmaurice et al., 2011.)

The regression parameters in a marginal model describe features of the mean response in the population and how these features relate to the covariates. This interpretation is not altered by the assumptions made about the nature or the magnitude of the lack of within-subject independence.

10.2.2 Conditional Models

In Chapter 9 we saw how the incorporation of random effects for individuals introduces correlations among the repeated measurements at the population level. In this section we describe briefly how a GLM approach can be applied to longitudinal data with a non-normal response by allowing some of the regression coefficients in a model to vary randomly from one individual to another, leading to what are called *generalized linear mixed effects models.* As a simple illustration of such a model we will consider a logistic regression model for longitudinal data with a binary response. Details are given in Technical Section 10.2.

Technical Section 10.2: Conditional Models

In detail, we assume data in which Y_{ij} is the value of a binary response for individual i at say time t_j. The logistic regression model (see Section 5.2 in Chapter 5) for the response is now written as

$$\text{logit}\left[Pr(Y_{ij}=1|u_i)\right]=\beta_0+\beta_1 t_j+u_i$$

where u_i is a random effect assumed to be normally distributed with zero mean and variance σ_u^2. The model is a simple logistic regression model with randomly varying intercepts and can be considered as a discrete data analogue of the random intercept model described in the previous chapter. The model allows for natural heterogeneity in individuals' propensity to respond positively, a propensity that persists in all the repeated binary responses for an individual. In this model the regression parameter β_1 represents the change in the log odds per unit change in time as in the usual logistic regression model, but now it is *conditional* on the random effect u_i. In other words, β_1 represents the change in the log odds per unit change in time for any given individual having an unobservable underlying propensity to respond positively, u_i. In other words, the regression parameter represents the influence of the covariate on a *specific subject's* mean response. We can illustrate the conditional nature of the model graphically by simulating the formerly specified logistic regression model; the result is shown in Figure 10.1.

In Figure 10.1, thin curves represent subject-specific relationships between the probability that the response equals one and a covariate t for the preceding model. The horizontal shifts are due to different values of the random intercept. The thick curve represents the population averaged relationship, formed by averaging the thin curves for each value of t. It is, in effect, the thick curve that would be estimated in a marginal model (see previous subsection). The population averaged regression parameters tend to be attenuated (closer to zero) relative to the subject-specific regression parameters. A marginal regression model does not address questions concerning heterogeneity between individuals.

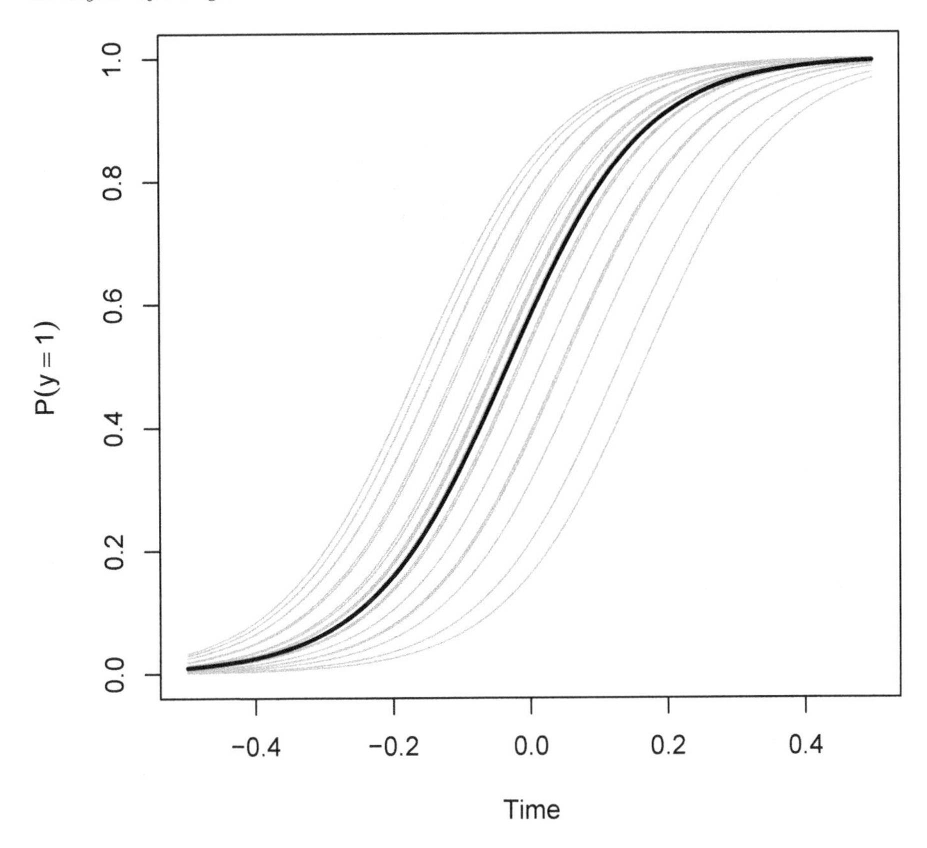

FIGURE 10.1
Simulation of random effects logistic regression model (Hothorn and Everitt, 2014, p. 267, used with permission).

Estimating the parameters in a generalized linear mixed effects model is undertaken by some form of maximum likelihood, but for details readers are referred to Fitzmaurice et al. (2011). The important point to reiterate here is that in conditional models the estimated regression coefficients have to be interpreted conditional on the random effects. The regression parameters in the model are said to be *subject-specific* and such effects will differ from the marginal or population averaged effects estimated using GEE, except when using an identity link function and a normal error distribution.

10.3 Using Generalized Estimating Equations to Fit Marginal Models

10.3.1 Beat the Blues Revisited

Although we have introduced generalized estimating equations as a method for analyzing longitudinal data where the response variable is non-normal, it can also be applied to data where the response can be assumed to follow a conditional normal distribution (conditioning being on the explanatory variables). Consequently, we first apply the method to the Beat the Blues data used in the previous chapter. Using generalized estimating equations approach we will fit two models, the first that assumes an independence structure for the working correlation matrix and then another model that assumes an exchangeable structure.

The results of fitting the independence model, giving both the empirical or robust (sandwich) estimators of the standard errors and the model based standard errors are shown in Table 10.3. Corresponding results from fitting the exchangeable model are given in Table 10.4.

Note how the model based and the robust estimates of the standard errors

TABLE 10.3
Results from Fitting Generalized Estimating Equations to the Beat the Blues Data Using the Normal Model with an Independent Correlation Structure

Estimates with Empirical (Robust) Standard Errors			
	Estimate	**Standard Error**	**z-Value**
Intercept	7.8831	2.1997	3.584
BDIpre	0.5724	0.0885	6.465
Time	−0.9608	0.1769	−5.432
Treatment	−3.3540	1.7139	−1.957
Drug	−3.5460	1.7307	−2.049
Length	1.7531	1.4195	1.235
Estimates with Model Based (Naive) Standard Errors			
	Estimate	**Standard Error**	**z-Value**
Intercept	7.8831	1.7805	4.427
BDIpre	0.5724	0.0549	10.433
Time	−0.9608	0.2326	−4.130
Treatment	−3.3540	1.0983	−3.054
Drug	−3.5460	1.1447	−3.098
Length	1.7531	1.1085	1.581

Note: Estimated Scale Parameter: 74.9; *BDIpre*, pretreatment Beck Depression Inventory.

TABLE 10.4
Results from Fitting Generalized Estimating Equations to the Beat the Blues Data Using the Normal Model with an Exchangeable Correlation Structure

Estimates with Empirical (Robust) Standard Errors			
	Estimate	**Standard Error**	**z-Value**
Intercept	5.8855	2.1071	2.793
BDIpre	0.6400	0.0793	8.069
Time	−0.7078	0.1539	−4.598
Treatment	−2.3360	1.6622	−1.405
Drug	−2.7743	1.6482	−1.683
Length	0.2085	1.4805	0.141
Estimates with Model Based (Naive) Standard Errors			
	Estimate	**Standard Error**	**z-Value**
Intercept	5.8855	2.3238	2.533
BDIpre	0.6400	0.0803	7.967
Time	−0.7078	0.1425	−4.966
Treatment	−2.3360	1.7262	−1.353
Drug	−2.7743	1.7840	−1.555
Length	0.2085	1.6918	0.123

Note: Estimated Scale Parameter: 77.1; *BDIpre*, pretreatment Beck Depression Inventory.

are considerably different for the independence structure (Table 10.3), but quite similar for the exchangeable structure (Table 10.4). This simply reflects that using an exchangeable working correlation matrix is more realistic for these data and that the standard errors resulting from this assumption are already quite reasonable to them without applying the "sandwich" procedure, i.e., the robust method of estimating the standard errors. In addition, if we compare the results under this assumed structure with those for the random intercept model given in Chapter 9 (see Table 9.9) we see that they are almost identical, since the random intercept model also implies an exchangeable structure for the correlations of the repeated measurements.

The single estimated parameter for the working correlation matrix from the GEE procedure is 0.692, quite similar to the estimated intra-class correlation coefficient from the random intercept model, i.e., $51.44/(25.27+51.44) = 0.671$ (see Table 9.9).

10.3.2 Respiratory Illness

As always, it is good practice to begin the exploration of any data set with one or more hopefully informative graphic(s). Here a simple bar-chart of the proportion of positive responses by visit (including baseline) is useful. The plot is shown in Figure 10.2 and suggests, informally at least, that the active

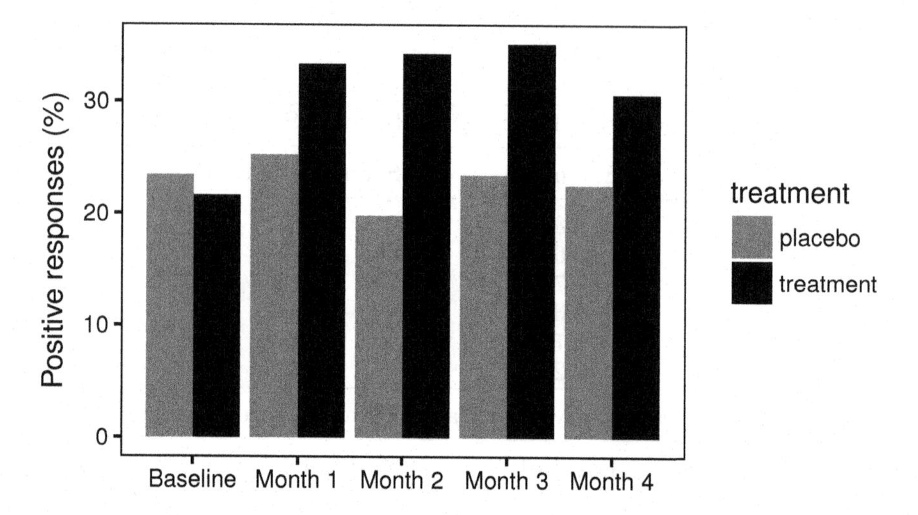

FIGURE 10.2
Proportion of positive responses by visit (including baseline) in the respiratory data.

treatment increases the proportion of positive responses at each post-baseline visit compared to the placebo.

First we will fit a marginal model that includes only the covariate age. As we are dealing with a binary response we shall assume that the probability of a good respiratory response is related to age by a logit link function and so the logistic regression model will be

$$\log\left\{\frac{Pr(Y_{ij}=1)}{Pr(Y_{ij}=0)}\right\} = \beta_0 + \beta_1 \text{age}_i \tag{10.1}$$

where $Y_{ij} = 1$ if the ith individual has good respiratory status on visit V_j and $Y_{ij} = 0$ if respiratory status is poor. To compare with later results, we will begin by fitting a model that assumes that the pairwise repeated measurements are independent.

The results giving both the empirical or robust (sandwich) estimators of the standard errors and the model based standard errors are shown in Table 10.5. The estimated regression coefficient for age is -0.0120 but of more interest are the two estimates of the standard error of this parameter estimate; the sandwich estimate is 0.0117 and the model based estimate is 0.0071. Here we know that the independence model is unrealistic and the standard error estimate based on the unrealistic model is too optimistic about the precision of the parameter estimate.

The assumption of independence is usually unrealistic for repeated measures data, and hence we need to consider a marginal model that allows for

TABLE 10.5
Results from Fitting a Logistic Regression Model to the Respiratory Data with Age as the Only Covariate and Assuming Independence between the Repeated Measurements

Estimates with Empirical (Robust) Standard Errors			
	Estimate	**Standard Error**	**z-Value**
Intercept	0.6458	0.3957	1.6320
age	−0.0120	0.0117	−1.0239
Estimates with Model Based (Naive) Standard Errors			
	Estimate	**Standard Error**	**z-Value**
Intercept	0.6458	0.2553	2.5294
age	−0.0120	0.0071	−1.7016

departures from independence. We apply the procedure with an exchangeable structure for the working correlation matrix and again fit the model to the data with the single covariate, age. The results are shown in Table 10.6. Comparing the values of the empirical and model based estimates of the standard error of the age regression coefficient in this model we find that they are very similar (0.0117 and 0.0114) and also very similar to the empirical estimate in the previous "independence" model (0.0117).

Finally, we will fit a marginal model, again with the exchangeable correlation structure, but now having also as covariates time, baseline respiratory status, centre, treatment, and sex. Using an obvious nomenclature to label the covariate values for the ith individual and where time_j takes the value j

TABLE 10.6
Results from Fitting a Logistic Regression Model to the Respiratory Data with a Single Covariate, Age, and Using an Exchangeable Correlation Structure

Estimates with Empirical (Robust) Standard Errors			
	Estimate	**Standard Error**	**z-Value**
Intercept	0.6458	0.3957	1.6320
age	−0.0120	0.0117	−1.0239
Estimates with Model Based (Naive) Standard Errors			
	Estimate	**Standard Error**	**z-Value**
Intercept	0.6458	0.4104	1.5736
age	−0.0120	0.0114	−1.0586

for $j = 1, 2, 3, 4$, the logistic regression model we shall fit is

$$\log\left\{\frac{Pr(Y_{ij}=1)}{Pr(Y_{ij}=0)}\right\} = \beta_0 + \beta_1\text{age}_i$$
$$+ \beta_2\text{time}_j + \beta_3\text{baseline}_i + \beta_4\text{centre}_i + \beta_5\text{treatment}_i + \beta_6\text{sex}_i \quad (10.2)$$

The results are shown in Table 10.7. First we might compare the empirical estimates of the standard errors and those derived from the fitted model. If we do this we find that the two sets of standard error estimates are very similar, suggesting that the chosen non-independence structure adequately describes the departures from independence in the data. The covariate of most interest in this study is, of course, treatment, and the estimated regression coefficient for this covariate is 1.292 with a 95% confidence interval [0.62, 1.96] (using the model standard errors). Exponentiating the limits of the confidence interval leads to the conclusion that the odds in favor of a good respiratory response in the active treatment group is between about 2 and 7 times the corresponding odds in the placebo group. The estimated treatment effect from the marginal model describes how the average rates (expressed as odds) of a good respiratory response in the study population would increase if patients were treated with the active drug.

TABLE 10.7
Results from Fitting a Logistic Regression Model Specified in Equation (10.2) to the Respiratory Data Using an Exchangeable Correlation Structure

Estimates with Empirical (Robust) Standard Errors			
	Estimate	**Standard Error**	**z-Value**
Intercept	−0.7332	0.5190	−1.413
age	−0.0184	0.0130	−1.415
time	−0.0642	0.0815	−0.788
baseline	1.8778	0.3501	5.364
centre	0.6809	0.3568	1.909
treatment	1.2922	0.3505	3.686
sex	0.1308	0.4441	0.294
Estimates with Model Based (Naive) Standard Errors			
	Estimate	**Standard Error**	**z-Value**
Intercept	−0.7332	0.5178	−1.416
age	−0.0184	0.0126	−1.465
time	−0.0642	0.0810	−0.793
baseline	1.8778	0.3419	5.492
centre	0.6809	0.3396	2.005
treatment	1.2922	0.3356	3.850
sex	0.1308	0.4179	0.313

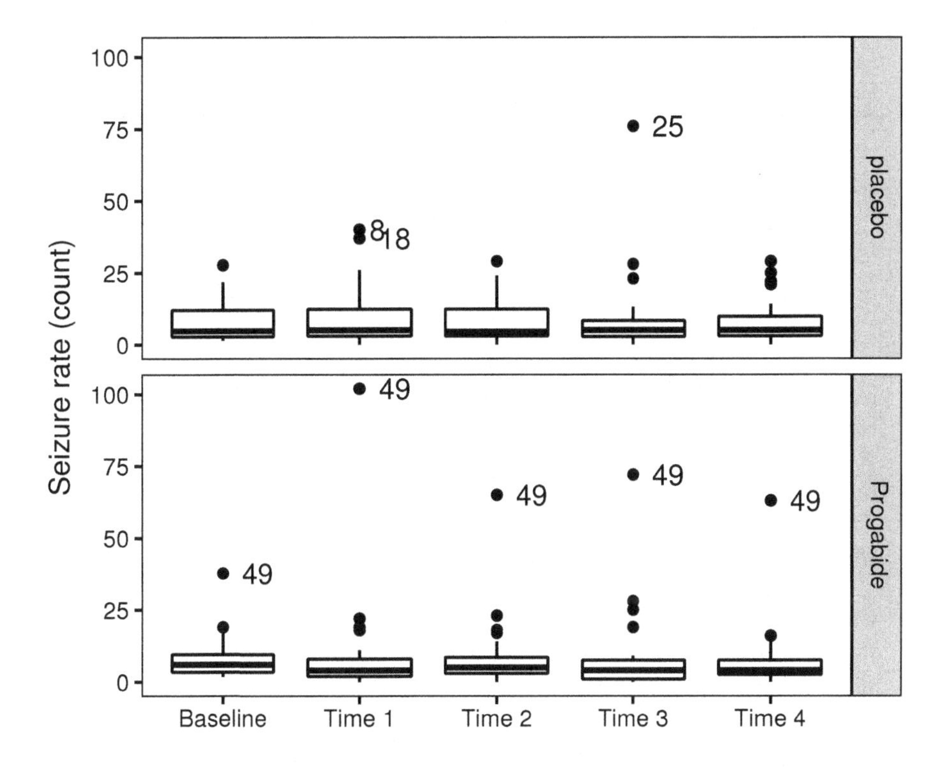

FIGURE 10.3
Box plots of seizure counts at each time point for each treatment group in the epilepsy data.

(Other correlation structures could be considered but this will be left as an exercise for the reader.)

10.3.3 Epilepsy

Again our starting point for dealing with these data is graphical, namely the box-plots of the number of epileptic seizures before and after treatment separately for the two treatment groups (the before count is the two-week average). The graph is shown in Figure 10.3. There is little convincing evidence for a treatment effect from this graph but there is evidence that some of the patients have some very large seizure rates, particularly patient 49. This patient could have a unreasonably large influence on any analysis of the data and should perhaps be considered for removal prior to any such analysis. Here, however, we shall model *all* the data *including* the observations with very high seizure rates. Readers are encouraged to repeat the analysis with at

TABLE 10.8
Means and Variances of Seizure Rates at Baseline and for the Post-Treatment Times in the Epilepsy Data

	Mean	**Variance**
Baseline	7.81	45.1
Time 1	8.95	220.1
Time 2	8.36	103.8
Time 3	8.44	200.2
Time 4	7.31	93.1

least observation number 49 removed and compare their results with the ones that follow.

It is also useful to look at the means and variances of seizure rates at baseline and at the post treatment times. The results are given in Table 10.8. The variances are far larger than the corresponding means—a point that we will return to in the next section.

Now we shall move on to fit a number of marginal models to these data. Count data are usually modelled as Poisson random variables using a log link function and that is what we shall do here. Using all available covariates then the model for the mean response is therefore

$$\log\left[\mathrm{E}(N_{ij})\right] = \beta_0 + \beta_1 \mathrm{time}_j + \beta_2 \mathrm{age}_i + \beta_3 \log(\mathrm{baseline}_i) + \beta_4 \mathrm{treatment}_i \quad (10.3)$$

where N_{ij} is the observed number of seizures for individual i in time period time_j for $j = 1, 2, 3, 4$. We use log(baseline) rather than baseline itself so that the exponentiated regression coefficient for the covariate represents the effect of the number of baseline seizures on subsequent seizure rates. We shall assume an exchangeable correlational structure.

The descriptive statistics given in Table 10.8 show that the variances are substantially greater than the corresponding means. As a result, the Poisson assumption that the means and variances are the same is not appropriate for these data; this overdispersion will be accounted for by allowing for the scale parameter to be estimated and not fixed at one.

The results are shown in Table 10.9. The model based standard errors and the corresponding empirical values are very similar. There is no evidence of a treatment effect. Clearly the baseline seizure rate influences the subsequent seizure rates and age also has an effect. The exponentiated confidence interval limits for the age effect using the robust standard errors in Table 10.9 indicate that the seizure rate for an increase in age of one year is about 1.001 to 1.04 times that of the younger age. Here it might be more useful to give the corresponding confidence interval for a ten year age difference; this can be calculated simply as $[\exp(10\times0.02-1.96\times0.0098), \exp(10\times0.02+1.96\times0.0098)]$, i.e., [1.20, 1.25].

TABLE 10.9
Results from Fitting the Model Specified by Equation (10.3) to the Epilepsy Data Using an Exchangeable Correlational Structure

Estimates with Empirical (Robust) Standard Errors			
	Estimate	**Standard Error**	**z-Value**
Intercept	−0.8753	0.4734	−1.849
time	−0.0587	0.0350	−1.678
age	0.0200	0.0098	2.042
log(baseline)	1.2247	0.1557	7.867
treatment	−0.0242	0.1911	−0.127
Estimates with Model Based (Naive) Standard Errors			
	Estimate	**Standard Error**	**z-Value**
Intercept	−0.8753	0.5295	−1.653
time	−0.0587	0.0346	−1.698
age	0.0200	0.0125	1.599
log(baseline)	1.2247	0.1055	11.603
treatment	−0.0242	0.1553	−0.156

Note: Estimated Scale Parameter: 4.82.

The scale parameter is estimated to be 4.82 indicating the overdispersion, relative to that predicted by Poisson variability, in these data.

10.4 Using Generalized Linear Mixed Effects Models to Fit Conditional Models

10.4.1 Respiratory Illness

In this subsection we will fit the following conditional model to the respiratory data:

$$\log\left\{\frac{Pr(Y_{ij}=1|u_i)}{Pr(Y_{ij}=0|u_i)}\right\} = \beta_0 + \beta_1\text{age}_i + \beta_2\text{time}_j + \beta_3\text{baseline}_i + \beta_4\text{center}_i + \beta_5\text{treatment}_i + \beta_6\text{sex}_i + u_i \quad (10.4)$$

where u_i is a random effect assumed to be normally distributed with zero mean and variance σ_u^2.

The results are shown in Table 10.10. Concentrating on the estimated treatment effect from the random effects model, namely 2.163 with estimated

TABLE 10.10
Results from Fitting the Logistic Regression Model Specified in Equation (10.4) to the Respiratory Data

		Random Effects		
$\hat{\sigma}_u^2 = 3.89$				
		Fixed Effects		
	Estimate	**Standard Error**	**z-Value**	**Pr(>|z|)**
Intercept	−1.3958	0.8383	−1.67	0.0959
age	−0.0255	0.0201	−1.26	0.2061
time	−0.1013	0.1252	−0.81	0.4182
baseline	3.0783	0.6027	5.11	3.3e-07
centre	1.0467	0.5478	1.91	0.0561
treatment	2.1632	0.5564	3.89	0.0001
sex	0.2025	0.6727	0.30	0.7634

Note: Log-likelihood: −214.

standard error 0.556, leading to a 95% confidence interval for the odds ratio of approximately [3, 26]. The variance of the random effects in the model is estimated to be 3.89. In this model, the treatment effect describes the effect of treatment on any patient's probability of a positive respiratory response.

So the answer to the question "how effective is the active drug" depends on whether the scientific interest is in the impact of the drug on the study population or on a randomly sampled individual from that population. Here the answer to both questions is that the drug is effective.

10.4.2 Epilepsy

We begin by fitting a random intercept model, namely

$$\log\left[\mathrm{E}(N_{ij}|u_i)\right] = \beta_0 + \beta_1 \mathrm{time}_j + \beta_2 \mathrm{age}_i + \beta_3 \log(\mathrm{baseline}_i) + \beta_4 \mathrm{treatment}_i + u_i \quad (10.5)$$

where u_i is a random effect assumed to be normally distributed with zero mean and variance σ_u^2.

The results are shown in Table 10.11. The random effects in the model are estimated to have variance 0.268. Of most interest here is that the treatment effect is marginally significant at the 5% level. An approximate 95% confidence interval for the effect of treatment on seizure count is [0.54, 0.98]. The seizure count with progabide is estimated to be between just over 50% to 98% of the seizure rate on the placebo; this describes the effect of treatment on a specific patient's seizure count. The treatment effect in the marginal model

TABLE 10.11
Results from Fitting the Model in Equation (10.5) to the Epilepsy Data

Random Effects				
$\hat{\sigma}_u^2 = 0.268$				
Fixed Effects				
	Estimate	**Standard Error**	**z-Value**	**Pr(>|z|)**
Intercept	−0.1753	0.4387	−0.40	0.6894
time	−0.0587	0.0202	−2.91	0.0036
age	0.0107	0.0122	0.88	0.3780
log(baseline)	1.0253	0.1012	10.14	<2e-16
treatment	−0.3200	0.1506	−2.12	0.0336

Note: Log-likelihood: −667.

fitted to these data in the previous section describe the effect of treatment in the population of patients assigned to placebo versus progabide.

Finally, we will fit a model that includes random effects for intercept and slope, i.e., the following model:

$$\log\left[\mathrm{E}(N_{ij}|\mathbf{u})\right] = \beta_0 + (\beta_1 + u_{i_2})\mathrm{time}_j + \beta_2\mathrm{age}_i + \beta_3\log(\mathrm{baseline}_i) + \beta_4\mathrm{treatment}_i + u_{i_1} \quad (10.6)$$

where $\mathbf{u}' = [u_{i_1}, u_{i_2}]$ and the random effects are assumed to have a bivariate normal distribution with zero mean and a covariance matrix $\boldsymbol{\Sigma}_\mathbf{u}$ given by

$$\boldsymbol{\Sigma}_\mathbf{u} = \begin{bmatrix} \sigma_{u1} & \sigma_{u1,u2} \\ \sigma_{u1,u2} & \sigma_{u2} \end{bmatrix} \quad (10.7)$$

Here, the model is a log-linear regression model with randomly varying intercept and slopes used to describe the heterogeneity among individuals in both their baseline seizure level and in the expected number of seizures over time.

The results are shown in Table 10.12. The parameter estimates in the table are very similar to those in Table 10.11. If we compare the log-likelihood values for the two models we see that the decrease when fitting the random intercept and random slope effects is 10 for the addition of two parameters, the variance of the slope random effect and the covariance of the two random effects. Testing the decrease as a chi-squared variables with 2 degrees of freedom suggests that the more complicated model gives a far better fit here.

The random effects for the intercept have estimated variance of 0.413 and for the slope random effects the corresponding value is 0.021; the correlation of the two types of random effects is −0.60.

TABLE 10.12
Results from Fitting the Model Specified in Equations (10.6) and (10.7) to the Epilepsy Data

	Random Effects[a]			
$\hat{\sigma}^2_{u_1} = 0.4134$, $\hat{\sigma}^2_{u_2} = 0.0213$				
	Fixed Effects			
	Estimate	**Standard Error**	**z-Value**	**Pr(>|z|)**
Intercept	−0.1877	0.4435	−0.42	0.672
time	−0.0514	0.0326	−1.57	0.116
age	0.0100	0.0123	0.81	0.417
log(baseline)	1.0252	0.1009	10.17	<2e-16
treatment	−0.3167	0.1502	−2.11	0.035

Note: Log-likelihood: −657.
[a] Estimated correlation between the u_1 and u_2 random effects is −0.60.

10.5 Summary

- In this chapter the generalized linear model (GLM, Chapter 5) has been extended to deal with longitudinal data in two different ways, marginal models and conditional or generalized linear mixed effects models.

- Marginal models are used to make inferences about population means on some transformed scale, for example, the logit or log scale, by modeling the mean conditional on the covariates but not on unobserved random effects.

- The model for the mean and the model to account for lack of pairwise independence in the repeated measurements are specified separately.

- In generalized linear mixed effects models, by contrast, random effects are used to model heterogeneity in some of the regression coefficients (e.g., slopes and intercepts) but conditional on these random effects the repeated measurements for an individual are independent.

- The regression coefficients in these models have subject-specific effects that describe changes in an individual's mean response and how these changes are related to covariates.

- Conditional models are of most use when the aim of the investigator is to make inferences about individuals rather than the study population.

- Though less unified than the methods available for the analysis of longitudinal data with normally distributed responses, the generalized estimating equations and the generalized linear mixed effects approaches provide powerful and flexible tools to analyse longitudinal data with binary and other non-normal response variables.

- More details are given in Fitzmaurice et al. (2011) who also give an excellent and more detailed account of marginal and conditional models for longitudinal data that given in this chapter which is, of necessity, relatively brief.

10.6 Exercises

10.1 For the epilepsy data investigate what Poisson models are most suitable when subject 49 is excluded from the analysis.

10.2 Investigate the use of other correlational structures than the independence and exchangeable structures used in the text, for both the respiratory and the epilepsy data.

* 10.3 The data shown in Table 10.13 were collected in a follow-up study of women patients with schizophrenia (Davis, 2002). The binary response recorded at 0, 2, 6, 8, and 10 months after hospitalization was thought disorder (absent or present). The single covariate is the factor indicating whether the patient had suffered early or late onset of her condition (age of onset less than 20 years or age of onset 20 years or above). The question of interest is whether the course of the illness differs between patients with early and late onset. Investigate this question using the generalized estimating equations approach.

TABLE 10.13
Clinical Trial Data from Patients Suffering from Schizophrenia (Only the Data of the First Four Patients Shown)

Subject	Onset	Disorder	Month
1	$<$ 20 yrs	present	0
1	$<$ 20 yrs	present	2
1	$<$ 20 yrs	absent	6
1	$<$ 20 yrs	absent	8
1	$<$ 20 yrs	absent	10
2	$>$ 20 yrs	absent	0
2	$>$ 20 yrs	absent	2
2	$>$ 20 yrs	absent	6
2	$>$ 20 yrs	absent	8
2	$>$ 20 yrs	absent	10
3	$<$ 20 yrs	present	0
3	$<$ 20 yrs	present	2
3	$<$ 20 yrs	absent	6
3	$<$ 20 yrs	absent	8
3	$<$ 20 yrs	absent	10
4	$<$ 20 yrs	absent	0
4	$<$ 20 yrs	absent	2
4	$<$ 20 yrs	absent	6
4	$<$ 20 yrs	absent	8
4	$<$ 20 yrs	absent	10

11

Missing Values

11.1 Introduction

Any well-designed study in the behavioral sciences should aim to draw a representative sample from the study population by following a well-defined sampling plan and a detailed protocol. But even the best laid plans can often go a little wrong and at the end of the study some of the data that should have been collected are *missing*. In a sample survey, for example, some individuals may have refused to respond or have not been contactable, or some of the participants have failed to answer particular items in a questionnaire. And as we have already seen in previous chapters, in longitudinal studies, data are often missing because subjects drop out prior to the end of the study.

Missing data can sometimes arise by design. For example, suppose one objective in a study of obesity is the estimate the distribution of a measure Y_1 of body fat in the population and correlate it with other factors. As Y_1 is expensive to measure it can only be obtained for a limited sample, but a crude proxy measure Y_2, for example, body mass index, can be obtained for a much larger sample. A useful design is to measure Y_2 and several covariates for a large sample and Y_1, Y_2 and the same covariates for a smaller subsample. The subsample allows predictions of the missing values of Y_1 to be generated for the larger sample using one or other of the methods we shall discuss later in the chapter thus yielding more efficient estimates than are possible from the subsample alone (this example is taken from Little, 2005).

Ignoring the situation when missing data are deliberate by design, the most important approach to the potential problems that missing data can cause is for the investigator to do his or her very best to avoid missing values in the first case. But despite the very best efforts of the investigator, some of the intended data will often be missing after data collection. In most research studies the intent of the majority of statistical analyses is to make valid inferences regarding a population of interest from the sample data at hand. Missing data threatens this goal if it is missing in a manner which makes the sample different than the population from which it was drawn, that is, if the missing data creates a biased sample. Therefore, it is important to respond to a missing data problem in a manner which, as far as possible, avoids this problem. It needs to be understood that once data are missing, it is impossible not to deal with them because once data are missing, any subsequent procedure applied to the data set represents a response in some form to the missing data problem. In this chapter a number of ways of dealing with missing values will be discussed but as pointed out by Little (2005) a basic (but often hidden) assumption with all these methods is that missingness of a particular value hides a "true" underlying value that is meaningful for analysis. This apparently obvious point is, however, not always the case. For example, consider a longitudinal analysis of CD4 counts in a clinical trial with patients suffering from AIDS. For patients who leave the study because they move to a different location, it makes sense to consider the CD4 counts that would have been recorded had they remained in the study. For subjects who die during the course of a study, it is less clear whether it is reasonable to consider CD4 counts after time of death as missing values. In such a case it may be preferable to treat death as a primary outcome and restrict analysis of CD4 counts to patients who are alive.

Dealing with missing values is a thriving area of current research amongst statisticians. In this chapter we restrict ourselves to a relatively brief overview of the possibilities along with a detailed analysis of an example.

11.2 Missing Data Mechanisms

The first question that needs to considered when deciding how best to deal with missing values is how such values may affect subsequent analyses on the data. To understand the problems that missing values can cause, the best way to begin is to consider a classification of missing-data mechanisms because the type of mechanism involved has implications for which approaches to analysis are suitable and which are not. The missing-data mechanism concerns the reasons why values are missing, in particular whether these reasons relate to recorded (non-missing) values for a subject. A useful classification of mechanisms, first introduced by Rubin (1976), involves three types of missing-data mechanism:

1. *Missing completely at random* (MCAR). The probability that a value is missing does not depend on either the observed or missing values of the response; so the classification of missing values as MCAR implies that $Pr(\text{missing} \mid \text{observed, unobserved}) = Pr(\text{missing})$. The probability of a missing value is the same for all cases implying that the causes of the missing data are unrelated to the data. Consequently, the observed (non-missing) values effectively constitute a simple random sample of the values for all the subjects. Possible examples include missing laboratory measurements because of a dropped test-tube (if it was not dropped because of the knowledge of any measurement), the accidental death of a participant in a study a participant moving to another area and in a longitudinal study a patient may miss an intended measurement occasion simply because they went shopping instead.

When data are MCAR, missing values are no different than non-missing in terms of their statistical properties and so in terms of the analyses to be performed the only real penalty in failing to account for the missing data is loss of power; MCAR causes least problems for data analysis. Sadly, for the data analyst MCAR is a relatively strong assumption and unrealistic for many (perhaps most) data sets.

2. *Missing at random* (MAR). This occurs when missingness depends only on observed values and not on values that are missing so that $Pr(\text{missing} \mid \text{observed, unobserved}) = Pr(\text{missing} \mid \text{observed})$. Some examples may help to clarify this mechanism. The first given by van Buuren (2018) involves the use of a weighing machine which produces more missing values when placed on a soft surface rather than a hard surface. Consequently, data are not MCAR. But if surface type is observed and it is realistic to assume MCAR within the type of surface, then the data are MAR.

A second example describing this type of missing-data mechanism is provided by Murray and Findlay (1988). The example arises from a study of hypertensive drugs in which the outcome measure was diastolic blood pressure. The protocol of the study specified that the participant was to be removed from the study when his/her blood pressure got too large. Here blood pressure at the time of dropout was observed before the participant dropped out, so although the missing-data mechanism is not MCAR because it depends on the values of blood pressure, it is MAR, because missingness depends only on the observed part of the data.

A further example of a MAR mechanism is provided by Heitjan (1997), and involves a study in which the response measure is body mass index (BMI). Suppose that the measure is missing because subjects who had high body mass index values at earlier visits avoided being measured at later visits out of embarrassment, regardless of whether they had gained or lost weight in the intervening period. The missing values here are MAR but not MCAR.

In all these examples, methods applied to the data that assumed MCAR might give misleading results (see later discussion). But as the missing data depend on known values and thus are described fully by variables observed in the data set, accounting for the values which "cause" the missing data will produce unbiased results in an analysis. (This type of missing value is also called *ignorable* because conclusions based on likelihood methods are not affected by MAR data, for example the models considered in chapters 9 and 10.)

3. *Missing not at random* (MNAR). This happens when neither MCAR or MAR holds. Here the probability of being missing depends on reasons that are not known, i.e., missingness depends on the unrecorded missing values. Consider again the BMI example introduced above; suppose subjects were more likely to avoid being measured (thus leading to a missing value) if they had put on extra weight since the last visit, then the data are MNAR. A further example is provided by a participant dropping out of a longitudinal study when his/her blood pressure became very high and this value is not

observed, or when their pain becomes intolerable and the associated pain value is not recorded. Dealing with data containing missing values that result from this type of missing-data mechanism is difficult. The correct analyses for such data must estimate the dependence of the missingness probability on the missing values. Models and software that attempt this are available (see, for example, Diggle and Kenward, 1994) but their use is not routine and, in addition, it must be remembered that the associated parameter estimates can be unreliable. Other approaches to handle MNAR are to find more data about the causes for the missingness, or to carry out sensitivity analyses to see how results alter under various assumptions about the possible causes.

11.3 Dealing with Missing Values

There are three major approaches to dealing with missing values:

1. Discard incomplete cases and analyze the remainder.
2. Impute or fill-in missing values by some procedure and then analyze the filled-in data.
3. Analyze the incomplete data by a method that does not require a complete (that is, rectangular) data set. This is the approach used when using maximum likelihood to estimate parameters in longitudinal data models as described in chapters 9 and 10, and leads to valid inferences on the parameters in the model being applied when the missing-data mechanism is MAR.

In this chapter the emphasis will be on imputation but before describing some of the methods available, we need to say a little about the first possibility in the list above.

A common (perhaps too common) approach to dealing with missing data in a study is *complete-case analysis*, where incomplete cases (cases with any missing value) are discarded and standard analysis methods applied to the remaining complete cases. In many statistical packages, this so-called *list-wise deletion* is the default approach. The major (perhaps only) advantage of complete-case analysis is convenience but in many cases this convenience is gained at a cost. When the missing data are MCAR the complete cases are a random subsample of the original sample and complete-case analysis provides valid inferences and unbiased estimates of means, variances and, for example, regression parameters, but when there is a substantial proportion of incomplete cases the method can be very inefficient and lead to a reduction of statistical power.

If the missing data are not MCAR then the complete cases are a biased sample and complete-case analysis can be misleading to a degree depending on the amount of missing data and the size of the departure from MCAR, producing biased estimates of means and other quantities of interest. Little and Rubin (2002), for example, show that the bias in the estimated mean increases with the difference between means of the observed and missing cases and with the proportion of missing data.

A simple alternative to complete-case analysis that is often used is *available-case analysis.* This is a method that uses all the cases available for estimating each quantity of interest. So, for example, all the cases that have recorded value for a pair of variables would be used to estimate the correlation between the two variables. Clearly available-case analysis makes more use of the available information in the data than the complete-case approach. But the method is not problem free; the sample base changes from, in our example, correlation to correlation, and there is no guarantee that the resulting correlation matrix is positive definite. In addition, the available-case approach creates potential problems when the missing-data mechanism is not MCAR.

The possible serious drawbacks of using listwise deletion or complete case analysis are discussed in Schafer and Graham (2002).

11.4 Imputing Missing Values

An ancient (almost) and still (too) often used technique for handling missing data is to replace them with the sample mean of the relevant variable, an approach known as *mean imputation.* This results in a complete data set so that standard methods of analysis can be applied. As pointed out by van Buuren (2018) "mean imputation is a fast and simple fix for missing data." But the method is poor because it will lead to underestimated variances, disturb the relationship between variables, and bias the estimate of the mean when the data are not MCAR. Van Buuren (2018) suggests that mean imputation should perhaps only be used as a rapid fix when a handful of values are missing, but that in general it should not be used. With little effort the more acceptable methods can be applied.

An improvement is *conditional mean imputation*, in which each missing value is replaced by an estimate of its conditional mean given the values of the non-missing values in the data, found via the prediction equation that results from the regression on the recorded values of a variable on the recorded values of the other variables in the data set. Although conditional mean imputation yields best predictions of the missing values in the sense of mean squared error, it leads to distorted estimates of quantities that are not linear in the data, for example, percentiles, variances, and correlations. Other improved methods are available but such *single imputation*, i.e., imputing one value for each missing datum by whatever method, fails to satisfy statistical objectives concerning the validity of resulting inferences based on the filled-in data. Because a single imputed value cannot reflect any of the uncertainty about the true underlying value, analyses that treat imputed values just like observed values systematically underestimate uncertainty (see Barnard et al., 2005). Consequently, imputing a single value for each missing datum and then analyzing the filled-in data using standard techniques for complete data will result in standard error estimates that are too small, confidence intervals that fail to attain their nominal coverage, and p-values that are too significant.

The problems of single imputation are largely overcome by the use of *multiple imputation* which is an approach to the missing values problem that allows the investigator to obtain valid assessments of uncertainty. The basic idea of multiple imputation is to impute two or more times for the missing data using independent draws of the missing values from a distribution that is appropriate under assumptions made about the data and the missing-data mechanism. The resulting multiple data sets are then each analyzed using the standard method appropriate for answering the questions of interest about the data and the analyses are then combined in a simple way that reflects the extra uncertainty due to having imputed rather than all the planned data being recorded. Multiple imputations can be created under a number of different models, and details are given in Rubin and Schenker (1991). But essentially the theoretical motivation for multiple imputation is Bayesian and for keen readers Technical Section 11.1 gives a very brief account that follows Barnard et al. (2005).

Technical Section 11.1: Multiple Imputation

We begin by letting Q be the population quantity of interest. If all the data have been observed then estimates of and inferences for Q would have been based on the complete-data posterior density $p(Q|Y_{\text{obs}}, Y_{\text{mis}})$. But because Y_{mis} is not observed, inferences, etc., have to be based on the actual posterior density, $p(Q|Y_{\text{obs}})$, which can be written as

$$p(Q|Y_{\text{obs}}) = \int p(Q|Y_{\text{obs}}, Y_{\text{mis}})\, p(Q|Y_{\text{mis}}, Y_{\text{obs}})\, dY_{\text{mis}}$$

The equation above shows that the actual posterior density of Q can be obtained by averaging the complete posterior density over the posterior predictive distribution of Y_{mis}. In principle, multiple imputations are repeated independent draws from $p(Q|Y_{\text{mis}}, Y_{\text{obs}})$. So multiple imputation allows approximating the equation above by separately analyzing each data set completed by imputation and then combining the results of the separate analyses. Details of suitable algorithms are given in Schafer (1997) and van Buuren (2018).

The question of how many imputations (m) is an obvious one that needs to be considered. In most cases a value for m between 3 and 10 is suggested.

Intuitively this seems rather small but Rubin (2004) shows that unless the rate of missing information is very high there is, in most cases, little advantage to producing and analyzing more than a few imputed data sets. White, Royston and Wood (2011) give a conservative rule of thumb that m should be set equal to the percentage of incomplete cases based on the argument that repeat analyses yield the same result. And van Buuren (2018) suggests that the substantive conclusions are unlikely to change as the result of raising m beyond $m = 5$.

11.5 Analyzing Multiply Imputed Data

From the analysis of each data set we need to look at the estimates of the quantity of interest and the standard errors of the estimates and then combine these in some way to produce estimates resulting from the multiple imputation. How this is done is described in Technical Section 11.2.

Technical Section 11.2: Combined Estimates and Standard Errors in Multiple Imputation

Let $\hat{Q}_i$ be the estimate of the parameter of interest obtained from the ith data set from the multiple imputation process and S_i be its corresponding standard error. The combined estimate of the quantity of interest is

$$\bar{Q} = \frac{1}{m}\sum_{i=1}^{m}\hat{Q}_i$$

To find the combined standard error involves first calculating the within-imputation variance,

$$\bar{S} = \frac{1}{m}\sum_{i=1}^{m}S_i$$

followed by the between-imputation variance,

$$B = \frac{1}{m-1}\sum_{i=1}^{m}(\hat{Q}_i - \bar{Q})^2$$

and then the required total variance can be found from

$$T = \bar{S} + (1 + \frac{1}{m})B$$

This total variance is made up of two components. The first which preserves the natural variability, $\bar{S}$, is simply the average of the variance estimates for each imputed data set and is analogous to the variance that would be suitable if we did not need to account for missing data. The second component, B, estimates uncertainty caused by missing data by measuring how the point estimates vary from data set to data set.

The overall standard error is simply the square root of T. A significance test for Q and a confidence interval are found from the usual test statistic, $(Q -$ hypothesized value of $Q)/\sqrt{T}$, the value of which is referred to a Student's t-distribution. The question arises however as to what is the appropriate value for the degrees of freedom of the test, say v_0? Rubin (2004) suggests that the answer to this question is given by

$$v_0 = (m-1)(1+1/r)^2$$

where

$$r = \frac{B + B/m}{\bar{S}}$$

But Barnard and Rubin (1999) noted that using this value of v_0 can produce values that are larger than the degrees of freedom in the complete data, a result which they considered "clearly inappropriate". Consequently they developed an adapted version that does not lead to the same problem. Barnard and Rubin's revised value for the degrees of freedom of the t-test in which we are interested is v_1 given by

$$v_1 = \frac{v_0 v_2}{v_0 + v_2}$$

where

$$v_2 = \frac{n(n-1)(1-\lambda)}{n+2}$$

and

$$\lambda = \frac{r}{\sqrt{r^2+1}}$$

The quantity v_1 is always less than or equal to the degrees of freedom of the test applied to the hypothetically complete data. (For more details see Schafer, 1997 and van Buuren, 2018).

11.6 Example of the Application of Multiple Imputation

The data in Table 11.1 involves the average amount of time per week in minutes a subject spends cleaning his car (all male subjects), the age of the subject, and a measure of extroversion. Only the time measurements are complete, while some data are missing from the two other variables. Here, we may well assume that they are missing at random (MAR).

TABLE 11.1
Car Wash Data that Includes Missing Values

Time	Age	Extro
46		40
79		
63	26	
42	34	43
44		41
59	26	42
27	25	
40		30
30	45	35
61	31	48
50	28	37
65	34	
70	44	
20		28
32		34
65	28	44
30	50	33
48	23	
20		
56		47

With these data we will estimate a linear regression model, where the time is the response variable and the two other variables are the explanatory variables. This will allow a brief comparison of

1. Analysis after removing subjects with any missing values,
2. Analysis using the mean imputation,
3. Multiple imputation approach.

We have met these data before in Chapter 2 and Chapter 4, where we created graphs of the data and analysed the dependencies with regression analysis. There we worked with the complete data, but here we have discarded some values of the data on purpose, in order to demonstrate how to cope with the problem of the missingness that is so often met in practice.

11.6.1 Complete-Case Analysis

We begin with a complete-case analysis, that is, implicitly remove all subjects with any missing values from the data. In this particular case, it means that only *five complete observations* will be used in the estimation. In practice, there should be good reasons to be worried, if a great part of the (small) data set is lost. However, that might not always be easy to notice, as the

TABLE 11.2
Results of Fitting the Multiple Linear Regression Model with the Two Explanatory Variables, Age and Extroversion, to the Car Wash Data Using Complete-Case Analysis

	Estimate	Standard Error	t-Value	Pr(>|t|)
Intercept	42.8029	38.6640	1.107	0.3304
Age	−0.9939	0.3979	−2.498	0.0669
Extro	0.9855	0.6849	1.439	0.2236

Note: Residual standard error: 6.3962 on 4 DF; multiple R-squared: 0.8708; F-statistic: 13.49 on 2 and 4 DF; p-value: 0.01668.

missingness may be scattered around multiple variables. Attention should be paid towards the number of (complete) observations retained in different, explorative analyses. Also the degrees of freedom of some analyses and tests might reveal surprisingly small numbers, compared to the total number of observations in the data at hand.

With those few observations that are left from the "listwise deletion", the model can still be estimated, and the results are displayed in Table 11.2. Both explanatory variables are statistically non-significant predictors of the Time, but the p-value of Age is quite close to the 5% level. The F-test is significant, however, and the R-square comes close to 90%.

11.6.2 Mean Imputation

The results of the same model applied to the mean imputed data are displayed in Table 11.3. Now, Age is clearly non-significant, but the Extroversion is significant, so the setting and interpretation change quite completely. However, the F-test gives a p-value that is over 5%, and the R-square drops to less than 30%. Now there are full degrees of freedom, but the mean imputation is only a single imputation and hence not very reliable. Perhaps more questions than answers remain after this "quick-and-dirty" trial, as one can not rely very much on these type of procedures.

TABLE 11.3
Results of Fitting the Multiple Linear Regression Model with the Two Explanatory Variables, Age and Extroversion, to the Car Wash Data Using Mean Imputation

	Estimate	Standard Error	t-Value	Pr(>|t|)
Intercept	−20.8419	37.9368	−0.549	0.5899
Age	−0.0541	0.5452	−0.099	0.9221
Extro	1.8119	0.7340	2.469	0.0245

Note: Residual standard error: 15.36 on 17 DF; multiple R-squared: 0.2877; F-statistic: 3.433 on 2 and 17 DF; p-value: 0.05593.

TABLE 11.4
Results of Fitting the Multiple Linear Regression Model with the Two Explanatory Variables, Age and Extroversion, to the Car Wash Data Using Multiple Imputation

	Estimate	Standard Error	Statistic	DF	p-Value
Intercept	−13.0507	45.9645	−0.284	4.6820	0.7817
Age	−0.3032	0.4478	−0.677	11.1316	0.5122
Extro	1.7959	0.9101	1.973	3.5875	0.0738

Note: Pooled estimates of 5 imputations.

11.6.3 Multiple Imputation

The third version of the model is based on a multiple imputation of the data set, and the results are displayed in Table 11.4. Instead of one set of estimates and their standard errors, we obtain five sets: one from each imputed data set. There are standard ways of pooling these estimates and their standard errors together and then summarizing them as average values of those multiple ones (see Section 11.5). So, based on the pooled estimates of the five imputations, both Age and Extroversion seem to be non-significant predictors of the Time.

As we have seen through this simplified example, the conclusions are different in all these three versions of the same linear regression model. The complete-case analysis and mean imputation may give any kind of results, bringing in the inference such uncertainty that is difficult or impossible to evaluate. Multiple imputation uses the available information in the most optimal way, and hence gives a more reliable basis for making inferences under the uncertainty caused by the missingness in the original data set.

11.7 Beat the Blues Revisited (Again)

To illustrate the use of multiple imputation in a more complex example than the one used in the previous section we shall, at the risk of boring our readers, return to the Beat the Blues (BtB) one more time, fitting models previously considered in Chapter 9 but here using both mean and multiple imputation.

We begin by considering a simple linear regression model for the data although it unrealistically assumes that the repeated measures of the Beck Depression Inventory (BDI) are independent. The results from the mean-imputed BtB data are given in Table 11.5. It appears that all covariates, including the Treatment, are now significant, as all the standard errors are lower than those in Table 9.8. This is most probably a consequence of replacing a large number of values with the simple mean of the BDI measurements. There are,

TABLE 11.5
Results from Fitting a Multiple Linear Regression Model to Mean Imputed BtB Data Assuming the Repeated Measurements of BDI are Independent

	Estimate	Standard Error	t-Value	Pr(>|t|)
Intercept	10.49720	1.41511	7.418	7.37e-13
BDIpre	0.35634	0.03866	9.217	<2e-16
Time	−0.68109	0.17778	−3.831	0.000148
Treatment	−2.61112	0.83968	−3.110	0.002009
Drug	−1.80777	0.87095	−2.076	0.038576
Length	2.31409	0.81991	2.822	0.005008

Note: Multiple R-squared: 0.269; F-statistic: 29 on 5 and 394 DF; *p*-value: <2.2e-16; *BDIpre*, pretreatment Beck Depression Inventory.

altogether, 120 more observations in the analysis after the imputation, as can be seen by comparing the degrees of freedoms of the F-statistics of these tables.

Next we fit the same model but now using multiple imputation with five imputations. The resulting pooled estimates are shown in Table 11.6. The covariate Drug is now shown to have a non-significant effect but Treatment and Length remain significant despite the increase in the associated standard error.

We know, of course, from the results given in Chapter 9 that the independence model is unrealistic for these data. So we will now fit a random intercept model using both mean imputation and multiple imputation with five imputations. The results from mean imputation are shown in Table 11.7. The standard errors are now larger than those given in Table 11.5 as a result of the use of a more realistic model that does not assume independence. But comparing the standard errors in Table 11.7 with those in Table 9.9 we see that they are smaller, reflecting the inadequacy of simple mean imputation.

Finally, we apply the multiple imputation to the BtB data and estimate the random intercept model again. The results of the pooled estimates are given

TABLE 11.6
Results from Fitting a Multiple Linear Regression Model to BtB Data Imputed Using Multiple Imputation

	Estimate	Standard Error	Statistic	DF	*p*-Value
Intercept	11.23758	2.05664	5.464	26.403	1.24e-07
BDIpre	0.34233	0.04663	7.342	223.776	3.88e-12
Time	−0.95620	0.21466	−4.455	219.500	0.000013
Treatment	−3.57255	1.52330	−2.345	10.571	0.019888
Drug	−2.32207	1.20800	−1.921	36.679	0.056046
Length	4.03393	1.07151	3.765	65.900	0.000213

Note: Pooled estimates of 5 imputations; *BDIpre*, pretreatment Beck Depression Inventory.

TABLE 11.7
Results from Fitting a Random Intercept Model to Mean Imputed BtB Data

Random Effects			
$\hat{\sigma}_u^2 = 31.75$, $\hat{\sigma}^2 = 32.59$			
Fixed Effects			
	Estimate	**Standard Error**	**t-Value**
Intercept	10.49720	1.86230	5.637
BDIpre	0.35634	0.06143	5.801
Time	−0.68109	0.12766	−5.335
Treatment	−2.61112	1.33417	−1.957
Drug	−1.80777	1.38386	−1.306
Length	2.31409	1.30276	1.776

Note: BDIpre, pretreatment Beck Depression Inventory.

in Table 11.8. In pooling the estimates (see Section 11.5), only the estimates of the fixed effects can be considered. Now, the standard errors are somewhat larger, but so are the coefficient estimates, and therefore the Treatment and Length are still significant, while the Drug is not. The covariates Time and the pretreatment BDI measurement (BDIpre) are clearly significant in all the models that we have considered. Also, the signs of the coefficient estimates are similar in each model, which gives some evidence of stability between the models. However, the uncertainty caused by the missingness and especially related to the significance of the treatment effect should be noted. In practice, these results suggest perhaps that multiple imputation is a tool that should not be ignored when fitting models to data sets that suffer from missing values.

TABLE 11.8
Results from Fitting a Random Intercept Model to BtB Data Imputed Using Multiple Imputation

Fixed Effects					
	Estimate	**Standard Error**	**Statistic**	**DF**	***p*-Value**
Intercept	11.23758	2.33200	4.819	41.559	0.000002
BDIpre	0.34233	0.06367	5.377	315.617	1.48e-07
Time	−0.95620	0.18200	−5.228	159.764	3.12e-07
Treatment	−3.57255	1.79076	−1.995	19.709	0.046903
Drug	−2.32207	1.55412	−1.494	84.846	0.136138
Length	4.03393	1.41182	2.857	145.246	0.004557

Note: Pooled estimates of 5 imputations; only fixed effects; *BDIpre*, pretreatment Beck Depression Inventory.

11.8 Summary

- For researchers in the behavioral sciences, it is an inescapable fact that the human subjects used in their studies often make life difficult by missing appointments.
- Consequently missing values are an ever-present possibility in studies in the behavioral sciences although everything possible should be done to avoid them.
- When data contain missing values, multiple imputation can be used to provide valid inferences for parameter estimates from the incomplete data.
- If carefully handled, multiple imputation can cope with missing data in all types of variables.
- In this chapter we have given only a brief account of dealing with missing values; a detailed account is available in the issue of *Statistical Methods in Medical Research* entitled *Multiple Imputation: Current Perspectives* (Volume 16, 2007) and in van Buuren (2018). See also Laaksonen (2018).

11.9 Exercises

11.1 The data in Table 11.9 show subjective estimates of the lengths (to the nearest 0.1 inch) of 15 pieces of string as assessed by three raters. Also given is the accurately measures length of the string. Missing values occur when a rater did not make an estimate in the five seconds allowed.

Use a multiple regression model to determine which rater gives the best prediction of the measured length, using first only complete observations and then both mean and multiple imputation.

11.2 With the same data, find the correlation matrix of the measured length and the three estimated lengths using

1. The complete case approach (i.e., the listwise deletion)
2. Mean value imputation
3. Multiple imputation

TABLE 11.9
Subjective Estimates of the Lengths (to the Nearest 0.1 Inch) of 15 Pieces of String as Assessed by Three Raters

String	Length	Graham	Brian	David
1	6.3	5.0	4.8	6.0
2	4.1	3.2	3.1	3.5
3	5.1		3.8	4.5
4	5.0	4.5	4.1	
5	5.7			5.0
6	3.3	2.5	2.8	2.6
7	1.3	1.7	1.4	
8	5.8	4.8	4.2	5.5
9	2.8			2.1
10	6.7	5.2		
11	1.5	1.2	1.1	1.2
12	2.1	1.8	1.6	1.8
13	4.6	3.4	4.1	3.8
14	7.6	6.0	6.3	6.5
15	2.5		1.6	2.0

* 11.3 The data in Table 11.10 given in Davis (2002) arise from the Iowa Cochlear Implant Project to compare the effectiveness of two types of cochlear implants in profoundly and bilaterally deaf patients. In one group of 23 subjects, the "Type A" implant was used. A second group of 21 subjects received the "Type B" implant. In both groups the electrode array was surgically implanted five to six weeks prior to electrical connection to the external speech processor. A sentence test was administered at 1, 9, 18, and 30 months after connection. The outcome variable of interest at each time point was the percentage of correct scores.

Assuming normality for the outcome variable, fit a random intercept model to the data using both mean imputation and multiple imputation with five imputations to the data and compare your results. Is there any evidence of a difference between the two implant types? Investigate how the multiple imputation results change when using different numbers of imputations.

TABLE 11.10
The Effectiveness of Two Types of Cochlear Implants in Profoundly and Bilaterally Deaf Patients

		Month			
Group	**ID**	**1**	**9**	**18**	**30**
1	1	28.57	53.00	57.83	59.22
1	2		13.00	21.00	26.50
1	3	60.37	86.41		
1	4	33.87	55.50	61.06	
1	5	1.61	0.69		
1	6	26.04	61.98	67.28	
1	7		59.00	66.80	83.20
1	8	11.29	38.02		
1	9	0.00	0.00	0.00	2.76
1	10		35.10	37.79	54.80
1	11	16.00	33.00	45.39	40.09
1	12	40.55	50.69	41.70	52.07
1	13	3.90	11.06	4.15	14.90
1	14	1.80	2.30	2.53	2.53
1	15	0.00	17.74	44.70	48.85
1	16	64.75	84.50	92.40	
1	17	38.25	81.57	89.63	
1	18	67.50	91.47	92.86	
1	19	45.62	58.00		
1	20	0.00	0.00	37.00	
1	21	51.15	66.13		
1	22	0.00	48.16		
1	23	0.00	0.92		
2	1		0.00	0.90	1.61
2	2	0.00	0.00	0.00	
2	3	0.00	0.00		
2	4	8.76	24.42		
2	5	0.00	20.79	27.42	31.80
2	6	2.30	12.67	28.80	24.42
2	7	12.90	28.34		
2	8		45.50	43.32	36.80
2	9	68.00	96.08	97.47	99.00
2	10	20.28	41.01	51.15	61.98
2	11	65.90	81.30	71.20	70.00
2	12	0.00	8.76	16.59	14.75
2	13	0.00	0.00	0.00	0.00
2	14	9.22	14.98	9.86	
2	15	11.29	44.47	62.90	68.20
2	16	30.88	29.72		
2	17	29.72	41.40	64.00	
2	18	0.00	43.55	48.16	
2	19	0.00	0.00		
2	20	8.76	60.00		

12

Multivariate Data and Multivariate Analysis

12.1 Introduction

In this chapter and the following five longer chapters (Chapters 13 to 17) we will be concerned with what is most often termed *multivariate data*. Such data arise when researchers measure several variables on each individual in their study, and where all variables need to be examined simultaneously in order both to uncover whatever "patterns" or "structure" the data may contain and understand the key features of the data. All the variables in this type of "multivariate data" are considered random unlike in the type of "multivariate data" discussed in Chapters 3 to 11 where only the response variable is considered random and where the regression models described seek to uncover the relationship between the response and the remaining variables.

The techniques described in this and the remaining chapters of the book are those that generally fall under the umbrella of *multivariate analysis* and they are largely, although not exclusively, descriptive rather than inferential, that have in common the aim to display or extract any "signal" in the data in the presence of noise and, in a very general sense, to discover what the data may be trying to tell us.

The starting point of most multivariate analysis methods is the data matrix $\mathbf{X}$ introduced in the Preface, that is

$$\mathbf{X} = \begin{bmatrix} x_{11} & x_{12} & \cdots & x_{1q} \\ x_{21} & x_{22} & \cdots & x_{2q} \\ \vdots & \vdots & \ddots & \vdots \\ x_{n1} & x_{n2} & \cdots & x_{nq} \end{bmatrix}$$

where n is the number of units in the sample, q is the number of variables measured on each unit, and x_{ij} denotes the value of the jth variable for the ith individual; an individual in this context may be something other than a human being. The variables will often be at different levels in the measurement hierarchy described in Chapter 1, for example, some categorical, some interval, and some ratio measurements. Further, often some variable values will be missing, and so, the problems associated with missing data mentioned in Chapter 1 and again in Chapter 11 may assume importance.

12.2 The Initial Analysis of Multivariate Data

The main techniques for analyzing multivariate data are those to be described in the subsequent chapters, but an initial graphical and numerical description can often be very helpful in gaining some insight into the data and in interpreting the results from later analyses. We will illustrate what can be done with an initial analysis using first the very simple multivariate data set shown in Table 12.1. The data consists of chest, hip, and waist measurements (in inches) of 20 individuals.

12.2.1 Summary Statistics for Multivariate Data

In order to provide a numerical summary for a multivariate data set, we need to produce summary statistics for each of the variables separately and also calculate appropriate statistics that summarize the relationships between the variables. For the former, we generally use means and variances (assuming that we are dealing with continuous variables), and for the latter, we usually take pairs of variables at a time and look at their covariances or correlations. Population and sample versions of all these quantities are defined in Technical Section 12.1.

TABLE 12.1
Chest, Waist, and Hip Measurements of 20 Individuals

Individual	Chest	Waist	Hips
1	34	30	32
2	37	32	37
3	38	30	36
4	36	33	39
5	38	29	33
6	43	32	38
7	40	33	42
8	38	30	40
9	40	30	37
10	41	32	39
11	36	24	35
12	36	25	37
13	34	24	37
14	33	22	34
15	36	26	38
16	37	26	37
17	34	25	38
18	36	26	37
19	38	28	40
20	35	23	35

Technical Section 12.1: Numerical Summary Statistics for Multivariate Data

For q variables, the population mean vector is usually represented as $\boldsymbol{\mu}' = [\mu_1, \mu_2, \ldots, \mu_q]$, where $\mu_i = E(x_i)$ is the population mean (or expected value as denoted by the E operator) of the ith variable. An estimate of $\boldsymbol{\mu}'$ based on n q-dimensional observations is $\bar{x}' = [\bar{x}_1, \bar{x}_2, \ldots, \bar{x}_q]$, where $\bar{x}_i$ is the sample mean of the variable x_i.

The vector of population variances can be represented by $\boldsymbol{\sigma}' = [\sigma_1^2, \sigma_2^2, \ldots, \sigma_q^2]$, where $\sigma_i^2 = E(x_i - \mu_i)^2$. An estimate of $\boldsymbol{\sigma}'$ based on n q-dimensional observations is $\mathbf{s}' = [s_1^2, s_2^2, \ldots, s_q^2]$, where s_i^2 is the sample variance of x_i.

The population covariance of two variables x_i and x_j is defined by

$$\text{Cov}\,(x_i, x_j) = E\,(x_i - \mu_i)\,(x_j - \mu_j)$$

If $i = j$, we note that the covariance of the variable with itself is simply its variance, and therefore, there is no real need to define variances and covariances independently in the multivariate case. The covariance of x_i and x_j is usually denoted by σ_{ij} (so, the variance of the variable x_i is often denoted by σ_{ii} rather than σ_i^2).

With q variables, $x_1, x_2, \ldots, x_q$, there are q variances and $q(q-1)/2$ covariances. In general, these quantities are arranged in a $q \times q$ symmetric matrix $\boldsymbol{\Sigma}$ where

$$\boldsymbol{\Sigma} = \begin{bmatrix} \sigma_{11} & \sigma_{12} & \cdots & \sigma_{1q} \\ \sigma_{21} & \sigma_{22} & \cdots & \sigma_{2q} \\ \vdots & \vdots & \ddots & \vdots \\ \sigma_{q1} & \sigma_{q2} & \cdots & \sigma_{qq} \end{bmatrix}$$

Note that $\sigma_{ij} = \sigma_{ji}$. This matrix is generally known as the *variance-covariance matrix* or simply the *covariance matrix*. The matrix $\boldsymbol{\Sigma}$ is estimated by the matrix $\mathbf{S}$, given by

$$\mathbf{S} = \frac{\sum_{i=1}^{n} (\mathbf{x}_i - \bar{\mathbf{x}})\,(\mathbf{x}_i - \bar{\mathbf{x}})'}{(n-1)}$$

where $\mathbf{x}_i' = [\mathbf{x}_{i1}, \mathbf{x}_{i2}, \ldots, \mathbf{x}_{iq}]$ is the vector of observations for the ith individual. The diagonal of $\mathbf{S}$ contains the variances of each variable. The covariance is often difficult to interpret because it depends on the units in which the two variables are measured; consequently, it is often standardized by dividing by the product of the standard deviations of the two variables to give a quantity called the correlation coefficient, ρ_{ij}, where

$$\rho_{ij} = \frac{\sigma_{ij}}{\sqrt{\sigma_{ii}\sigma_{jj}}}$$

The correlation coefficient lies between −1 and +1 and gives a measure of the linear relationship between the variables x_i and x_j. It is positive if high values of x_i are associated with high values of x_j, and negative if high values of x_i are associated with low values of x_j. With q variables, there are $q(q-1)/2$ distinct correlations, which may be arranged in a $q \times q$ matrix whose diagonal elements are unity.

For sample data, the correlation matrix contains the usual estimates of the ρ values, namely, Pearson's correlation coefficient, and is generally denoted by **R**. The matrix may be written in terms of the sample covariance matrix **S** as follows:

$$\mathbf{R} = \mathbf{D}^{-1/2}\mathbf{S}\mathbf{D}^{-1/2}$$

where $\mathbf{D}^{-1/2} = \text{diag}\,(1/\text{s}_\text{i})$.

In most situations, we will be dealing with covariance and correlation matrices of full rank q, so that both matrices will be nonsingular (i.e., invertible).

For the body measurements data, the numerical summary statistics described earlier are as follows:

Means:

Chest	Waist	Hips
37.00	28.00	37.05

Variances:

Chest	Waist	Hips
6.63	12.53	5.94

Covariance matrix:

	Chest	Waist	Hips
Chest	6.63	6.37	3.00
Waist	6.37	12.53	3.58
Hips	3.00	3.58	5.94

Correlation matrix:

	Chest	Waist	Hips
Chest	1.00	0.70	0.48
Waist	0.70	1.00	0.41
Hips	0.48	0.41	1.00

We see that the waist measurements have the largest variance, with chest and hip measurements having very similar variances. Waist and hip measurements have the largest correlation with a value of 0.70.

12.2.2 Graphical Descriptions of the Body Measurement Data

As always, numerical summaries need to be interpreted alongside appropriate graphics, and so, in Figure 12.1, we give the boxplots of each of the three body measurements, and then, in Figure 12.2, a scatterplot matrix of the three variables is shown with a histogram of each measurement placed on the diagonal. The boxplot for chest measurements shows a mild degree of skewness and one potential outlier. The scatterplot matrix is more interesting, with the panel showing the scatterplot of waist and hip measurements suggesting the possibility that the data may consist of two separate groups of observations, a possibility underlined by the bimodality of the histogram for waist measurements. The two-group possibility could be investigated further by applying cluster analysis (see Chapter 17) to the data, but here, it is not too taxing to come up with an explanation for the possible two-group structure, which is that there are men and women in the sample. (If there really are distinct groups of observations in the data, the previously calculated summary statistics for the whole sample may not give an accurate description of the separate groups.)

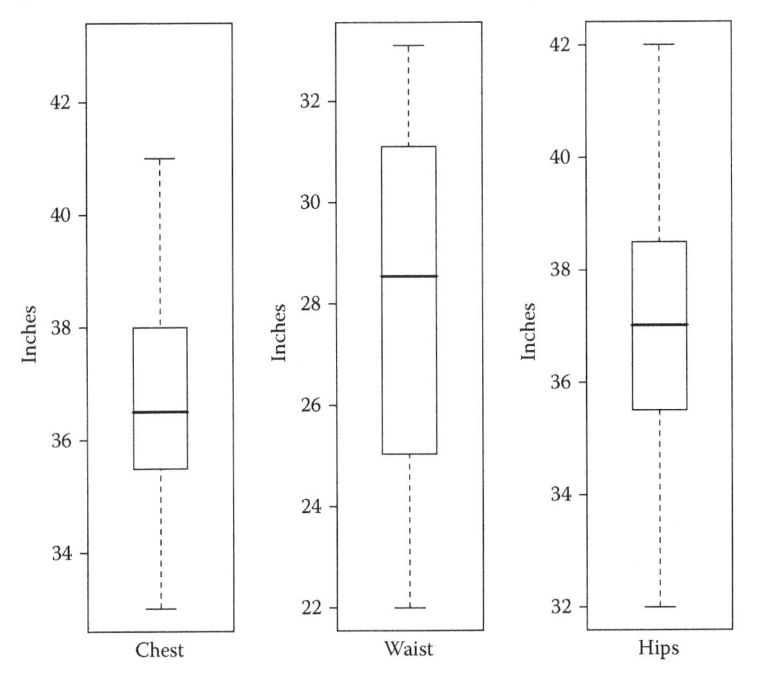

FIGURE 12.1
Boxplots of body measurements.

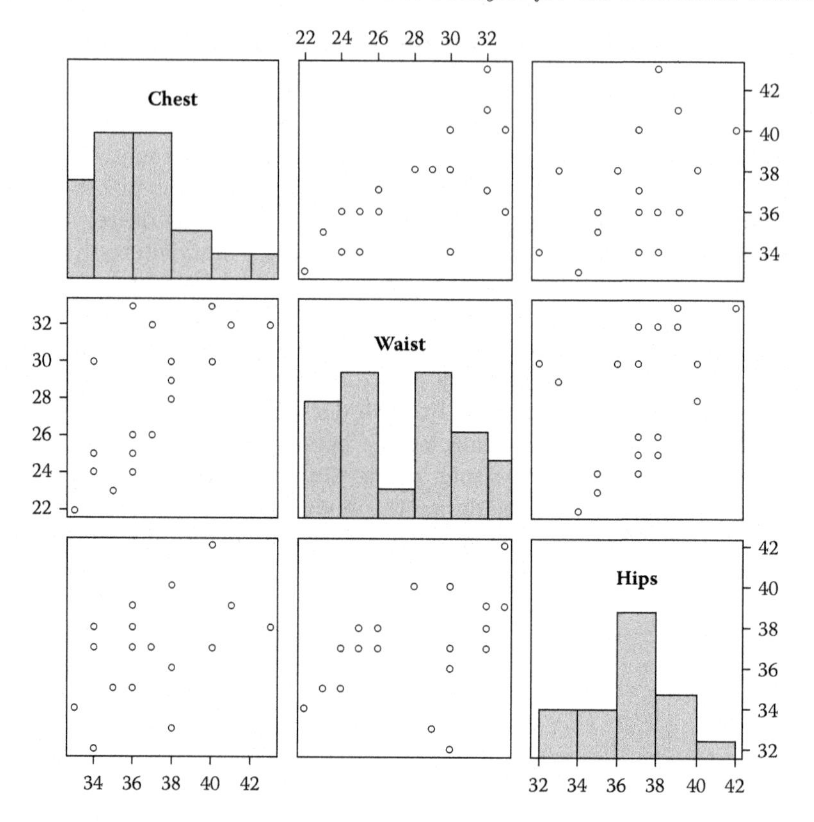

FIGURE 12.2
Scatterplot matrix of body measurements.

12.3 The Multivariate Normal Probability Density Function

As we have seen in earlier chapters, the normal probability density function is the basis of the inferences derived from most multivariable techniques. In multivariate analysis, it is the multivariate normal density function that has a similar role, although many multivariate analyses are carried out in the spirit of data exploration, where questions of statistical significance are of minor or no importance. Nevertheless, researchers in behavioral sciences dealing with the complexities of multivariate data may on occasion need to know a little about the multivariate density function and, in particular, how to assess whether or not a set of multivariate data can be assumed to have this probability density function. So, in Technical Section 12.2, we define the multivariate normal density and describe some of its properties.

Technical Section 12.2: Multivariate Normal Density Function

For a vector of q random variables $\mathbf{x}' = [x_1, x_2, \ldots, x_q]$, the multivariate normal density function takes the form

$$f(\mathbf{x}) = (2\pi)^{-q/2}|\boldsymbol{\Sigma}|^{-1/2}\exp\left\{-\frac{1}{2}(\mathbf{x}-\boldsymbol{\mu})'\boldsymbol{\Sigma}^{-1}(\mathbf{x}-\boldsymbol{\mu})\right\}, -\infty < x_i < \infty$$

where $\boldsymbol{\Sigma}$ is the population covariance matrix of the variables, and $\boldsymbol{\mu}$ is the vector of population mean values of the variables. The simplest example of the multivariate normal density function is the bivariate normal density with $q = 2$; this can be written explicitly as

$$f(x_1, x_2) = \frac{1}{2\pi\sigma_1\sigma_2\sqrt{(1-\rho^2)}} \times$$

$$\exp\left\{\frac{-1}{2(1-\rho^2)}\left[\left(\frac{x_1-\mu_1}{\sigma_1}\right)^2 - 2\rho\left(\frac{x_1-\mu_1}{\sigma_1}\right)\left(\frac{x_2-\mu_2}{\sigma_2}\right)+\left(\frac{x_2-\mu_2}{\sigma_2}\right)^2\right]\right\}$$

where μ_1 and μ_2 are the population means of the two variables, σ_1^2 and σ_2^2 are the population variances, and ρ is the population correlation between the two variables. Figure 12.3 shows an example of a bivariate normal density function with both means equal to 0, both variances equal to 1, and correlation equal to 0.5.

The population mean vector and the population covariance matrix of a multivariate density function are estimated from a sample of multivariate observations as described in Technical Section 12.1.

One property of a multivariate normal density function that is worth mentioning here is that linear combinations of the variables, that is,

$$y = a_1x_1 + a_2x_2 + \cdots + a_qx_q$$

where $a_1, a_2, \ldots, a_q$ is a set of scalars, are themselves normally distributed with mean $\mathbf{a}'\boldsymbol{\mu}$ and variance $\mathbf{a}'\boldsymbol{\Sigma}\mathbf{a}$ where $\mathbf{a}' = [a_1, a_2, \ldots, a_q]$. Linear combinations of variables will be of importance in later chapters.

For many multivariate methods to be described in later chapters, the assumption of multivariate normality is often not critical to the results of the analysis. But there may be occasions when testing for multivariate normality may be of interest. A start can be made perhaps by testing each separate variable for univariate normality using, say, a probability plot, as described in Chapter 2. Such plots can be helpful, but unfortunately, marginal multivariate normality does not necessarily imply multivariate normality. An alternative (additional)

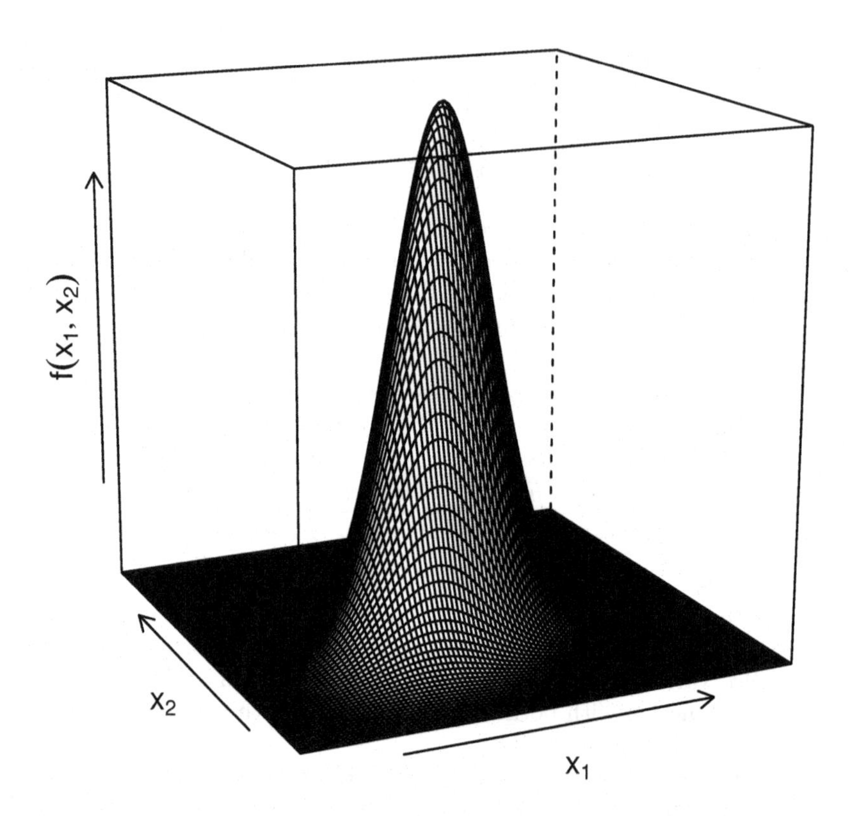

FIGURE 12.3
Bivariate normal density function with means equal to 0, variances equal to 1, and correlation equal to 0.5.

approach is to convert the multivariate observations $\mathbf{x}_1, \mathbf{x}_2, \ldots, \mathbf{x}_n$ into a set of generalized distances d_i^2, giving a measure of the distance of each particular observation from the mean vector of the complete sample $\bar{\mathbf{x}}$. The generalized distances (also called Mahalanobis distances) d_i^2 are calculated as

$$d_i^2 = (\mathbf{x}_i - \bar{\mathbf{x}})' \, \mathbf{S}^{-1} \, (\mathbf{x}_i - \bar{\mathbf{x}})$$

where $\mathbf{S}$ is the sample covariance matrix. This distance measure takes into account the different variances of the variables and the covariances of pairs of variables. If the observations do arise from a multivariate normal distribution, then the generalized distances have, approximately, a chi-squared distribution with q degrees of freedom. So, plotting the ordered distances against the corresponding quantiles of the appropriate chi-squared distribution should lead to a straight line through the origin.

12.3.1 Assessing Multivariate Data for Normality

We will now assess the body measurements data for normality, although, because there are only 20 observations in the sample, there is really too little information to come to any convincing conclusion. Figure 12.4 shows separate probability plots for each measurement; in the plots, there appears to be no evidence of any departures from linearity.

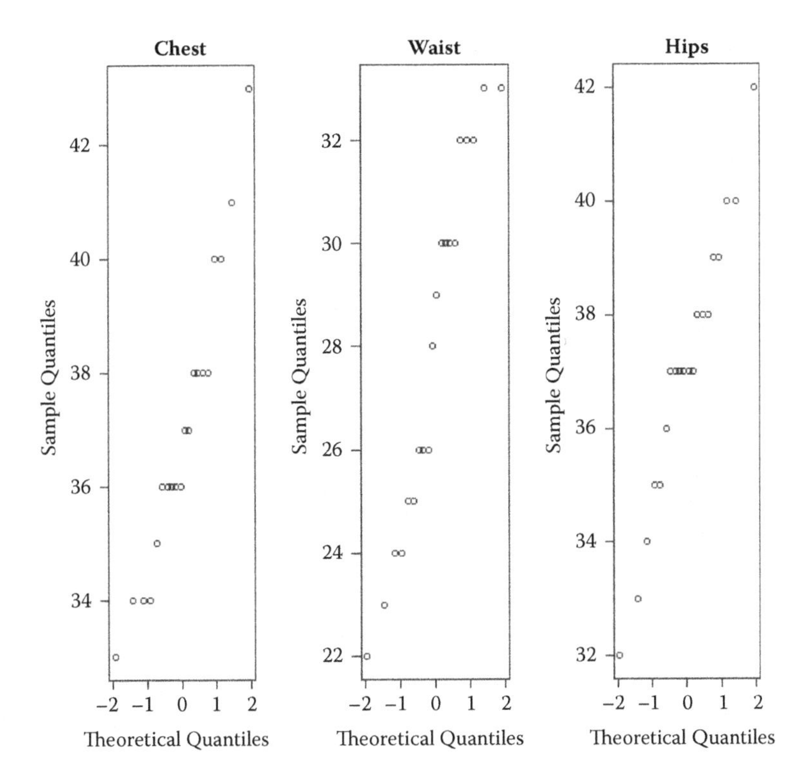

FIGURE 12.4
Normal probability plots of chest, waist, and hip measurements.

The chi-square plot of the 20 generalized distances in Figure 12.5 does seem to deviate a little from linearity, but with so few observations, it is hard to be certain.

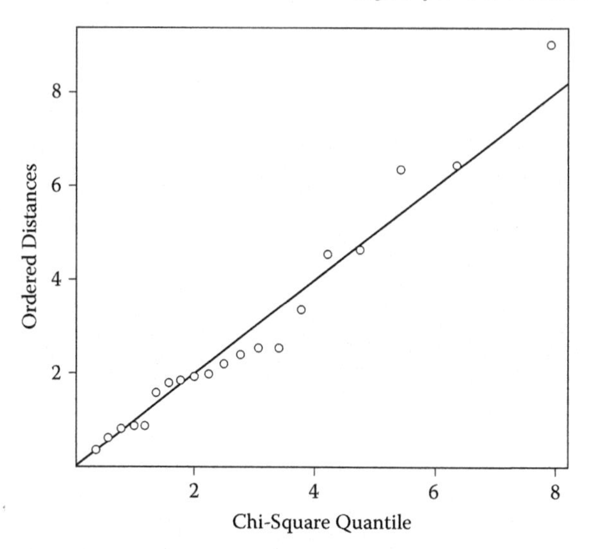

FIGURE 12.5
Chi-square plot of generalized distances for the body measurements data.

Now let us look at a larger example of a multivariate data set collected in a health survey of 103 paint sprayers in a car assembly plant. Six variables were recorded on each individual:

Hemo: Hemoglobin concentration
PCV: Packed cell volume
WBC: White blood cell count
Lympho: Lymphocyte count
Neutro: Neutrophil count
Lead: Serum lead concentration

Data for the first five paint sprayers are shown in Table 12.2. We begin by looking at separate normal probability plots of each of the six variables; the plots are given in Figure 12.6. The plots for WBC, Lympho, and Lead show some deviation for linearity that would suggest that the six variables

TABLE 12.2
Part of the Data on Paint Sprayers

Case	Hemo	PCV	WBC	Lympho	Neutro	Lead
1	13.4	39	4100	14	25	17
2	14.6	46	5000	15	30	20
3	13.5	42	4500	19	21	18
4	15.0	46	4600	23	16	18
5	14.6	44	5100	17	31	19

do not have a multivariate normal density. The chi-square plot of generalized distances in Figure 12.7 appears to confirm this because there is considerable departure from linearity in this plot although this is primarily due to a relatively few observations. Here, it is of interest to see what happens when we take a log transformation of all the variables and then look at the chi-square plot of the transformed data, given in Figure 12.8. In this plot, it looks like just six observations deviate from the linearity required to give the data a multivariate normal density seal of approval. Identifying these outliers and creating a chi-square plot for the remaining log-transformed observations might be useful (see Exercise 12.2).

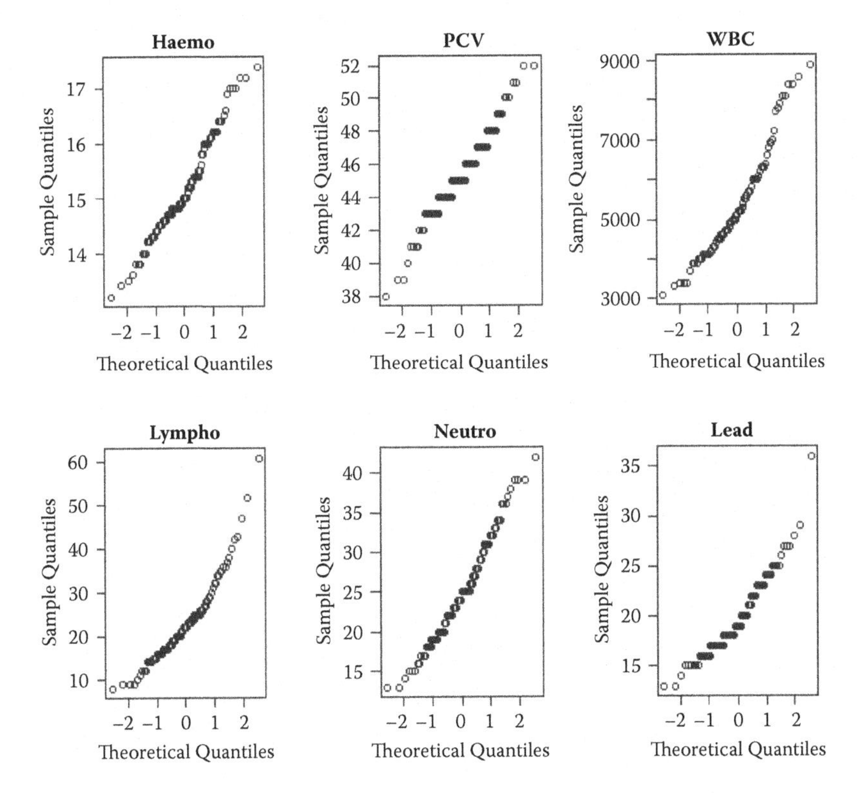

FIGURE 12.6
Normal probability plots for the six variables in the data on paint sprayers.

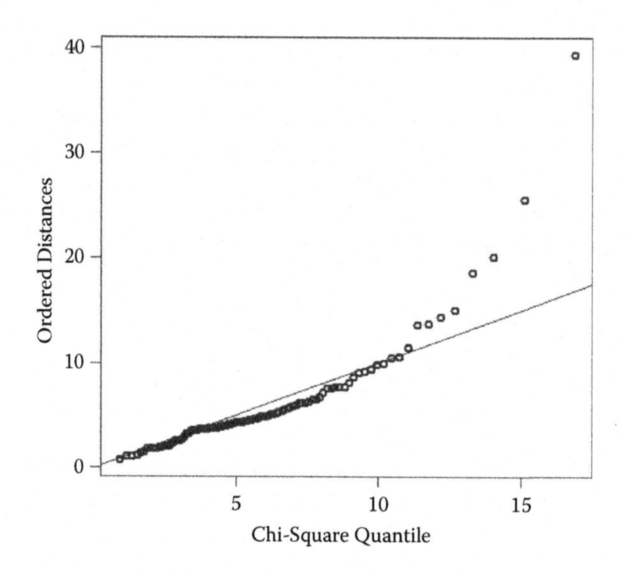

FIGURE 12.7
Chi-square plot of generalized distances for paint-sprayer data.

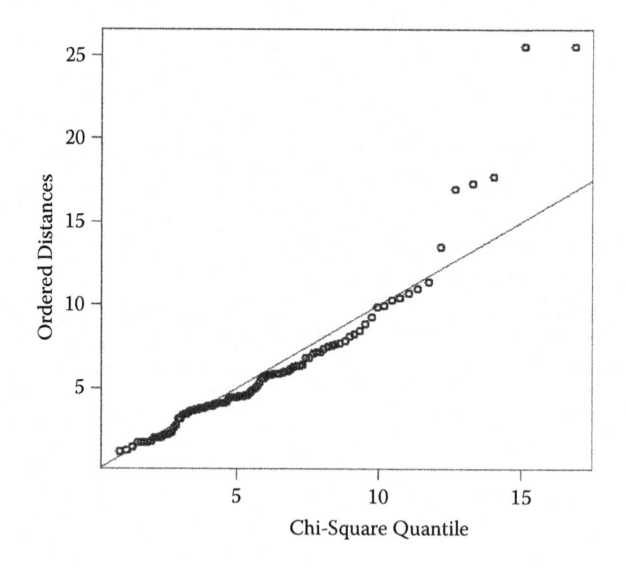

FIGURE 12.8
Chi-square plot of paint sprayers data after a log transformation of all the six variables.

12.4 Summary

- Multivariate data arise when researchers measure several variables on each individual in their sample and there is no response variable.
- Although in some cases it may make sense to isolate each variable and study it separately, in the main it does not. In most instances the variables are related in such a way that, when analyzed in isolation, they may often fail to reveal the full structure of the data. With the great majority of multivariate data sets, all the variables need to be examined simultaneously in order to uncover the patterns and key features in the data.
- Several multivariate techniques are largely "exploratory" in nature and, in many cases, a set of multivariate data may not arise as a sample from some populations. Consequently, questions of inference become less important.
- Where inference for multivariate data is an issue, it is usually based on the assumption that the sample data arise from a population in which the variables have a multivariate normal density function. In such cases, it may be worth assessing the assumption.

12.5 Exercises

12.1 For the body measurements data, using the scatterplot matrix of the data as a guide, try to identify the men and the women in the sample and then find the means, variances, and covariance and correlation matrices of the groups you identify.

12.2 For the paint-sprayer data, use some suitable graphics to identify any observations that you think are outliers. Construct chi-square plots of the generalized distances after removing the outlying observations both for the raw data and log-transformed data.

12.3 The data give the life expectancies in different countries by age and by sex. Find numerical summaries separately for men and women, and construct suitable graphics for an initial examination of the data.

13

Principal Components Analysis

13.1 Introduction

One of the problems with many sets of multivariate data is that there are simply too many variables to make the application of, say, some of the graphical techniques described in Chapter 2 successful in providing an informative initial assessment of the data. Further, having too many variables may cause problems for other statistical techniques that the researcher may want to apply to the data. The possible problem of too many variables is sometimes known as the curse of dimensionality. Clearly, the scatterplots, scatterplot matrices, and other graphics that might be applied to multivariate data for an initial assessment are likely to be more useful when the number of variables in the data, the dimensionality of the data, is relatively small rather than large. This brings us to *principal components analysis* (PCA), a multivariate technique with the central aim of reducing the dimensionality of a multivariate data set while retaining as much as possible of the variation present in it. This aim is achieved by transforming to a new set of variables the principal components that are uncorrelated and that are ordered, so that the first few of them account for most of the variation in all the original variables. In the best of all possible worlds, the result of a PCA would be the creation of a small number of new variables that can be used as surrogates for the originally large number of variables and, consequently, that provide a simpler basis for, say, graphing or summarizing the data and also, perhaps, when undertaking further multivariate analyses of the data.

13.2 Principal Components Analysis (PCA)

The basic goal of PCA is to describe variation in a set of correlated variables, $x_1, x_2, \ldots, x_q$, in terms of a new set of uncorrelated variables, $y_1, y_2, \ldots, y_q$, each of which is a linear combination of the x variables. The new variables are derived in decreasing order of "importance" in the sense that y_1 accounts for as much as possible of the variation in the original data among all linear combinations of $x_1, x_2, \ldots, x_q$. Then, y_2 is chosen to account for as much as possible of the remaining variation, subject to being uncorrelated with y_1, and so on. The new variables defined by this process, $y_1, y_2, \ldots, y_q$, are the principal components. Principal components analysis was first suggested by Pearson (1901) and independently by Hotelling (1933).

The general hope of PCA is that the first few components will account for a substantial proportion of the variation in the original variables $x_1, x_2, \ldots, x_q$, and can, consequently, be used to provide a convenient lower-dimensional summary of these variables that might prove useful for a variety of reasons. Consider, for example, a set of data consisting of examination scores for several different subjects for each of a number of students. One question of interest might be how best to construct an informative index of overall examination performance. One obvious possibility would be to take the mean score for each student, although, if the possible or observed range of examination scores varied from subject to subject, it might be more sensible to weight the scores in some way before calculating the average or, alternatively, standardize the results for the separate examinations before attempting to combine them. In this way, it might be possible to spread the students out further and so obtain a better ranking. The same result could often be achieved by applying the principal components to the observed examination results and using the students' scores on the first principal component to provide a measure of examination success that maximally discriminated between them.

A further possible application for PCA arises in the field of economics, where complex data are often summarized by some kind of index number, for example, indices of prices, wage rates, cost of living, and so on. When assessing changes in prices over time, the economist will wish to allow for the fact that prices of some commodities are more variable than others or that the prices of some of the commodities are considered more important than others; in each case, the index will need to be weighted accordingly. In such examples, the first principal component can often satisfy the investigator's requirements.

However, it is not always the first principal component that is of most interest to a researcher. A taxonomist, for example, when investigating variation in morphological measurements on animals for which all the pairwise correlations are likely to be positive, will often be more concerned with the second and subsequent components since these might provide a convenient description of aspects of an animal's "shape"; this will often be of more interest to the researcher than aspects of an animal's "size," which here, because of the positive correlations, will be reflected in the first principal component. For essentially the same reasons, the first principal component derived from, say, clinical psychiatric scores on patients may only provide an index of the severity of symptoms, and it is the remaining components that will give the psychiatrist important information about the "pattern" of symptoms.

The principal components are most commonly (and properly) used as a means of constructing an informative graphical representation of the data (see examples later in the chapter) or as inputs to some other analysis. One example of the latter is provided by regression analysis; principal components may be useful here when

- There are too many explanatory variables relative to the number of observations.
- The explanatory variables are highly correlated.

Both situations lead to problems when applying regression techniques—problems that may be overcome by replacing the original explanatory variables with the first few principal component variables derived from them. An example will be given later, and other applications of the technique are described in Rencher and Christensen (2012).

A further example when the results from a PCA may be useful is in the application of multivariate analysis of variance (see Chapter 18), when there are too many original variables to ensure that the technique can be used with reasonable power. In such cases, the first few principal components might be used to provide a smaller number of variables for analysis.

In the behavioral sciences, particularly psychology, the principal components are often considered an end in themselves, and researchers may then try to interpret them in a similar fashion to the factors in an exploratory factor analysis (see Chapter 15). We shall make some comments about this practice later in that chapter.

13.3 Finding the Sample Principal Components

PCA is overwhelmingly an exploratory technique for multivariate data. Although there are inferential methods available for using the sample principal components derived from a random sample of individuals from a population to test hypotheses about population principal components (see Jolliffe, 2002), they are very rarely to be seen in the accounts of PCA analysis that appear in the literature. Quintessentially, PCA is an aid in helping us understand the sample data. We use this observation as the rationale for describing only sample principal components in this chapter.

The first principal component of the observations is that linear combination of the original variables whose sample variance is the greatest among all possible such linear combinations. The second principal component is defined as that linear combination of the original variables that accounts for a maximal proportion of the remaining variances, subject to being uncorrelated with the first principal component. Subsequent components are defined similarly. The question now arises as to how the coefficients specifying the linear combinations of the original variables defining each component are found. This question is answered in Technical Section 13.1.

Technical Section 13.1: Extracting Principal Components

The first principal component of the observations, y_1, is the linear combination

$$y_1 = a_{11}x_1 + a_{12}x_2 + \cdots + a_{1q}x_q$$

whose sample variance is the greatest among all such linear combinations. Because the variance of y_1 could be increased without limit simply by increasing the values of the coefficients $a_{11}, a_{12}, \ldots, a_{1q}$ (which we will write as the vector $\mathbf{a}_1$), a restriction must be placed on these coefficients. As we shall see later, a sensible constraint is to require that the sum of squares of the coefficients should take the value 1, although other constraints are possible, and any multiple of the vector $\mathbf{a}_1$ produces basically the same component. To find the coefficients defining the first principal component, we need to choose the elements of the vector $\mathbf{a}_1$ so as to maximize the variance of y_1 subject to the sum-of-squares constraint, which can be written as $\mathbf{a}_1'\mathbf{a}_1 = 1$. The sample variance of y_1, which is a linear function of the x variables, is given by (see Chapter 12)

$$\text{var}(y_1) = \mathbf{a}_1'\mathbf{S}\mathbf{a}_1$$

where $\mathbf{S}$ is the $q \times q$ sample covariance matrix of the x variables. To maximize a function of several variables subject to one or more constraints, the method of Lagrange multipliers is used. Full algebraic details are given in Morrison (1990) and Jolliffe (2002), and we will not give them here. (The algebra of an example with $q = 2$ is, however, given in Section 13.5.) We simply state that the Lagrange multiplier approach leads to the solution that $\mathbf{a}_1$ is what is called an *eigenvector* or *characteristic vector* of the sample covariance matrix $\mathbf{S}$, and that it is the eigenvector corresponding to the largest of what are called the *eigenvalues* or *characteristic roots* of $\mathbf{S}$. (Eigenvalues of $\mathbf{S}$ and the corresponding eigenvectors are found by numerical algorithms, the details of which are not necessary to know to understand PCA.)

The second principal component y_2 is defined to be the linear combination

$$y_2 = a_{21}x_1 + a_{22}x_2 + \cdots + a_{2q}x_q$$

that is,

$$y_2 = \mathbf{a}_2'\mathbf{x}, \text{ where } \mathbf{a}' = [a_{21}, a_{22}, \ldots, a_{2q}] \text{ and } \mathbf{x}' = [x_1, x_2, \ldots, x_q]$$

which has the greatest variance subject to the following two conditions:

$$\mathbf{a}_2'\mathbf{a}_2 = 1$$
$$\mathbf{a}_2'\mathbf{a}_1 = 0$$

(The second condition specifies that y_1 and y_2 are uncorrelated.)

Similarly, the jth principal component is the linear combination $y_j = \mathbf{a}_j'\mathbf{x}$, which has the greatest variance subject to the conditions

$$\begin{aligned} \mathbf{a}_j'\mathbf{a}_j &= 1 \\ \mathbf{a}_j'\mathbf{a}_i &= 0 \qquad (i < j) \end{aligned}$$

Application of the Lagrange multiplier technique demonstrates that the vector of coefficients defining the jth principal component, that is, $\mathbf{a}_j$, is the eigenvector of $\mathbf{S}$ associated with its jth largest eigenvalue. If the q eigenvalues of $\mathbf{S}$ are denoted by $\lambda_1, \lambda_2, \ldots, \lambda_q$, then by requiring that $\mathbf{a}_i'\mathbf{a}_i = 1$, it can be shown that the variance of the ith principal component is given by λ_i. The total variance of the q principal components will equal the total variance of the original variables so that

$$\sum_{i=1}^{q} \lambda_i = s_1^2 + s_2^2 + \cdots + s_q^2$$

where s_i^2 is the sample variance of x_i. We can write this more concisely as

$$\sum_{i=1}^{q} \lambda_i = \text{trace}(\mathbf{S})$$

Consequently, the jth principal component accounts for a proportion of the total variation of the original data, where

$$P_j = \frac{\lambda_j}{\text{trace}(\mathbf{S})}$$

The first m principal components, where $m < q$, account for a proportion of the total variation in the original data, where

$$P^{(m)} = \frac{\sum_{i=1}^{m} \lambda_i}{\text{trace}(\mathbf{S})}$$

In geometrical terms, it is easy to show that the first principal component defines the line of best fit (in the least-squares sense) to the q-dimensional observations in the sample. These observations may therefore be represented in one dimension by taking their projection onto this line, that is, finding their first principal component score. If the observations happen to be collinear in q dimensions, this representation would completely account for the variation in the data, and the sample covariance matrix would have only one nonzero eigenvalue. In practice, of course, such collinearity is extremely unlikely, and an improved representation would be given by projecting the q-dimensional observations onto the space of the best fit, this being defined by the first two principal components. Similarly, the first m components give the best fit in m dimensions. If the observations fit exactly into a space of m dimensions, it would be indicated by the presence of $q - m$ zero eigenvalues of the covariance matrix. This would imply the presence of $q - m$ linear relationships between the variables. Such constraints are sometimes referred to as *structural relationships*. In practice, in the vast majority of applications of PCA, all the eigenvalues of the covariance matrix will be nonzero.

13.4 Should Principal Components be Extracted from the Covariance or the Correlation Matrix?

The account of principal components given above has them extracted from the covariance matrix of the data. However, imagine a set of multivariate data in which the variables $x_1, x_2, \ldots, x_q$ are of completely different types, for example, length, temperature, blood pressure, anxiety rating, etc. With such a data set, the structure of the principal components derived from the covariance matrix will depend upon the essentially arbitrary choice of units of measurement; for example, changing lengths from centimeters to inches will alter the derived components. Additionally, if there are large differences between the variances of the original variables, then the ones whose variances are the largest will tend to dominate the early components. This difficulty is overcome in practice by extracting the components from the correlation matrix **R**. Extracting the components as the eigenvectors of **R** is equivalent to calculating the principal components from the original variables after each has been standardized to have unit variance. It should be noted, however, that there is rarely any simple correspondence between the components derived from **S** and those derived from **R**. In addition, choosing to work with **R** rather than **S** involves a definite, but possibly arbitrary, decision to make the variables "equally important."

To demonstrate how the principal components of the covariance matrix of a data set can differ from the components extracted from the data's correlation

TABLE 13.1
Correlations of Blood Chemistry Variables and Their Standard Deviations

	rBlood	Plate	wBlood	Neut.	Lymph	Bilir.	Sodium	Potass.
rBlood	1.000	0.290	0.202	−0.055	−0.105	−0.252	−0.229	0.058
Plate	0.290	1.000	0.415	0.285	−0.376	−0.349	−0.164	−0.129
wBlood	0.202	0.415	1.000	0.419	−0.521	−0.441	−0.145	−0.076
Neut.	−0.055	0.285	0.419	1.000	−0.877	−0.076	0.023	−0.131
Lymph	−0.105	−0.376	−0.521	−0.877	1.000	0.206	0.034	0.151
Bilir.	−0.252	−0.349	−0.441	−0.076	0.206	1.000	0.192	0.077
Sodium	−0.229	−0.164	−0.145	0.023	0.034	0.192	1.000	0.423
Potass.	0.058	−0.129	−0.076	−0.131	0.151	0.077	0.423	1.000
Std Dev	0.371	41.253	1.935	0.077	0.071	4.037	2.732	0.297

matrix, we will use the example given in Jolliffe (2002). The data in this example consist of eight blood chemistry variables measured on 72 patients in a clinical trial. The correlation matrix of the data, together with the standard deviations of each of the eight variables, is given in Table 13.1; there are considerable differences between these standard deviations. We can apply PCA to both the covariance and correlation matrix of the data (the covariance matrix is not given, but it can be easily calculated from the correlation matrix and the standard deviations—see Chapter 12). The details of the principal components of the covariance matrix are given in Table 13.2, and those of the

TABLE 13.2
Principal Components of the Covariance Matrix of Blood Chemistry Data

	Variances, etc. of the Components							
	Comp.1	**Comp.2**	**Comp.3**	**Comp.4**	**Comp.5**	**Comp.6**	**Comp.7**	**Comp.8**
Variance	1704.68	15.06	6.98	2.64	0.13	0.07	0.00	0.00
Proportion of variance	0.986	0.0087	0.00404	0.00153	0.000	0.000	0.000	0.000
Cumulative proportion	0.986	0.9943	0.99836	0.99989	1.000	1.000	1.000	1.000
	Component Loadings							
rBlood	0.000	0.000	0.000	0.000	0.943	0.329	0.000	0.000
Plate	−0.999	0.000	0.000	0.000	0.000	0.000	0.000	0.000
wBlood	0.000	−0.192	0.000	−0.981	0.000	0.000	0.000	0.000
Neut.	0.000	0.000	0.000	0.000	0.000	0.000	0.758	0.650
Lymph	0.000	0.000	0.000	0.000	0.000	0.000	−0.649	0.760
Bilir.	0.000	0.961	0.195	−0.191	0.000	0.000	0.000	0.000
Sodium	0.000	0.193	−0.979	0.000	0.000	0.000	0.000	0.000
Potass.	0.000	0.000	0.000	0.000	0.329	−0.942	0.000	0.000

TABLE 13.3
Principal Components of the Correlation Matrix of Blood Chemistry Data

	Variances, etc. of the Components							
	Comp.1	**Comp.2**	**Comp.3**	**Comp.4**	**Comp.5**	**Comp.6**	**Comp.7**	**Comp.8**
Variance	2.792	1.532	1.249	0.778	0.622	0.489	0.436	0.102
Proportion of variance	0.349	0.191	0.156	0.0973	0.0777	0.0611	0.0545	0.0128
Cumulative proportion	0.349	0.540	0.697	0.7939	0.8716	0.9327	0.9872	1.0000
	Component Loadings							
rBlood	−0.194	0.417	0.400	0.652	0.175	−0.363	0.176	0.102
Plate	−0.400	0.154	0.168	0.000	−0.848	0.230	−0.110	0.000
wBlood	−0.459	0.000	0.168	−0.274	0.251	0.403	0.677	0.000
Neut.	−0.430	−0.472	−0.171	0.169	0.118	0.000	−0.237	0.678
Lymph	0.494	0.360	0.000	−0.180	−0.139	0.136	0.157	0.724
Bilir.	0.319	−0.320	−0.277	0.633	−0.162	0.384	0.377	0.000
Sodium	0.177	−0.535	0.410	−0.163	−0.299	−0.513	0.367	0.000
Potass.	0.171	−0.245	0.709	0.000	0.198	0.469	−0.376	0.000

correlation matrix in Table 13.3 (in both tables, very small values have been set to 0).

Examining the results in Tables 13.2 and 13.3, we see that each of the principal components of the covariance matrix is largely dominated by a single variable, whereas those for the correlation matrix have moderate-sized coefficients on several of the variables. In addition, the first component of the covariance matrix accounts for almost 99% of the total variance of the observed variables. The components of the covariance matrix are completely dominated by the fact that the variance of the plate variable is 100 times larger than the variance of any of the other seven variables. Consequently, the principal components from the covariance matrix simply reflect the order of sizes of the variances of the observed variables. The results from the correlation matrix tell us, in particular, that a weighted contrast of the first four and last four variables is the linear function with the largest variance. This example illustrates that, when variables are on very different scales or have very different variances, a PCA of the data should be performed on the correlation matrix, not on the covariance matrix.

13.5 Principal Components of Bivariate Data with Correlation Coefficient r

Before we move on to look at some practical examples of the application of PCA, it will be helpful to look in a little more detail at the mathematics of the method in one very simple case. We will do this in Technical Section 13.2, using bivariate data in which the two variables x_1 and x_2 have correlation coefficient r.

Technical Section 13.2: Principal Components of Bivariate Data

Suppose we have just two variables x_1 and x_2, measured on a sample of individuals, with sample correlation matrix given by

$$\mathbf{R} = \begin{bmatrix} 1.0 & r \\ r & 1.0 \end{bmatrix}$$

In order to find the principal components of the data, we need to find the eigenvalues and eigenvectors of $\mathbf{R}$. The eigenvalues are roots of the equation

$$|\mathbf{R} - \lambda\mathbf{I}| = 0$$

where the vertical lines indicate the determinant of the matrix enclosed by them. This leads to the following quadratic equation in λ:

$$(1-\lambda)^2 - r^2 = 0$$

which has roots (eigenvalues) $\lambda_1 = 1+r$, $\lambda_2 = 1-r$. The first component has variance $1+r$, and the second has variance $1-r$. Note that the sum of the eigenvalues is 2, equal to trace($\mathbf{R}$), that is, the sum of the elements on the main diagonal. The eigenvector corresponding to λ_1 is obtained by solving the equation

$$\mathbf{R}\mathbf{a}_1 = \lambda_1 \mathbf{a}_1$$

This leads to the equations

$$a_{11} + r a_{12} = (1+r)a_{11}$$
$$r a_{11} + a_{12} = (1+r)a_{12}$$

The two equations are identical, and both reduce to $a_{11} = a_{12}$. If we now introduce the normalization constraint $\mathbf{a}_1'\mathbf{a}_1 = 1$, we find that

$$a_{11} = a_{12} = \frac{1}{\sqrt{2}}$$

Similarly, we find the elements of the second eigenvector as $a_{21} = 1/\sqrt{2}$ and $a_{22} = -1/\sqrt{2}$. The two principal components are then given by

$$y_1 = \frac{1}{\sqrt{2}}(x_1 + x_2)$$

and

$$y_2 = \frac{1}{\sqrt{2}}(x_1 - x_2)$$

Notice that if $r < 0$, the order of the eigenvalues, and hence that of the principal components, is reversed; if $r = 0$, the eigenvalues are both equal to 1, and any two solutions at right angles could be chosen to represent the two components.

Three further points:

- There is an arbitrary sign in the choice of the elements of $\mathbf{a}_i$; it is customary to choose a_{i1} to be positive.
- The components do not depend on r, although the proportion of variance explained by each does change with r. As r tends to 1, the proportion of variance accounted for by y_1, namely, $(1+r)/2$, also tends to 1.
- When $r = 1$ the points all lie on a straight line, and the variation in the data is unidimensional.

13.6 Rescaling the Principal Components

The coefficients defining the principal components derived as described in Section 13.5 are often rescaled so that they are correlations or covariances between the original variables and the derived components. The rescaled coefficients are often more useful in interpreting a PCA. The covariance of variable x_i with component y_j is given by

$$\text{Cov}(x_i, y_j) = \lambda_j a_{ji}$$

The correlation of variables x_i with component y_j is therefore

$$r_{x_i,y_j} = \frac{\lambda_j a_{ji}}{\sqrt{\text{Var}(x_i)\,\text{Var}(y_j)}} = \frac{\lambda_j a_{ji}}{s_i\sqrt{\lambda_j}} = \frac{a_{ji}\sqrt{\lambda_j}}{s_i}$$

If the components are extracted from the correlation matrix rather than the covariance matrix, the correlation between variable and component becomes

$$r_{x_i,y_j} = a_{ji}\sqrt{\lambda_j}$$

because in this case the standard deviation s_i is 1. (Although for convenience we have used the same nomenclature for the eigenvalues and the eigenvectors extracted from the covariance matrix or the correlation matrix, they will, of course, not be equal.)

The rescaled coefficients from a PCA of a correlation matrix are analogous to factor loadings as we shall see in Chapter 15. It is often these rescaled coefficients that are presented as the results of a PCA and used in interpretation.

13.7 How the Principal Components Predict the Observed Covariance Matrix

In Technical Section 13.3, we will look at how the principal components reproduce the observed covariance or correlation matrix from which they were extracted.

Technical Section 13.3: How Principal Components Reproduce the Sample Covariance Matrix

To begin, let the initial vectors $\mathbf{a}_1, \mathbf{a}_2, \ldots, \mathbf{a}_q$ that define the principal components be used to form a $q \times q$ matrix $\mathbf{A} = [\mathbf{a}_1, \ldots, \mathbf{a}_q]$; these are vectors extracted from the covariance matrix $\mathbf{S}$ and scaled so that

$\mathbf{a}_i'\mathbf{a}_i = 1$. Arrange the eigenvalues $\lambda_1, \ldots, \lambda_q$ along the main diagonal of a diagonal matrix $\mathbf{\Lambda}$. Then, it can be shown that the covariance matrix of the observed variables $x_1, x_2, \ldots, x_q$ is given by

$$\mathbf{S} = \mathbf{A\Lambda A}'$$

Rescaling the vectors $\mathbf{a}_1, \mathbf{a}_2, \ldots, \mathbf{a}_q$ so that the sum of squares of their elements is equal to the corresponding eigenvalue, that is, calculating $\mathbf{a}_i^* = \lambda_i^{1/2}\mathbf{a}_i$, allows $\mathbf{S}$ to be written more simply as

$$\mathbf{S} = \mathbf{A}^*(\mathbf{A}^*)'$$

where $\mathbf{A}^* = \left[\mathbf{a}_1^*, \ldots, \mathbf{a}_q^*\right]$

If the matrix $\mathbf{A^*}_m$ is formed from, say, the first m components rather than from all q, then $\mathbf{A^*}_m(\mathbf{A^*}_m)'$ gives the predicted value of $\mathbf{S}$ based on these m components. It is often useful to calculate the predicted value based on the number of components considered to adequately describe the data. How this number might be chosen is considered in the next section.

13.8 Choosing the Number of Components

As described earlier, PCA is seen to be a technique for transforming a set of observed variables into a new set of variables that are uncorrelated with one another. The variation in the original q variables is only completely accounted for by all q principal components. The usefulness of these transformed variables, however, stems from their property of accounting for the variance in decreasing proportions. The first component, for example, accounts for the maximum amount of variation possible for any linear combination of the original variables. But, how useful is this artificial variate constructed from the observed variables? To answer this question, we would first need to know the proportion of the total variance of the original variables for which it accounted. If, for example, 80% of the variation in a multivariate data set involving six variables could be accounted for by a simple weighted average of the variable values, then almost all the variation can be expressed along a single continuum rather than in six-dimensional space. The PCA would have provided a highly parsimonious summary (reducing the dimensionality of the data from six to one) that might be useful in later analysis.

So, the question we need to ask is how many components are needed to provide an adequate summary of a given data set? A number of informal and more formal techniques are available. Here we shall concentrate on the former; examples of the use of formal inferential methods are given in Jolliffe (2002) and Rencher and Christensen (2012).

The most common of the relatively *ad hoc* procedures that have been suggested for deciding on the number of components to retain are the following:

- Retain just enough components to explain some specified, large percentage of the total variation of the original variables. Values between 70% and 90% are usually suggested, although smaller values might be appropriate as q or n, the sample size, increases.

- Exclude those principal components whose eigenvalues are less than the average, that is, $\sum_{i=1}^{q} \lambda_i/q$. Since $\sum_{i=1}^{q} \lambda_i = \text{trace}(\mathbf{S})$, the average eigenvalue is also the average variance of the original variables. This method then retains those components that account for more variance than the average for the observed variables.

- When the principal components are extracted from the correlation matrix, $\text{trace}(\mathbf{R}) = q$, and the average variance is, therefore, 1; so, applying the rule in the previous bullet point, components with eigenvalues less than 1 will be excluded. This rule was originally suggested by Kaiser (1958), but Jolliffe (1972), on the basis of a number of simulation studies, proposed that a more appropriate procedure would be to exclude components extracted from a correlation matrix whose associated eigenvalues are less than 0.7.

- Cattell (1965) suggested examination of the plot of λ_i against i, the so-called *scree diagram*. The number of components selected is the value of i corresponding to an "elbow" in the curve, that is, a change of slope from "steep" to "shallow." In fact, Cattell was more specific than this; he recommended looking for a point on the plot beyond which the scree diagram defines a more or less straight line, not necessarily horizontal. The first point on the straight line is then taken as the last component to be retained. Further, it should also be remembered that Cattell suggested the scree diagram in the context of factor analysis rather than as applied to PCA.

- A modification of the scree diagram described by Jolliffe (1989) is the log-eigenvalue diagram consisting of a plot of $\log(\lambda_i)$ against i.

Returning to the results of the PCA of the correlation matrix of the blood chemistry data given in Section 13.4, we find that the first four components account for nearly 80% of the total variance, but it takes a further two components to push this figure up to 90%. A cutoff of 1 for the eigenvalues leads to retaining three components, and with a cutoff of 0.7, four components are kept. Figure 13.1 shows the scree diagram and the log-eigenvalue diagram for the data. The former plot may suggest four components, although this is fairly subjective, and the latter seems to be of little help here because it appears to indicate retaining seven components, which is hardly much of a dimensionality reduction. The example illustrates that the proposed methods for deciding how many components to keep can (and often do) lead to different conclusions.

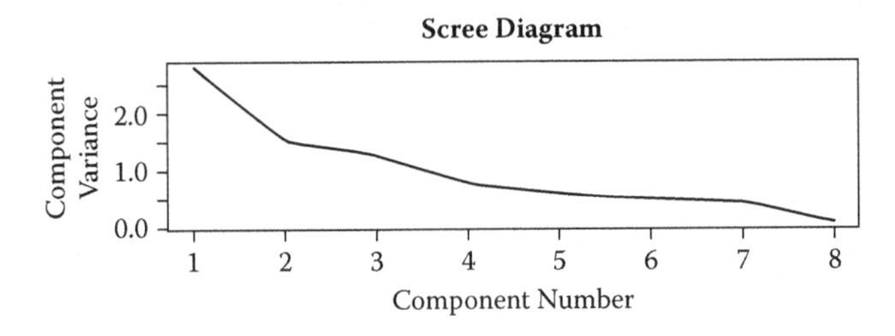

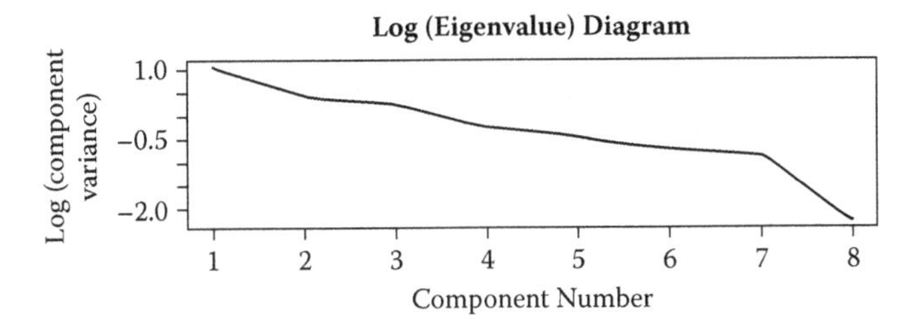

FIGURE 13.1
Scree diagram and log-eigenvalue diagram for the principal components of the correlation matrix of the blood chemistry data.

13.9 Calculating Principal Component Scores

If we decide that we need, say, m principal components to adequately represent our data (using one or other of the methods described in the previous section), then we will generally wish to calculate the scores on each of these components for each individual in our sample. How we do this is described in Technical Section 13.4.

Technical Section 13.4: Calculating Principal Component Scores

First, let us assume that we have derived the components from the covariance matrix **S**. The m principal component scores for individual i with original $q \times 1$ vector of variable values x_i are obtained as

$$
\begin{aligned}
y_{i1} &= \mathbf{a}_1'\mathbf{x}_i \\
y_{i2} &= \mathbf{a}_2'\mathbf{x}_i \\
&\vdots \\
y_{im} &= \mathbf{a}_m'\mathbf{x}_i
\end{aligned}
$$

If the components are derived from the correlation matrix, then $\mathbf{x}_i$ would contain the ith individual's standardized scores for each variable.

The principal component scores calculated as shown here have variances equal to λ_j for $j = 1, \ldots, m$. Many investigators might prefer to have scores with means equal to 0 and variances equal to 1. Such scores can be found as follows:

$$\mathbf{z} = \mathbf{\Lambda}_m^{-1}\mathbf{A}_m'\mathbf{x}$$

where $\mathbf{\Lambda}_m$ is an $m \times m$ diagonal matrix with $\lambda_1, \lambda_2, \ldots, \lambda_m$ on the main diagonal, $\mathbf{A}_m = [\mathbf{a}_1, \ldots, \mathbf{a}_m]$, and $\mathbf{x}$ is the $q \times 1$ vector of standardized scores. We should note here that the first m principal component scores are the same whether we retain all possible q components or just the first m. As we shall see in Chapter 15, this is not the case with the calculation of factor scores.

13.10 Some Examples of the Application of PCA

In this section, we will look at the application of PCA to a number of data sets, beginning with one involving only two variables as this allows us to illustrate graphically an important point about this type of analysis.

13.10.1 Head Size of Brothers

The data in Table 13.4 give the head lengths (in millimeters) for each of the first two adult sons in 25 families. The mean vector and covariance matrix of the data are

$$\bar{\mathbf{x}}' = [185.72, 183.84] \quad \mathbf{S} = \begin{bmatrix} 95.29 & 69.66 \\ 69.66 & 100.81 \end{bmatrix}$$

The principal components of these data extracted from their covariance matrix are

$$y_1 = 0.693x_1 + 0.721x_2 \quad y_2 = -0.721x_1 + 0.693x_2$$

with variances 167.77 and 28.33. The first principal component accounts for a proportion $167.77/(167.77 + 28.33) = 0.86$ of the total variance in the original variables. Note that the total variance of the principal components is 196.10, which, as expected, is equal to the total variance of the original variables found by adding the relevant terms in the covariance matrix above, that is, $95.29 + 100.81 = 196.10$.

How should the two derived components be interpreted? The first component is essentially the sum of the head lengths of the two sons, and the

TABLE 13.4
Head Lengths (in Millimeters) of First and Second Sons of 25 Families

Family	First Son	Second Son
1	191	179
2	195	201
3	181	185
4	183	188
5	176	171
6	208	192
7	189	190
8	197	189
9	188	197
10	192	187
11	179	186
12	183	174
13	174	185
14	190	195
15	188	187
16	163	161
17	195	183
18	186	173
19	181	182
20	175	165
21	192	185
22	174	178
23	176	176
24	197	200
25	190	187

second component is the difference in head lengths. Perhaps we can label the first component "size" and the second component "shape," but see the next subsection for some comments about trying to give principal components such labels.

To calculate an individual's score on a component, we simply multiply the variable values, subtract the appropriate mean by the loading for the variable, and add these values over all variables. We can illustrate this calculation using the data for the first family, in which the head length of the first son is 191 mm and the second son, 179 mm. The score for this family on the first principal component is calculated as

$$0.693 \times (191 - 185.72) + 0.721 \times (179 - 183.84) = 0.169$$

and on the second component the score is

$$-0.721 \times (191 - 185.72) + 0.693 \times (179 - 183.84) = -7.61$$

The variance of the first principal component scores will be 167.77, and that of the second principal component scores will be 28.33.

We can plot the data showing the axes corresponding to the principal components. The first axis passes through the mean of the data and has a slope 0.721/0.693, and the second axis also passes through the mean and has a slope −0.693/0.721. The plot is shown in Figure 13.2. This example illustrates that a PCA is essentially simply a rotation of the axes of the multivariate data scatter. Further, we can also plot the principal component scores to give Figure 13.3. (Note that in this figure, the range of the x-axis and the range of the y-axis have been made the same so that the larger variance of the first principal component is clearly shown.)

We can use the PCA of the head size data to demonstrate how the principal components reproduce the observed covariance matrix. We first need to rescale the principal components we have at this point by multiplying them by the square roots of their respective variances to give the new components:

$$y_1 = 12.952[0.693x_1 + 0.721x_2], \quad \text{that is ,} \quad y_1 = 8.976x_1 + 9.338x_2$$

and

$$y_2 = 5.323[-0.721x_1 + 0.693x_2], \quad \text{that is ,} \quad y_2 = -3.837x_1 + 3.688x_2$$

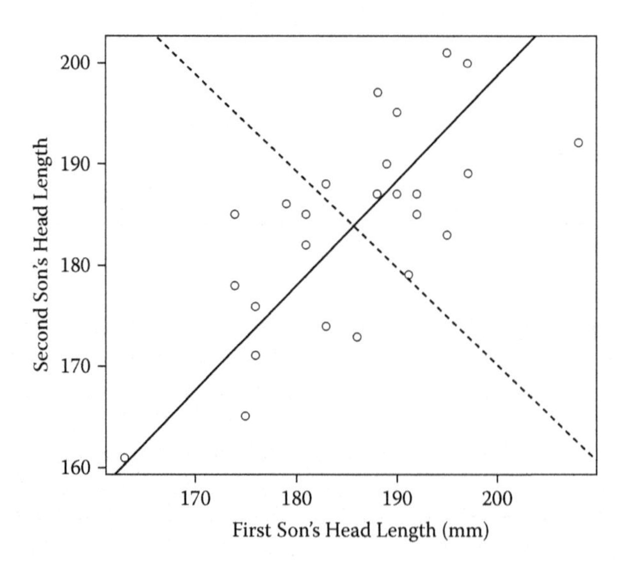

FIGURE 13.2
Head lengths of first and second sons of 25 families, showing axes corresponding to the principal components of the sample covariance matrix of the data.

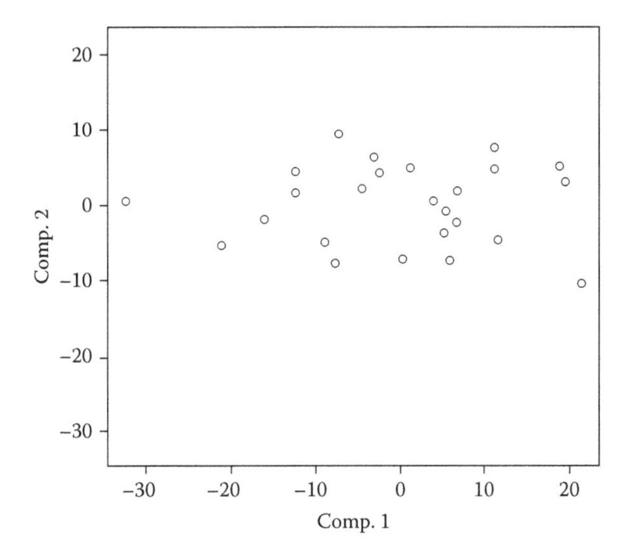

FIGURE 13.3
Plot of the first two principal components score for the head size data.

leading to the matrix $\mathbf{A}^*{}_2$ as defined in Section 13.7:

$$\mathbf{A}^*{}_2 = \begin{bmatrix} 8.976 & -3.837 \\ 9.338 & 3.688 \end{bmatrix}$$

Multiplying this matrix by its transpose should recreate the covariance matrix of the head length data; performing the matrix multiplication shows that it does recreate $\mathbf{S}$:

$$\mathbf{A}^*{}_2(\mathbf{A}^*{}_2)' = \begin{bmatrix} 95.29 & 69.66 \\ 69.66 & 100.81 \end{bmatrix}$$

As an exercise, readers might like to find the predicted covariance matrix using only the first component.

The head size example has been useful for discussing some aspects of PCA, but it is not, of course, typical of multivariate data sets encountered in behavioral research, where many more than two variables will be recorded for each individual in a study. In the following two subsections, we consider some more realistic examples.

13.10.2 Crime Rates in the United States

The *Statistical Abstract of the USA* (1988) gives rates of different types of crime per 100,000 residents of 50 states of the United States plus the District of Columbia for the year 1986. The data for the first five states are given

TABLE 13.5
Crime Rates in the United States

State	Murder	Rape	Robbery	Assault	Burglary	Theft	Vehicle
ME	2.0	14.8	28	102	803	2347	164
NH	2.2	21.5	24	92	755	2208	228
VT	2.0	21.8	22	103	949	2697	181
MA	3.6	29.7	193	331	1071	2189	906
RI	3.5	21.4	119	192	1294	2568	705

in Table 13.5. We shall use PCA to explore these data, but to start, it will be useful to look at the scatterplot matrix of the seven types of crime, and this is given in Figure 13.4. The plot shows that the relationships between crime rates are of varying strengths and that there are a number of outlying states in some of the panels, for example, the one for rape and murder rates. We shall ignore the outlier problem in the following analysis, but readers are encouraged to see how the results that follow are altered if the outliers are removed (see Exercise 13.1). In this example, the variables are all on the same scale, crimes per 100,000 of the resident population of a state, but if we look at the variances of each crime rate,

Murder	Rape	Robbery	Assault	Burglary	Theft	Vehicle
23.20	212.31	18993.37	22004.31	177912.83	582812.84	50007.37

we see that they are very different, and the results from a PCA on the unstandardized variables would be swamped by those variables with the largest variances. Consequently, we will apply the analysis to the standardized variables, that is, components will be extracted from the correlation rather than the covariance matrix.

The results of the PCA on the crime rate data are shown in Table 13.6. A scree plot of variances of principal components is shown in Figure 13.5. Only the variance of the first component is greater than 1, although that of the second component is very close to 1. The scree plot suggests that perhaps two components might be adequate to describe these data. Many users of PCA search for an interpretation of the derived components that allow them to be "labeled" in some sense. This requires examining the coefficients defining each component; in Table 13.6, the coefficients are scaled so that their sums of squares equal 1; "—" indicates near-zero values. (Remember also that the signs of the coefficients are arbitrary in the sense that the minus signs and positive signs could be reversed without altering the structure or the interpretation of the components.) Examining the coefficients defining the principal components in Table 13.6, we see that the first component might be regarded as some index of overall crime rate in a state, with states that have larger crime

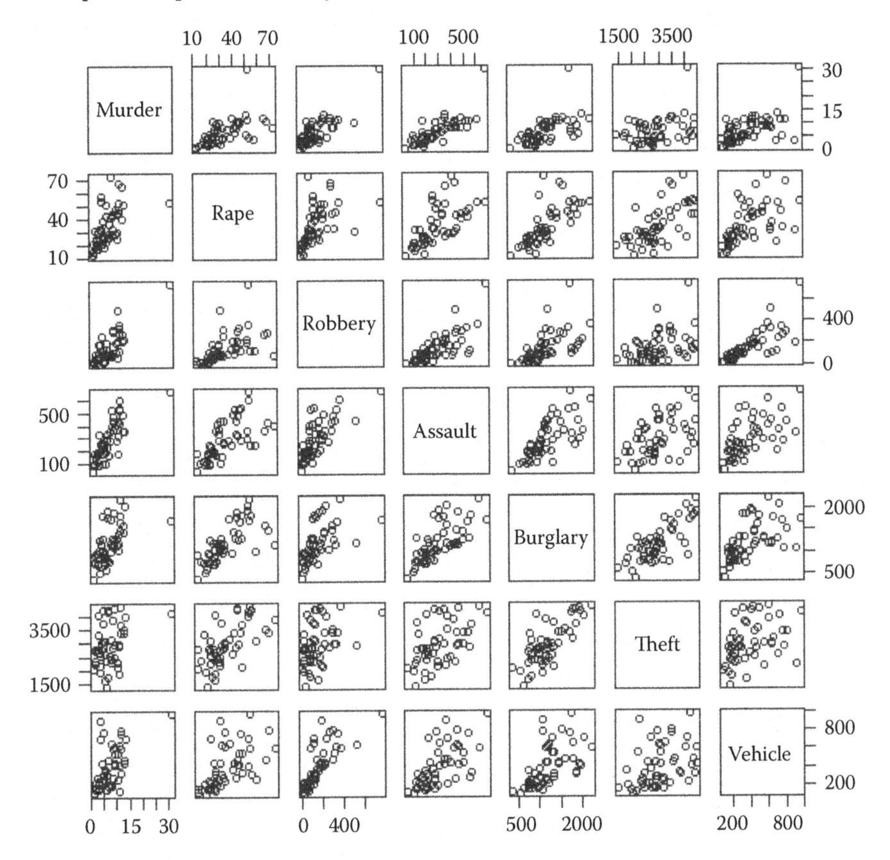

FIGURE 13.4
Scatterplot matrix of crime rate data.

rates having larger negative scores on this component (negative because of the minus signs attached to each loading); perhaps we could label this component "dangerousness." Labeling the second component is more difficult; in this component, the coefficients have a mixture of positive and negative signs, and the component appears to contrast "property crimes," that is, larceny and burglary, with crimes against the person, that is, robbery and murder. A not very inspired label might be "property versus person." The other components have very small variances, and we shall not try to interpret them.

Attempting to label components in this way is not without its critics; the following quotation from Marriott (1974) should act as a salutary warning about the dangers of overinterpretation:

> It must be emphasized that no mathematical method is, or could be, designed to give physically meaningful results. If a mathematical expression of this sort has an obvious physical meaning, it must be attributed to a lucky change, or to the fact that the data have a strongly marked structure

TABLE 13.6
Results from a PCA of the Correlation Matrix of Crime Rate Data

	Comp.1	Comp.2	Comp.3	Comp.4	Comp.5	Comp.6	Comp.7
Importance of Components							
Variance	4.69	0.99	0.46	0.34	0.24	0.18	0.09
Proportion of Variance	0.67	0.14	0.07	0.05	0.03	0.03	0.01
Cumulative Proportion	0.67	0.81	0.88	0.93	0.96	0.99	1.00
Coefficients Defining Components							
Murder	−0.381	−0.350	−0.538	—	−0.274	0.370	0.480
Rape	−0.377	0.279	—	−0.830	−0.250	—	−0.151
Robbery	−0.391	−0.420	0.131	0.275	−0.387	—	−0.651
Assault	−0.410	−0.124	−0.335	—	0.564	−0.620	—
Burglary	−0.394	0.367	—	0.162	0.466	0.622	−0.283
Theft	−0.321	0.628	—	0.449	−0.388	−0.282	0.256
Vehicle	−0.366	−0.282	0.758	—	0.163	—	0.422

Note: — indicates a coefficient that is almost zero.

that shows up in analysis. Even in the latter case, quite small sampling fluctuations can upset the interpretation; for example, the first two principal components may appear in reverse order or may become confused

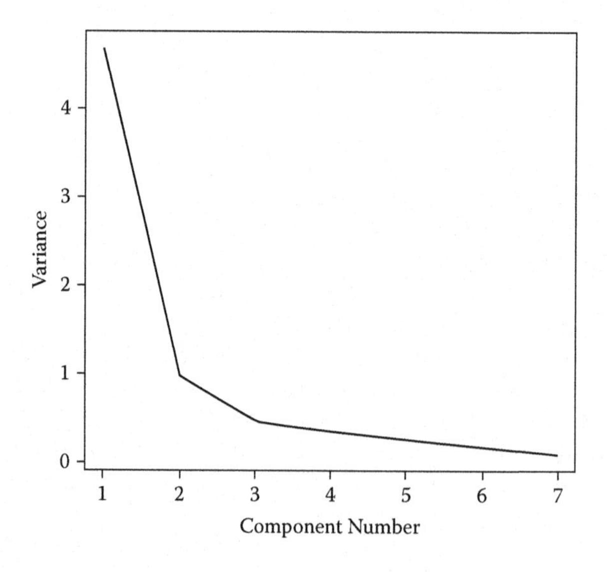

FIGURE 13.5
Scree plots of variances of principal components in the correlation matrix of crime rate data.

altogether. Reification then requires considerable skill and experience if it is to give a true picture of the physical meaning of the data.

Perhaps a more suitable use for the principal components of the crime rate data is as the basis of various graphical displays of cities. In fact, this is often the most useful aspect of a PCA and as a means to providing informative "views" of multivariate data, PCA has the advantage of making it less urgent or tempting to try to interpret and label the components. The first few component scores provide a low-dimensional "map" of the observations in which the Euclidean distances between the points representing the individuals best approximate in some sense the Euclidean distances between the individuals based on the original variables (for details, see the next chapter).

A plot of the crime rate data in the space of the first two principal components showing state labels is given in Figure 13.6. Clearly, DC is not a place to live in by choice! On the second component, WA has a high score because it has a low murder and robbery rate and relatively high burglary and larceny rates; contrast NY, which has relatively high murder and robbery rates and relatively low burglary and larceny rates and has a low score on the second component. Because the first two components account for over 80% of the variance in the crime rates, the two-dimensional plot in Figure 13.6 gives a very accurate description of the original seven-variable data. This claim is

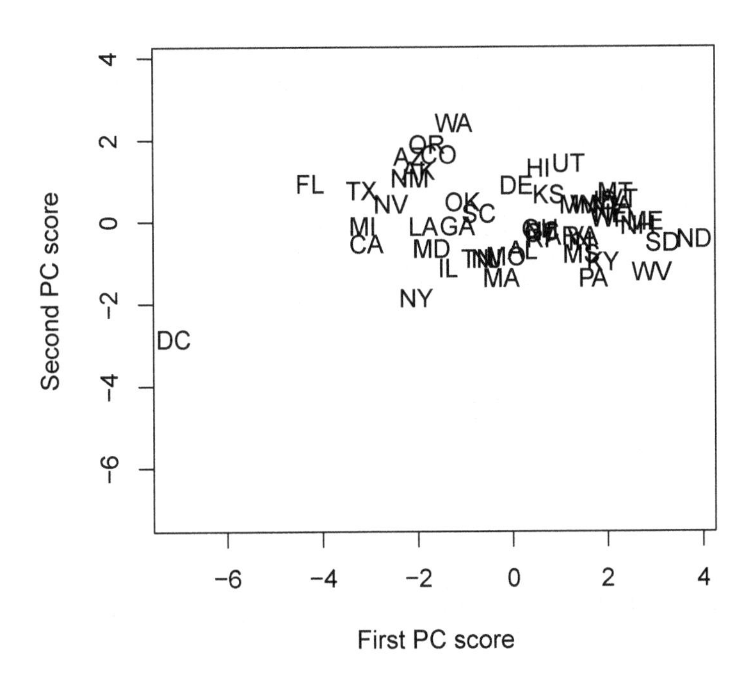

FIGURE 13.6
Plot of the crime rate data in the space of the first two principal components showing state labels.

TABLE 13.7
Observed Correlation Matrix for Crime Rate Data Compared to Matrix Constructed from the Two-Component Solution

	Observed						
	Murder	**Rape**	**Robbery**	**Assault**	**Burglary**	**Theft**	**Vehicle**
Murder	1.000	0.578	0.804	0.781	0.581	0.361	0.573
Rape	0.578	1.000	0.530	0.659	0.721	0.635	0.569
Robbery	0.804	0.530	1.000	0.740	0.551	0.400	0.786
Assault	0.781	0.659	0.740	1.000	0.710	0.512	0.638
Burglary	0.581	0.721	0.551	0.710	1.000	0.764	0.579
Theft	0.361	0.635	0.400	0.512	0.764	1.000	0.386
Vehicle	0.573	0.569	0.786	0.638	0.579	0.386	1.000
	Predicted						
Murder	0.801	0.578	0.843	0.775	0.578	0.357	0.752
Rape	0.578	0.745	0.576	0.691	0.799	0.741	0.571
Robbery	0.843	0.576	0.889	0.802	0.571	0.328	0.788
Assault	0.775	0.691	0.802	0.803	0.713	0.540	0.739
Burglary	0.578	0.799	0.571	0.713	0.862	0.821	0.576
Theft	0.357	0.741	0.328	0.540	0.821	0.872	0.377
Vehicle	0.752	0.571	0.788	0.739	0.576	0.377	0.708

backed up by comparing the observed correlation matrix with the correlation matrix "predicted" by the two-component solution after it has been rescaled, as described in Section 13.7. The two matrices are shown in Table 13.7. Corresponding elements of the two matrices are quite similar.

13.10.3 Drug Usage by American College Students

The majority of adult and adolescent Americans regularly use psychoactive substances and often do so for a substantial proportion of their lifetime. Various forms of licit and illicit psychoactive substance use are prevalent, suggesting that patterns of psychoactive substance taking are a major part of the individual's behavioral repertory and have pervasive implications on the performance of other behaviors. In an investigation of these phenomena, Huba et al. (1981) collected data on drug usage rates for 1634 students in the seventh to ninth grades in 11 schools in the greater metropolitan area of Los Angeles. Each participant completed a questionnaire about the number of times a particular substance had ever been used. The substances asked about were as follows:

1. Cigarettes
2. Beer
3. Wine
4. Liquor
5. Cocaine
6. Tranquilizers
7. Drugstore medications used to get "high"
8. Heroin and other opiates
9. Marijuana
10. Hashish
11. Inhalants (glue, gasoline)
12. Hallucinogenics (LSD, psilocybin, mescaline)
13. Amphetamine stimulants

Responses were recorded on a five-point scale:

1. Never tried
2. Only once
3. A few times
4. Many times
5. Regularly

The correlations between the usage rates of the 13 substances are shown in Table 13.8.

Applying a PCA to this correlation matrix gives the results shown in Table 13.9. A scree plot of the variances of the components is shown in Figure 13.7. The first two components have variances greater than 1, and the third has a variance slightly lower than 1. The scree plot suggests that perhaps three components need to be used to describe the correlations for these data, although these three components account for only 57% of the variance in drug usage rates. The first component is clearly a measure of overall drug usage, as might be expected since all the coefficients have relatively similar numerical values and all have the same sign. The second component contrasts "legal" with "illegal" substances (with the exception of marijuana, which has the same sign as the legal substances). This component might be seen as contrasting "soft" and "hard" drug usage. So, after overall usage has been accounted for, the main source of variation is between the different patterns of consumption of soft and hard drugs. The third component is essentially a contrast of drugstore and inhalant substance usage on the one hand, with marijuana, hashish, and amphetamine usage on the other.

We will return to the drug usage data in Chapter 16.

TABLE 13.8
Correlation Matrix for Drug Usage Data

	Drug	1	2	3	4	5	6	7	8	9	10	11	12	13
1	Cigarettes	1												
2	Beer	0.447	1											
3	Wine	0.442	0.619	1										
4	Liquor	0.435	0.604	0.583	1									
5	Cocaine	0.114	0.068	0.053	0.115	1								
6	Tranquilizers	0.203	0.146	0.139	0.258	0.349	1							
7	Drugstore	0.091	0.103	0.110	0.122	0.209	0.221	1						
8	Heroin	0.082	0.063	0.066	0.097	0.321	0.355	0.201	1					
9	Marijuana	0.513	0.445	0.365	0.482	0.186	0.315	0.150	0.154	1				
10	Hashish	0.304	0.318	0.240	0.368	0.303	0.377	0.163	0.219	0.534	1			
11	Inhalants	0.245	0.203	0.183	0.255	0.272	0.323	0.310	0.288	0.301	0.302	1		
12	Hallucinogenics	0.101	0.088	0.074	0.139	0.279	0.367	0.232	0.320	0.204	0.368	0.304	1	
13	Amphetamine	0.245	0.199	0.184	0.293	0.278	0.545	0.232	0.314	0.394	0.467	0.392	0.511	1

TABLE 13.9
Results of Applying PCA to the Correlation Matrix of Drug Usage Rates

	Variances of Components												
	C1	**C2**	**C3**	**C4**	**C5**	**C6**	**C7**	**C8**	**C9**	**C10**	**C11**	**C12**	**C13**
Variance	4.38	2.05	0.96	0.81	0.76	0.69	0.63	0.62	0.57	0.40	0.39	0.38	0.36
Proportion of Variance	0.34	0.16	0.07	0.06	0.059	0.053	0.050	0.048	0.044	0.031	0.030	0.029	0.027
Cumulative Proportion	0.34	0.49	0.57	0.63	0.69	0.74	0.79	0.84	0.88	0.91	0.94	0.97	1.00

	Coefficients Defining Components												
Var	**C1**	**C2**	**C3**	**C4**	**C5**	**C6**	**C7**	**C8**	**C9**	**C10**	**C11**	**C12**	**C13**
1	−0.280	−0.283	—	—	−0.300	−0.387	−0.124	0.137	0.655	−0.139	−0.136	−0.169	−0.263
2	−0.287	−0.394	0.120	—	0.187	0.161	0.114	—	—	—	—	0.695	−0.410
3	−0.267	−0.393	0.207	−0.139	0.309	0.141	—	—	0.107	−0.421	0.210	−0.188	0.564
4	−0.318	−0.322	—	—	0.181	0.142	—	−0.164	−0.214	0.563	−0.181	−0.519	−0.219
5	−0.208	0.290	—	−0.582	−0.432	0.416	0.185	−0.244	0.204	—	0.154	—	—
6	−0.293	0.262	−0.165	—	0.122	—	−0.629	−0.399	—	−0.124	−0.421	0.170	0.138
7	−0.176	0.190	0.723	0.372	−0.178	0.277	−0.309	0.253	—	—	—	—	—
8	−0.201	0.317	0.153	−0.534	0.327	−0.359	—	0.525	−0.169	—	—	—	—
9	−0.340	−0.160	−0.228	0.112	−0.365	−0.129	—	0.285	−0.149	0.418	0.154	0.285	0.502
10	−0.329	—	−0.352	0.125	−0.256	0.243	0.167	0.274	−0.400	−0.496	−0.187	−0.240	−0.152
11	−0.274	0.163	0.330	0.159	−0.152	−0.531	0.466	−0.417	−0.228	—	—	—	—
12	−0.245	0.327	−0.144	0.272	0.379	0.210	0.413	0.162	0.440	0.179	−0.308	—	0.159
13	−0.328	0.235	−0.235	0.267	0.203	—	−0.132	−0.177	—	—	0.733	—	−0.269

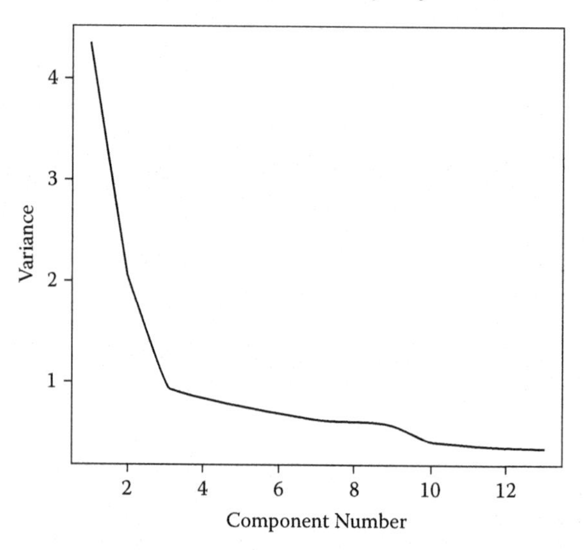

FIGURE 13.7
Scree plot of component variances for the drug usage data.

13.11 Using PCA to Select a Subset of the Variables

Although the first few principal component scores for each individual may provide a very useful summary for a set of multivariate data, all the original variables are needed in their computation. In many cases, an investigator might be happier with determining a subset of the original variables that contains, in some sense, virtually all the information contained in the complete set of these variables. In a series of papers, Jolliffe (1970, 1972, 1973) discusses a number of approaches to selecting subsets of variables, several of which are based on PCA. One such method is to first use one or another of the criteria for choosing the number of components described in Section 13.8; this number, say, m, is taken to indicate the effective dimensionality of the data and will be the size of the subset of the original variables to be retained. The variables are then chosen, one associated with each of the first m components and having the largest absolute coefficient value on the component but not already having been selected. If we use the eigenvalue criterion suggested by Jolliffe, namely, retain components with eigenvalues greater than 0.7 (assuming the correlation matrix is being used) for the drug usage data, then we keep the first five components. Examining the coefficients defining these components, we find that we are led to the following five variables:

1. Marijuana usage
2. Beer usage

3. Drugstore medication usage
4. Cocaine usage
5. Hallucinogenics usage

These variables could be used in future analyses of the data with little loss of information, compared to using all 13 of the original variables.

13.12 Summary

- PCA provides a way of reducing the complexity of multivariate data by reducing their dimensionality.
- The reduction in dimensionality that can often be achieved by a PCA is possible only if the original variables are correlated; if the original variables are independent of one another, a PCA cannot lead to any simplification.
- In most applications, variables will be on different scales, so components will need to be extracted from the correlation matrix of the data.
- In essence, PCA is simply a rotation of the axes of multivariate data scatter.
- The first few principal component scores can often be used to provide a convenient summary of a multivariate data set, particularly for looking at the data via simple scatterplots.
- Trying to give meaningful labels to components is often confusing and a waste of time.
- Two techniques that are related to principal components are multidimensional scaling and correspondence analysis; both topics are discussed in the next chapter.

13.13 Exercises

13.1 The crime rate data considered in the text contains a number of possible outliers. Reanalyze the data using principal components after removing the observations you consider to be outliers, and compare your results with those given in the text.

13.2 Find the principal components of the following correlation matrix and compare how the one- and two-component solutions reproduce the matrix.

$$\mathbf{R} = \begin{bmatrix} 1.0000 & & \\ 0.6579 & 1.0000 & \\ 0.0034 & -0.0738 & 1.0000 \end{bmatrix}$$

13.3 Macdonnell (1902) obtained measurements on seven physical characteristics for each of 300 criminals. The seven variables measured were (1) head length, (2) head breadth, (3) face breadth, (4) left finger length, (5) left forearm length, (6) left foot length, and (7) height. The correlation matrix calculated by Macdonnell is

	1	2	3	4	5	6	7
1	1.000						
2	0.402	1.000					
3	0.396	0.618	1.000				
4	0.301	0.150	0.321	1.000			
5	0.305	0.135	0.289	0.846	1.000		
6	0.339	0.206	0.363	0.759	0.797	1.000	
7	0.340	0.183	0.345	0.661	0.800	0.736	1.000

Find the principal components of this correlation matrix and interpret the results.

13.4 The data give prestige, income, education, and suicide rates for 36 occupations, originally given in Labovitz (1970). Undertake a PCA of the data and use the results to try to answer the question of whether there are several distinct types of jobs.

13.5 Rescale the coefficients defining the principal components of the crime rate data so that they represent correlations between the components and crime rates.

14

Multidimensional Scaling and Correspondence Analysis

14.1 Introduction

The concept of *distance* is at the heart of many methods for dealing with multivariate data. For example, in the scatterplots shown in Chapter 2 the relative "distances" between points are obviously central to making interpretations about the plots; this is clear without giving an explicit definition of distance but simply appealing to its everyday usage. But in this chapter, we will need to consider the distance concept in more detail as we consider methods which aim to give useful low-dimensional "maps" of particular types of data. We have, of course, already met one such technique, principal components analysis, in the previous chapter. As mentioned briefly there, this technique finds the required low-dimensional map in such a way that Euclidean distances between the observations in the space of the original variables are preserved as far as possible by the points representing the observations in the principal component plot. For the observations in the data matrix $\mathbf{X}$ (see Chapter 12), the Euclidean distance between observation i with variable values $x_{i1}, x_{i2}, \ldots, x_{iq}$, and observation j with variable values $x_{j1}, x_{j2}, \ldots, x_{jq}$ is defined as

$$d_{ij} = \sqrt{\sum_{k=1}^{q} (x_{ik} - x_{jk})^2}$$

Applying this formula to all pairs of observations results in an $n \times n$ Euclidean *distance matrix*. To illustrate the calculation of a distance matrix we shall use the data in Table 14.1 showing birth and death rates for seven countries; the Euclidean distance matrix for these data is shown in Table 14.2. The distance matrix is symmetric because $d_{ij} = d_{ji}$ and it has zeros on the main diagonal.

The distance matrix in Table 14.2 was derived from the raw data in Table 14.1 simply by applying the formula for Euclidean distance given above. But similar matrices can arise more directly. For example, Table 14.3 shows judgments about various brands of cola made by two subjects using a visual analogue scale with anchor points "same" (having a score of 0) and "different" (having a score of 100). In this example, the resulting rating for a pair of colas is labeled as a measure of *dissimilarity*—low values indicate that the two colas

TABLE 14.1
Birth and Death Rates for Seven Countries

Country	Birth rate	Death rate
Japan	8.4	9.9
Italy	8.6	10.2
Spain	9.4	8.6
United Kingdom	12.4	9.0
Finland	10.8	9.5
Cuba	11.2	7.6
United States	12.5	8.2

Note: Average data from 2010–2015 taken from the *World Population Prospects* by United Nations, Population Division (2017).

TABLE 14.2
Euclidean Distance Matrix Based on the Birth and Death Rates in Table 14.1

	Japan	Italy	Spain	UK	Finland	Cuba	US
Japan	0.00						
Italy	0.36	0.00					
Spain	1.64	1.79	0.00				
United Kingdom	4.10	3.98	3.03	0.00			
Finland	2.43	2.31	1.66	1.68	0.00		
Cuba	3.62	3.68	2.06	1.84	1.94	0.00	
United States	4.44	4.38	3.13	0.81	2.14	1.43	0.00

TABLE 14.3
Dissimilarity Data for All Pairs of 10 Colas for 2 Subjects

Subject 1:											Subject 2:										
	Cola Number											Cola Number									
	1	2	3	4	5	6	7	8	9	10		1	2	3	4	5	6	7	8	9	10
1	0										1	0									
2	16	0									2	20	0								
3	81	47	0								3	75	35	0							
4	56	32	71	0							4	60	31	80	0						
5	87	68	44	71	0						5	80	70	37	70	0					
6	60	35	21	98	34	0					6	55	40	20	89	30	0				
7	84	94	98	57	99	99	0				7	80	90	90	55	87	88	0			
8	50	87	79	73	19	92	45	0			8	45	80	77	75	25	86	40	0		
9	99	25	53	98	52	17	99	84	0		9	87	35	50	88	60	10	98	83	0	
10	16	92	90	83	79	44	24	18	98	0	10	12	90	96	89	75	40	27	14	90	0

are regarded as more alike than high values, and vice versa. A *similarity measure* would have been obtained had the anchor points been reversed, although similarities are often scaled to lie in the interval $[0, 1]$.

Distance matrices, dissimilarity and similarity matrices however they arise are generally referred to collectively as *proximity matrices.* It is the analysis of such matrices that is the subject of this chapter where the aims are to clarify, display, and possibly explain any structure or pattern not readily apparent in the collection of numerical values. In some areas, particularly psychology, the ultimate goal in the analysis of a set of proximities is more specifically the development of theories for explaining similarity judgments, or, in other words, finding an answer to the question "what makes things seem alike or different?" Here we will confine ourselves to discussing the methods for analysing proximity values without searching for the answer to this type of "holy grail" question.

The two classes of techniques to be described in this chapter are *multidimensional scaling* which is used essentially to represent an observed proximity matrix geometrically and *correspondence analysis* which is quintessentially a method for displaying the associations among a set of categorical variables in a type of scatterplot or map, thus allowing a visual examination of any structure or pattern in the data. Both these methods are exploratory in nature and they share more similarities than differences with each other, so they fit extremely well in the same chapter. In addition, both methods, especially correspondence analysis, have interesting connections to factor analysis that is the topic of the next chapter. We will start from multidimensional scaling, covering its basic principles and practices.

14.2 Multidimensional Scaling

A spatial representation of a proximity matrix consists of a set of q-dimensional coordinates representing each of the observations, chosen so that the distances (generally Euclidean but not always) between the points in the q-dimensional space match closely in some sense the observed proximities. Finding both the "best" fitting set of coordinates and the appropriate value of q (which we would like to be as low as possible) is the general aim of *multidimensional scaling* (MDS) techniques that have originated in psychometrics. The variety of methods that have been suggested differ largely in how agreement between fitted distances and observed proximities are measured, i.e., in their goodness-of-fit criteria. Here, we discuss only *classical scaling* and *nonmetric scaling.* Accounts of other methods are given in Everitt and Rabe-Hesketh (1997).

14.2.1 Classical Multidimensional Scaling

Classical multidimensional scaling seeks to represent a proximity matrix by a simple geometrical model or map. Such a model is characterized by a set of points $\mathbf{x}_1, \mathbf{x}_2, \ldots, \mathbf{x}_n$, in q dimensions, each point representing one of the stimuli of interest, and a measure of the distance between pairs of points. The objective of MDS is to determine both the dimensionality, q, of the model, and the n, q-dimensional coordinates $\mathbf{x}_1, \mathbf{x}_2, \ldots, \mathbf{x}_n$ so that the model gives a "good" fit for the observed proximities. Fit will often be judged by some numerical index that measures how well the proximities and the distances in the geometrical model match. In essence, this simply means that the larger an observed dissimilarity between two stimuli (or the smaller their similarity), the further apart should be the points representing them in the final geometrical model.

The question now arises as to how we estimate q, and the coordinate values $\mathbf{x}_1, \mathbf{x}_2, \ldots, \mathbf{x}_n$, from the observed proximity matrix? *Classical scaling* provides an answer to this question based on the early work of Young and Householder (1938) and the development of the method due to Torgerson (1952, 1958). To begin, we must note that there is no unique set of coordinate values that give rise to these distances, since they are unchanged by shifting the whole configuration of points from one place to another, or by rotation or reflection of the configuration. In other words, we cannot uniquely determine either the location or the orientation of the configuration. The location problem is usually overcome by placing the mean vector of the configuration at the origin. The orientation problem means that any configuration derived can be subjected to an arbitrary orthogonal transformation. Such transformations can often be used to facilitate the interpretation of solutions as will be seen later.

The essential details of classical scaling are given in Technical Section 14.1.

Technical Section 14.1: Classical Multidimensional Scaling

To begin our account of the method, we shall assume that the proximity matrix we are dealing with is a matrix of Euclidean distances $\mathbf{D}$ derived from a raw data matrix $\mathbf{X}$. In the beginning of the chapter, we saw how to calculate Euclidean distances from $\mathbf{X}$. Now, MDS is essentially concerned with the reverse problem: Given the distances (arrayed in the $n \times n$ matrix $\mathbf{D}$) how do we find $\mathbf{X}$? To begin, define an $n \times n$ inner product matrix $\mathbf{B}$ as follows:

$$\mathbf{B} = \mathbf{X}\mathbf{X}' \tag{14.1}$$

where the elements of $\mathbf{B}$ are given by

$$b_{ij} = \sum_{k=1}^{q} x_{ik} x_{jk} \tag{14.2}$$

It is easy to see that the squared Euclidean distances between the rows of $\mathbf{X}$ can be written in terms of the elements of $\mathbf{B}$ as

$$d_{ij}^2 = b_{ii} + b_{jj} - 2b_{ij} \tag{14.3}$$

If the bs could be found in terms of the ds in the equation above, then the required coordinate values could be derived by factoring $\mathbf{B}$ as in (14.1). No unique solution exists unless a location constraint is introduced. Usually the center of the points $\bar{\mathbf{x}}$ is set at the origin, so that $\sum_{i=1}^{n} x_{ik} = 0$ for all $k = 1, 2, \dots, q$. These constraints and the relationship given in (14.2) imply that the sum of the terms in any row of $\mathbf{B}$ must be zero. Consequently, summing the relationship given in (14.3) over i, over j, and finally over both i and j, leads to the following series of equations:

$$\sum_{i=1}^{n} d_{ij}^2 = \text{trace}(\mathbf{B}) + nb_{jj}$$

$$\sum_{j=1}^{n} d_{ij}^2 = \text{trace}(\mathbf{B}) + nb_{ii}$$

$$\sum_{i=1}^{n}\sum_{j=1}^{n} d_{ij}^2 = 2n \times \text{trace}(\mathbf{B})$$

where $\text{trace}(\mathbf{B}) = \sum_{i=1}^{n} b_{ii}$. The elements of $\mathbf{B}$ can now be found in terms of squared Euclidean distances as

$$b_{ij} = -\frac{1}{2}\left[d_{ij}^2 - d_{i.}^2 - d_{.j}^2 + d_{..}^2\right],$$

where

$$d_{i.}^2 = \frac{1}{n}\sum_{j=1}^{n} d_{ij}^2$$

$$d_{.j}^2 = \frac{1}{n}\sum_{i=1}^{n} d_{ij}^2$$

$$d_{..}^2 = \frac{1}{n^2}\sum_{i=1}^{n}\sum_{j=1}^{n} d_{ij}^2$$

Having now derived the elements of $\mathbf{B}$ in terms of Euclidean distances, it remains to factor it to give the coordinate values. In terms of its spectral decomposition, $\mathbf{B}$ can be written as

$$\mathbf{B} = \mathbf{V}\boldsymbol{\Lambda}\mathbf{V}',$$

where $\boldsymbol{\Lambda} = \text{diag}(\lambda_1, \lambda_2, \dots, \lambda_n)$ is the diagonal matrix of eigenvalues of $\mathbf{B}$ and $\mathbf{V}$ the corresponding matrix of eigenvectors, normalized so that the sum of squares of their elements is unity, that is, $\mathbf{V}'\mathbf{V} = \mathbf{I}_n$. The eigenvalues are assumed labeled such that $\lambda_1 \geq \lambda_2 \geq \cdots \geq \lambda_n$.

When the matrix of Euclidean distances $\mathbf{D}$ arises from an $n \times k$ matrix of full column rank, then the rank of $\mathbf{B}$ is k, so that the last $n - k$ of its eigenvalues will be zero. So $\mathbf{B}$ can be written as $\mathbf{B} = \mathbf{V}_1 \mathbf{\Lambda}_1 \mathbf{V}_1'$, where $\mathbf{V}_1$ contains the first k eigenvectors and $\mathbf{\Lambda}_1$ the k non-zero eigenvalues. The required coordinate values are thus $\mathbf{X} = \mathbf{V}_1 \mathbf{\Lambda}_1^{1/2}$, where $\mathbf{\Lambda}_1^{1/2} = \text{diag}(\sqrt{\lambda_1}, \sqrt{\lambda_2}, \ldots, \sqrt{\lambda_k})$.

Using all q dimensions would lead to complete recovery of the original Euclidean distance matrix. The best-fitting k-dimensional representation is given by the k eigenvectors of $\mathbf{B}$ corresponding to the k largest eigenvalues. The adequacy of the k-dimensional representation can be judged by the size of the criterion

$$P_k = \sum_{i=1}^{k} \lambda_i \Big/ \sum_{i=1}^{n-1} \lambda_i$$

Values of P_k of the order of 0.8 suggest a reasonable fit.

When the observed dissimilarity matrix is not Euclidean, the matrix $\mathbf{B}$ is not positive-definite. In such cases, some of the eigenvalues of $\mathbf{B}$ will be negative; correspondingly, some coordinate values will be complex numbers. If, however, $\mathbf{B}$ has only a small number of small negative eigenvalues, a useful representation of the proximity matrix may still be possible using the eigenvectors associated with the k largest positive eigenvalues. The adequacy of the resulting solution might be assessed using one of the following two criteria suggested by Mardia et al. (1979):

$$P_k^{(1)} = \sum_{i=1}^{k} |\lambda_i| \Big/ \sum_{i=1}^{n} |\lambda_i| \quad \text{or} \quad P_k^{(2)} = \sum_{i=1}^{k} \lambda_i^2 \Big/ \sum_{i=1}^{n} \lambda_i^2$$

Again, we would look for values above 0.8 to claim a "good" fit. Alternatively, Sibson (1979) recommends one of the following criteria for deciding on the number of dimensions to adequately represent the observed proximities:

1. *Trace criterion:* Choose the number of coordinates so that the sum of the *positive* eigenvalues is approximately equal to the sum of *all* the eigenvalues.

2. *Magnitude criterion:* Accept as genuinely positive only those eigenvalues whose magnitude substantially exceeds that of the largest negative eigenvalue.

If, however, the matrix $\mathbf{B}$ has a considerable number of large negative eigenvalues, classical scaling of the proximity matrix may be inadvisable and some other methods of scaling, for example nonmetric or least-squares scaling, might be better employed (for details, see Everitt and Rabe-Hesketh, 1997).

As first example of applying classical scaling, we will return to the distance matrix in Table 14.2. The two-dimensional solution is shown in Table 14.4 and we can use these results to produce a two-dimensional "map" of the data as shown in Figure 14.1. Of course, in this example the raw data consists of only two variables, so consequently the classical scaling solution achieves complete recovery of the observed distance matrix. For example, the Euclidean distance between Japan and Italy found from the relevant values given in Table 14.4 is calculated as

$$\sqrt{[-2.26 - (-2.19)]^2 + [0.03 - 0.39]^2} = 0.36,$$

which is the distance between these two countries derived for the raw data values in Table 14.1 (see Table 14.2).

TABLE 14.4
Two-Dimensional Solution from Classical MDS Applied to the Distance Matrix in Table 14.2

Country	Dimension 1	Dimension 2
Japan	−2.26	0.03
Italy	−2.19	0.39
Spain	−0.83	−0.78
United Kingdom	1.78	0.74
Finland	0.11	0.59
Cuba	1.21	−1.01
United States	2.18	0.05

14.2.2 Connection to Principal Components

It should be mentioned here that when the proximity matrix of interest contains Euclidean distances derived from a raw data matrix (as in the previous example of the seven countries), classical scaling can be shown to be equivalent to principal components analysis (PCA, see Chapter 13), with the derived coordinate values corresponding to the scores on the principal components derived from the covariance matrix. One result of this duality is that the classical scaling is often referred to as *principal coordinates analysis* (see Gower, 1966). The low-dimensional representation achieved by classical scaling for Euclidean distances (and that produced by PCA) is such that the function ϕ given by

$$\phi = \sum_{r,s}^{n} (d_{rs}^2 - \hat{d}_{rs}^2)$$

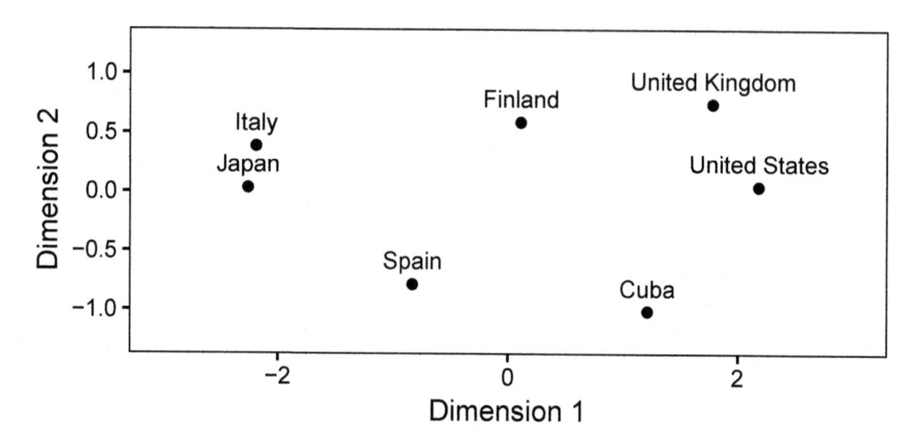

FIGURE 14.1
Resulting map from classical scaling of the seven countries' distance matrix in Table 14.2.

is minimized. In this expression, d_{rs} is the Euclidean distance between observations r and s in the original q-dimensional space, and $\hat{d}^2_{rs}$ is the corresponding distance in k-dimensional space ($k < q$) chosen for the classical scaling solution (equivalently the first k principal components).

For more details of the connections of MDS to other methods, see Groenen and van de Velden (2005).

14.2.3 Road Distances in Finland

It is almost *de rigueur* in any account of MDS to include an example involving road distances between towns or cities of a particular country, usually the country of origin of the author involved; see, for example, Kruskal and Wish (1978), Manly (1986), Krzanowski and Marriott (1994), Mustonen (1995), and Everitt and Rabe-Hesketh (1997). The present account is no exception and so Table 14.5 gives the road distances between 15 cities, towns etc. in Finland. Road distances are unlikely to be Euclidean (they are measured on the surface of a sphere for one thing) and, in fact, the eigenvalues of the matrix **B** (see Technical Section 14.1) arising from the distances in Table 14.6 include seven negative values (including one that is practically zero).

The first question that has to be addressed is how many coordinate values are needed to adequately represent the observed distance matrix? We could answer this question by using one or other of the fit criteria described in Technical Section 14.1, but here we shall simply use the two-dimensional solution and judge informally how this matches up to the usual map of Finland. This plot (after a convenient rotation to put Helsinki in the South) is shown in Figure 14.2. This diagram gives a good approximation to the map of Finland, as the mean absolute error of approximation between the 15 selected places

TABLE 14.5
Road Distances of 15 Selected Places in Finland

	Hel	**Joe**	**Jyv**	**Kil**	**Kok**	**Kot**	**Kuo**	**Kuu**	**Lap**	**Nuo**	**Oul**	**Rov**	**Tam**	**Tur**	**Vaa**
Helsinki	0														
Joensuu	438	0													
Jyväskylä	272	245	0												
Kilpisjärvi	1202	979	931	0											
Kokkola	491	429	241	789	0										
Kotka	134	343	244	1177	488	0									
Kuopio	383	136	144	878	315	316	0								
Kuusamo	802	459	551	620	411	735	417	0							
Lappeenranta	223	236	219	1144	460	108	264	682	0						
Nuorgam	1328	1045	1056	649	916	1303	1006	617	1271	0					
Oulu	612	393	339	592	196	585	286	212	551	719	0				
Rovaniemi	832	550	561	428	419	806	509	191	774	495	222	0			
Tampere	174	394	151	1079	324	243	293	702	275	1207	491	712	0		
Turku	166	549	304	1226	436	295	448	848	364	1353	633	856	153	0	
Vaasa	419	494	282	910	121	484	379	533	501	1035	318	541	244	331	0

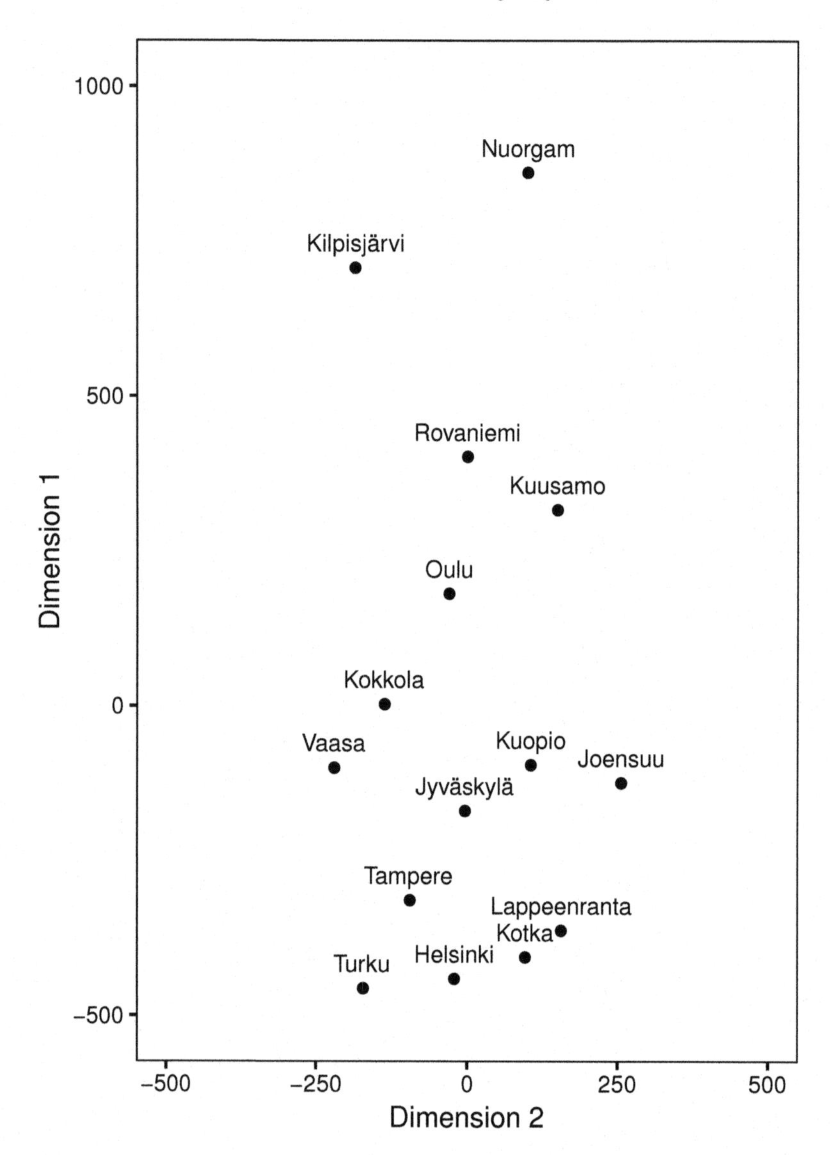

FIGURE 14.2
Resulting map from classical scaling of the Finnish road distances.

is only about 16 km. The case of the two northernmost places of these data, both located in rather extreme corners of Finland, reveals a huge exception, however, as the actual road distance between the villages of Nuorgam and Kilpisjärvi (based on these data) is 649 km, but the estimate based on the classical scaling is only about half of that (326 km). The reason for this error

TABLE 14.6
Eigenvalues and Eigenvectors Arising from Classical Scaling Applied to the Distance Matrix in Table 14.5

Eigenvalues	City/Town/Village	Dim 1	Dim 2
2395788	Helsinki	−440.7	−21.5
285542	Joensuu	−124.5	255.9
198140	Jyväskylä	−169.6	−3.8
56660	Kilpisjärvi	707.1	−186.0
16092	Kokkola	2.6	−137.4
15119	Kotka	−405.8	96.9
7806	Kuopio	−95.7	106.0
724	Kuusamo	316.6	150.3
0	Lappeenranta	−362.9	156.5
−995	Nuorgam	860.8	101.0
−5886	Oulu	180.5	−30.4
−13245	Rovaniemi	402.0	0.5
−21336	Tampere	−314.1	−95.3
−37055	Turku	−455.9	−172.5
−56468	Vaasa	−100.4	−220.2

should come quite clear by looking at the (true) map of Finland, for example, in Google Maps (see Figure 14.3).

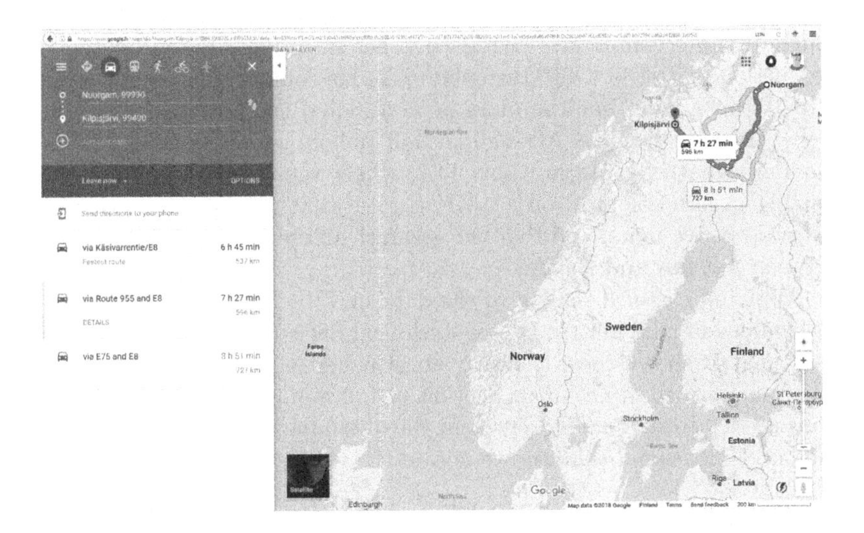

FIGURE 14.3
Routes and road distances between the villages of Nuorgam and Kilpisjärvi, Finland (source: Google Maps).

TABLE 14.7
Proximity Matrix of Ten Remarkable Classical Music Composers Selected and Compared by Olli Mustonen

	Bac	Hay	Moz	Bee	Sch	Bra	Sib	Deb	Bar	Šos
Bach	0									
Haydn	50	0								
Mozart	30	10	0							
Beethoven	20	15	20	0						
Schubert	40	30	25	10	0					
Brahms	40	70	40	20	15	0				
Sibelius	40	90	70	25	60	20	0			
Debussy	50	50	50	80	50	70	35	0		
Bartók	30	80	80	60	70	70	35	15	0	
Šostakovitš	30	40	50	40	60	70	20	40	20	0

14.2.4 Mapping Composers of Classical Music

Our final example of the use of classical scaling involves data on composers of classical music, taken with permission from Mustonen (1996, 156–159) and Mustonen (1995, 167–170). Seppo Mustonen (a Finnish professor of Statistics) asked his son Olli Mustonen (a Finnish pianist, conductor, and composer) to select ten remarkable composers from different era of classical music and compare those composers with each other intuitively based on their entire production and style. Olli Mustonen made his comparisons using a scale from 0 to 100 in a way that the more he considered the composers to differ, the higher the score he gave. After about half an hour's reflection, he presented the proximity matrix given in Table 14.7, where the selected composers appear roughly in chronological order. We can see that the scale was applied with intervals of five units, and that the greatest difference was 90 units, occurring between Sibelius and Haydn.

Applying classical scaling to the data in Table 14.7 leads to four negative eigenvalues for the matrix **B** (see Technical Section 14.1) and so the dissimilarity matrix shown there is clearly non-Euclidean. Here we will look at the fit criteria described in Technical Section 14.1 as a guide to the number of dimensions needed to adequately represent the dissimilarity values in Table 14.7. For the one-dimensional solution we obtain the values

$$P_1^{(1)} = 0.35 \text{ and } P_1^{(2)} = 0.58$$

while for the two-dimensional solution, the values obtained are

$$P_2^{(1)} = 0.58 \text{ and } P_2^{(2)} = 0.83$$

which would seem to suggest two dimensions (although the first one does not approach 0.8 before eight dimensions). Also both the alternative criteria (the

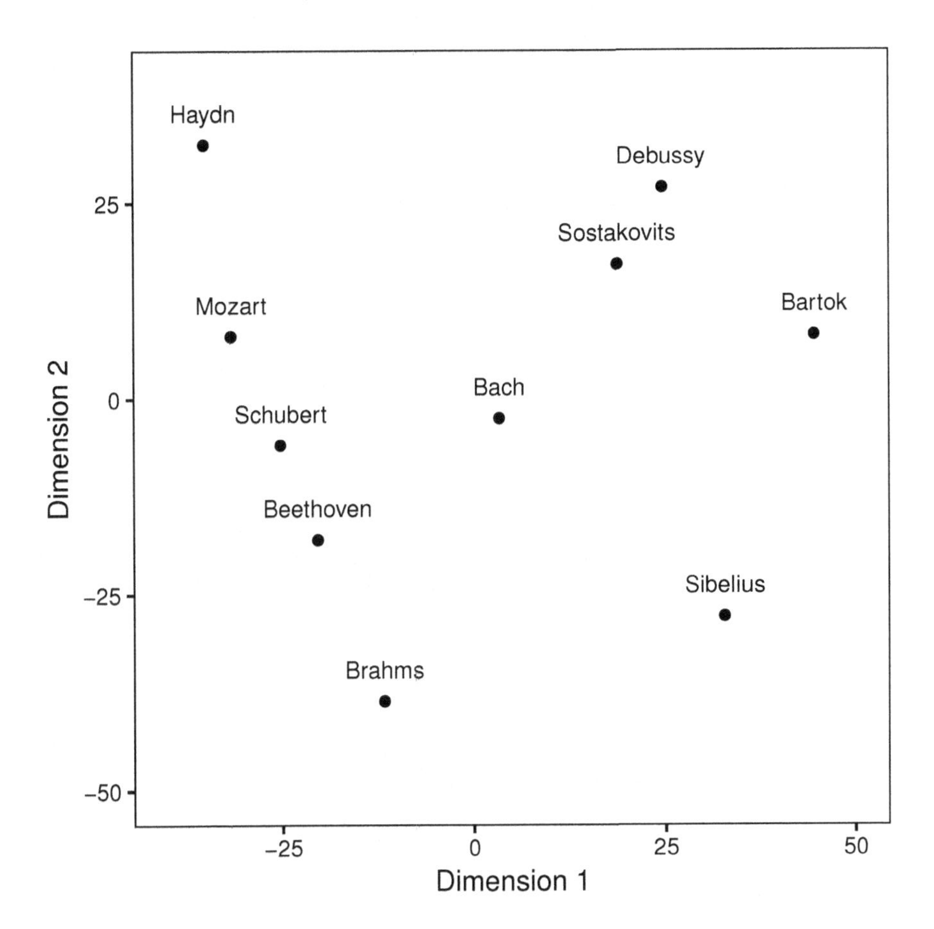

FIGURE 14.4
Resulting map from classical scaling of the classical composers.

trace and the magnitude) support the conclusion, so we shall proceed with two dimensions, following the original lines of interpretations of Mustonen (1996).

The resulting map of composers is shown in Figure 14.4. The first dimension (from left to right) appears to be related to time, with one significant exception: the "timeless" Bach is placed in the middle. The second dimension (from top to bottom) can be interpreted as a transition from "light" music to "heavy" music. Indeed, the *Viennese Classics* (Haydn, Mozart, Schubert, and Beethoven) form a logical chain, accompanied by Brahms, who, together with Sibelius, is located in the "heavyweight division". The modern composers (Debussy, Šostakovitš, and Bartók) seem to form a cluster of their own, and it is perfectly understandable that, of these composers, it is Šostakovitš who gets settled nearest to Bach. A rather lonely Sibelius is placed at a considerable distance from all other composers.

One feature of the proximity matrix in Table 14.7 which we should point to, involves the rating between Debussy and Beethoven which is the second largest in the whole matrix. According to the rater, Olli Mustonen, this rating was "coloured" by the fact that he was aware that Debussy hated Beethoven's music!

We shall close this subsection by noting that obviously musicians do not have a uniform perception of the relations between the composers (see Georges, 2017). The reader might easily make experiments in this respect and compare the resulting maps to the one we have presented here. For further analysis and interesting ways of studying the behavioral dimensions of music, including emotions, see, for example, Eerola (2011). And about the term "classic" in the context of music, take a look at Bennett (2009).

14.2.5 Nonmetric Multidimensional Scaling

In some psychological work and in market research, proximity matrices arise from asking human subjects to make judgments about the similarity or dissimilarity of objects or stimuli of interest. When collecting such data, the investigator may feel that realistically subjects are only able to give "ordinal" judgments; for example, when comparing a range of colours they might be able to specify with some confidence that one colour is brighter than another but would be far less confident if asked to put a value to how much brighter. Such considerations led, in the 1960s, to the search for a method of multidimensional scaling that uses only the *rank order* of the proximities to produce a spatial representation of them. In other words, a method was sought that would be invariant under *monotonic transformations* of the observed proximity matrix; i.e., the derived coordinates will remain the same if the numerical values of the observed proximities are changed but their rank order is not.

The essential details of nonmetric multidimensional scaling are given briefly in Technical Section 14.2.

Technical Section 14.2: Nonmetric Multidimensional Scaling

Nonmetric MDS was proposed in landmark papers by Shepard (1962a,b) and by Kruskal (1964a). The quintessential component of the method is that the coordinates in the spatial representation of the observed proximities give rise to fitted distances, d_{ij}, and that these distances are related to a set of numbers which we will call *disparities*, $\hat{d}_{ij}$, by the formula $d_{ij} = \hat{d}_{ij} + \epsilon_{ij}$, where the ϵ_{ij} are error terms representing errors of measurement plus distortion errors arising because the distances do not correspond to a configuration in the particular number of dimensions chosen. The disparities are monotonic with the observed proximities and, subject to this constraint, resemble the fitted distances as closely as possible.

Monotonic regression (see Barlow et al., 1972) is used to find the disparities, and then the required coordinates in the spatial representation of the observed dissimilarities, are found by minimizing a criterion known as *stress*, which is defined as

$$stress = \min \frac{\sum_{i<j}(\hat{d}_{ij} - d_{ij})^2}{\sum_{i<j} d_{ij}^2},$$

where the minimum is taken over the disparities $\hat{d}_{ij}$ such that $\hat{d}_{ij}$ is monotonic with the observed dissimilarities. In essence, *stress* represents the extent to which the rank order of the fitted distances disagrees with the rank order of the observed dissimilarities. The denominator in the formula for *stress* is chosen to make the final spatial representation invariant under changes of scale; that is, uniform stretching or shrinking.

An algorithm to minimize the *stress* criterion (see the Technical Section 14.2) and so find the coordinates of the required spatial representation is described in Kruskal (1964b). For each value of the number of dimensions, q, in the spatial configuration, the configuration that has the smallest *stress* is called the best-fitting configuration in q dimensions. For judging the goodness (or badness) of fit, Kruskal (1964a) gave the following verbal evaluation of various values of *stress* (typically expressed in percentages):

Stress (%)	**Goodness-of-fit**
20	Poor
10	Fair
5	Good
2.5	Excellent
0	Perfect

(The last one, "perfect," only occurs if the rank order of the fitted distances matches the rank order of the observed dissimilarities, an event, which is, of course, very rare in practise.)

14.2.6 Re-mapping Composers of Classical Music

We now apply the nonmetric MDS to the classical music composers' proximity matrix in Table 14.7. The resulting two-dimensional map is shown in Figure 14.5. There are small differences between this plot and that resulting from classical scaling (Figure 14.4), but in general the interpretation of each diagram is very similar. The value of *stress* (defined earlier) for Figure 14.5 is 5.5%.

Now let us look at the plot provided by coordinates one and three from the nonmetric scaling solution for the composers' data; this is shown in Figure 14.6. The third dimension seems to highlight the previously mentioned closeness of Šostakovitš and Bach. The same applies to Sibelius and Bartók.

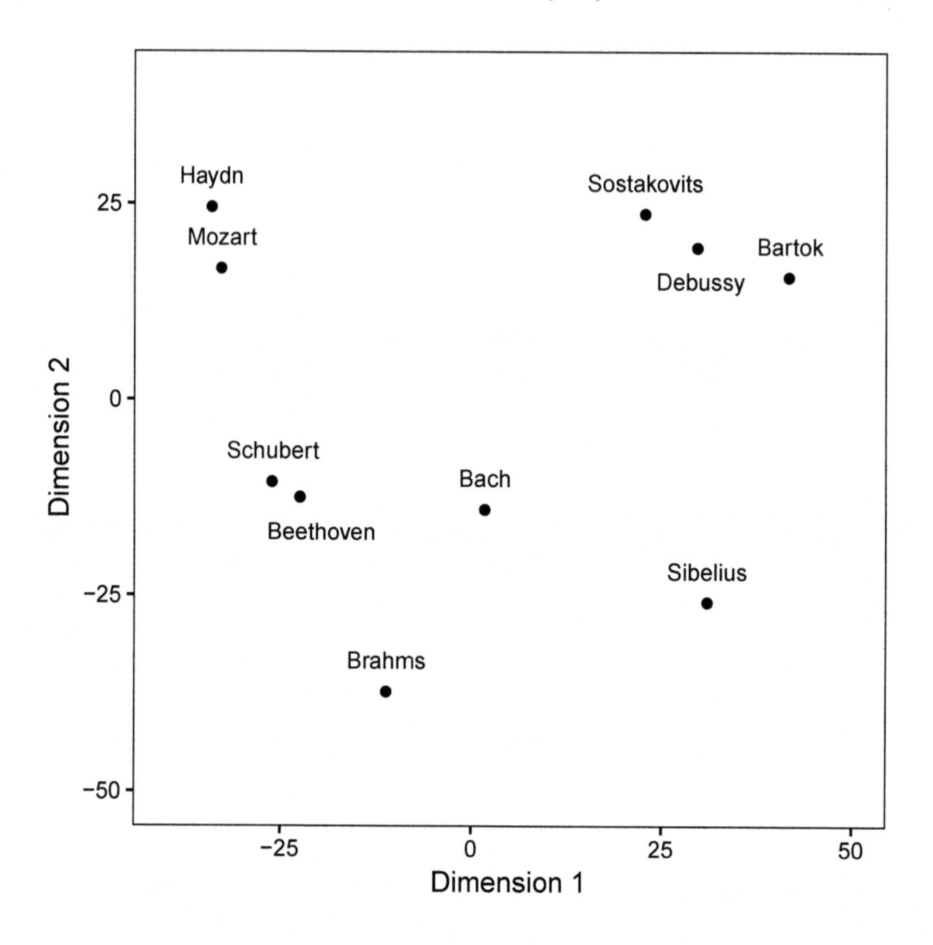

FIGURE 14.5
Resulting map from nonmetric scaling of the classical composers (dimensions 1 and 2).

A further feature of nonmetric scaling that might aid in evaluating solutions is the *Shepard diagram*; this involves a plot of the observed dissimilarities against the corresponding distances obtained from the scaling solution. In an ideal world, the plotted points would fall on a straight line corresponding to a perfect match between the rank order of the dissimilarities and that of the fitted distances. But the world is rarely ideal and such a perfect fit is extremely unlikely in practice. However, such a plot is useful in indicating how far from "perfection" is the scaling solution. The Shepard diagram for the two-dimensional solution from the application of nonmetric scaling to the data on composers is shown in Figure 14.7. Clearly the fit is not perfect but the step function shows that the departure from a straight line is relatively small; this reflects the small value of the stress associated with this solution.

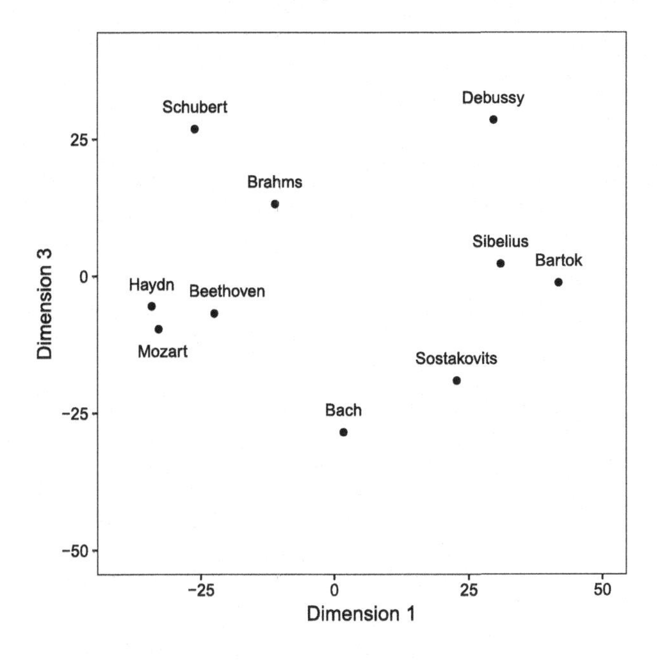

FIGURE 14.6
Resulting map from nonmetric scaling of the classical composers (dimensions 1 and 3).

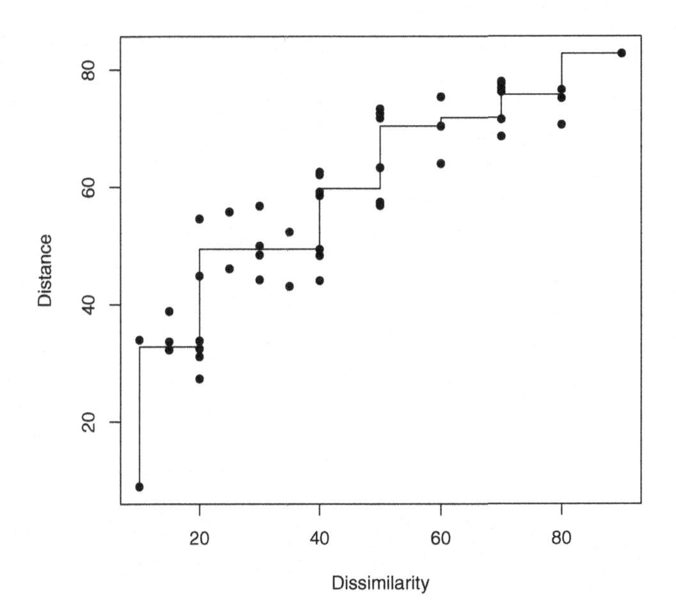

FIGURE 14.7
Shepard diagram for the classical composers' nonmetric scaling.

14.3 Correspondence Analysis

Correspondence analysis (CA) is a versatile method for visualizing and analyzing categorical data. As with many other multivariate methods, its history can be traced back to the 1930s and 1940s, to the publications of Fisher, among others. However, the essential development of the method took place in France in the 1960s, based on the visionary work of a French linguist and data analyst Jean-Paul Benzécri and his co-workers (Benzécri 1973). The wider dissemination of correspondence analysis outside France began with the books by Lebart et al. (1984), Greenacre (1984), and Benzécri (1992). The French tradition of correspondence analysis and other methods of exploratory multivariate analysis is continued in the recent book by Husson et al. (2017). A detailed bibliographical account of the history of correspondence analysis, spanning (or spinning!) the French, Japanese, and Dutch schools of data analysis, is included in Greenacre (2016). See also Blasius and Greenacre (2014).

Mathematically speaking, correspondence analysis can be regarded as either

- a method for decomposing the chi-squared statistic used to test for independence in a contingency table into components corresponding to different dimensions of the heterogeneity between its columns; or
- a method for simultaneously assigning a scale to the rows and a separate scale to the columns of a table in order to maximize the correlation between the two scales.

Quintessentially, correspondence analysis is a technique for displaying multivariate (often bivariate) categorical data graphically, by deriving coordinates to represent the categories of both the row and the column variables, which may then be plotted to display the pattern of association between the variables graphically.

The basic details of correspondence analysis are given in Technical Section 14.3.

Technical Section 14.3: Correspondence Analysis

Correspondence analysis (CA) is based on straightforward, classical results on matrix theory. The central result is the *singular value decomposition* (SVD), which is the basis of many multivariate statistical methods. We shall briefly summarize the theory behind correspondence analysis. For a more general account of the subject, see Greenacre (2016).

Let $\mathbf{N}$ denote an $I \times J$ data matrix, with positive row and column sums. Typically, the elements of $\mathbf{N}$ represent some form of count data, such as frequencies. Dividing $\mathbf{N}$ by its grand total, $n = \sum_i \sum_j n_{ij}$, converts $\mathbf{N}$ to the *correspondence matrix*, denoted by $\mathbf{P}$:

$$\mathbf{P} = \frac{1}{n}\mathbf{N}$$

The concept of a set of relative frequencies, or a *profile*, is fundamental to correspondence analysis. Profiles have special geometric features, because their elements add up to 1 (or 100%), and that is why graphical representations of data, especially scatterplots, are so central in the method.

The weights assigned to the profiles are called *row and column masses* and they are denoted by

$$r_i = \sum_{j=1}^{J} p_{ij} \quad \text{and} \quad c_i = \sum_{i=1}^{I} p_{ij}$$

or, in matrix form, as vectors

$$\mathbf{r} = \mathbf{P1} \quad \text{and} \quad \mathbf{c} = \mathbf{P'1}$$

where $\mathbf{1}$ is the vector of ones (either $J \times 1$ or $I \times 1$, to match the dimensions of $\mathbf{P}$ and $\mathbf{P}'$). The coordinates of the row and column profiles are obtained using the SVD as follows.

To begin, we calculate the matrix $\mathbf{S}$ of standardized residuals:

$$\mathbf{S} = \mathbf{D}_r^{-\frac{1}{2}}(\mathbf{P} - \mathbf{rc}')\mathbf{D}_c^{-\frac{1}{2}}$$

where $\mathbf{D}_r = \text{diag}(\mathbf{r})$ and $\mathbf{D}_c = \text{diag}(\mathbf{c})$. The crucial step is the SVD of the matrix $\mathbf{S}$:

$$\mathbf{S} = \mathbf{UD}_\alpha\mathbf{V}'$$

where $\mathbf{U}'\mathbf{U} = \mathbf{V}'\mathbf{V} = \mathbf{I}$ and $\mathbf{D}_\alpha$ is the diagonal matrix of (positive) singular values in descending order ($\alpha_1 \geq \alpha_2 \geq \cdots$) as all essential results are included in these three output matrices of the SVD.

First, from the results of the SVD, we obtain the *principal coordinates* of rows and columns as matrices $\mathbf{F}$ and $\mathbf{G}$, respectively:

$$\mathbf{F} = \mathbf{D}_r^{-\frac{1}{2}}\mathbf{UD}_\alpha \quad \text{and} \quad \mathbf{G} = \mathbf{D}_c^{-\frac{1}{2}}\mathbf{VD}_\alpha$$

We also obtain the eigenvalues or *principal inertias* λ_k:

$$\lambda_k = \alpha_k^2, \quad k = 1, 2, \ldots, K$$

where $K = \min(I - 1, J - 1)$.

The concept of *inertia* refers to the measure of variance in the categorical data. The principal inertia λ_k refers to the inertia accounted for by each of the principal axes $k = 1, 2, \ldots, K$ obtained from the SVD.

The square root of the principal inertia, i.e., the first singular value α_1, can be interpreted as the maximum correlation (also called *canonical correlation*) between the scale values assigned to the categories of the row and column variables.

The principal coordinates **F** and **G** refer to the profiles with respect to principal axes. In the so-called *symmetric map*, where the separate configurations of row profiles and column profiles are overlaid in a joint display, all points are displayed in principal coordinates. This is the most typical way of mapping the results of correspondence analysis as a *biplot*, a general graphical display for multivariate data, see Gabriel (1971), Gabriel and Odoroff (1990), and Greenacre (2010). Strictly speaking, the symmetric map is not a true biplot, but it can be interpreted in the biplot style, see Gabriel (2002).

In a symmetric map summarizing the results of correspondence analysis, the row-to-row and column-to-column Euclidean distances are approximate chi-squared distances between the respective profiles, but there is no specific row-to-column distance interpretation. For a contingency table with I rows and J columns, it can be shown that the chi-squared distances can be represented *exactly* in $\min(I-1, J-1)$ dimensions. When both I and J are greater than three, an exact two-dimensional representation of the chi-squared distances is not possible, so the question of the adequacy of the fit will need to be addressed. In some of these cases, more than two dimensions may be required to give an acceptable fit (again for details, see Greenacre, 2016).

The symmetric map is interpreted by examining the positions of the points representing the categories of the row and column variables. Assuming that a two-dimensional solution provides an adequate fit for the data, row points that are close together represent row categories that have similar profiles across columns. Accordingly, column points that are close together indicate columns with similar profiles down the rows. Finally, row points that lie close to column points represent a row/column combination that occurs more frequently in the table than would be expected if the row and column variables were independent. Conversely, row and column points that are distant from one another indicate a cell in the table where the count is lower than would be expected under independence.

Next we will consider a simple example of the application of correspondence analysis.

14.3.1 Simple Example of the Application of Correspondence Analysis

The data shown in Table 14.8 is concerned with the influence of a girl's age on her relationship with her boyfriend. In this table, each of 139 girls has been classified into one of three groups:

TABLE 14.8
The Influence of Age of Relationship Status with Boyfriends: Observed Frequencies along with Expected Frequencies (in Parentheses)

	Age group					
Relationship	**< 16**	**16–17**	**17–18**	**18–19**	**19–20**	**Total**
No boyfriend	21	21	14	13	8	77
(expected)	(17.2)	(18.3)	(13.3)	(17.2)	(11.1)	
Boyfriend: No sex	8	9	6	8	2	33
(expected)	(7.4)	(7.8)	(5.7)	(7.4)	(4.7)	
Boyfriend: Sex	2	3	4	10	10	29
(expected)	(6.5)	(6.9)	(5.0)	(6.5)	(4.2)	
Total	31	33	24	31	20	139

- No boyfriend,
- Boyfriend: No sex(ual intercourse),
- Boyfriend: Sex(ual intercourse).

In addition, the age of each girl was recorded and used to divide the girls into five age groups. According to the labels in Table 14.8 the groups might seem overlapping but to be precise, 16–17 means "16 but less than 17," and hence all the girls in the data are teenagers.

The standard analysis for such a contingency table would involve the calculation of the usual chi-square statistic for assessing independence of row and column variables. This involves first calculating the expected values under the assumption of independence—these are given in Table 14.8; details of their calculation and of the chi-square test itself are given in any introductory statistics text. For Table 14.8 the value of the statistic is 20.6 with eight degrees of freedom. The associated p-value is 0.008. Clearly the relationship groups and age are not independent, a result that will only be a surprise to our readers who have lived a rather sheltered life.

Now let us apply correspondence analysis to the data. The result is the two-dimensional map of the categories shown in Figure 14.8. The adequacy of the fit need not to be addressed, as the relationship variable includes only three categories and hence the chi-squared distances are represented exactly by the Euclidean distances in two dimensions. The principal inertia (defined earlier) is 0.141 (95.4%) for the first dimension and only 0.007 (4.6%) for the second dimension. It can be inferred that actually one dimension would be enough to account for the associations of the relationship data. Interpretation of the diagram involves examination of the positions of the points representing the row and column categories of Table 14.8.

Examining the plot in Figure 14.8, we see that it tells the age-old story of girls travelling through their teenage years, initially having no boyfriend,

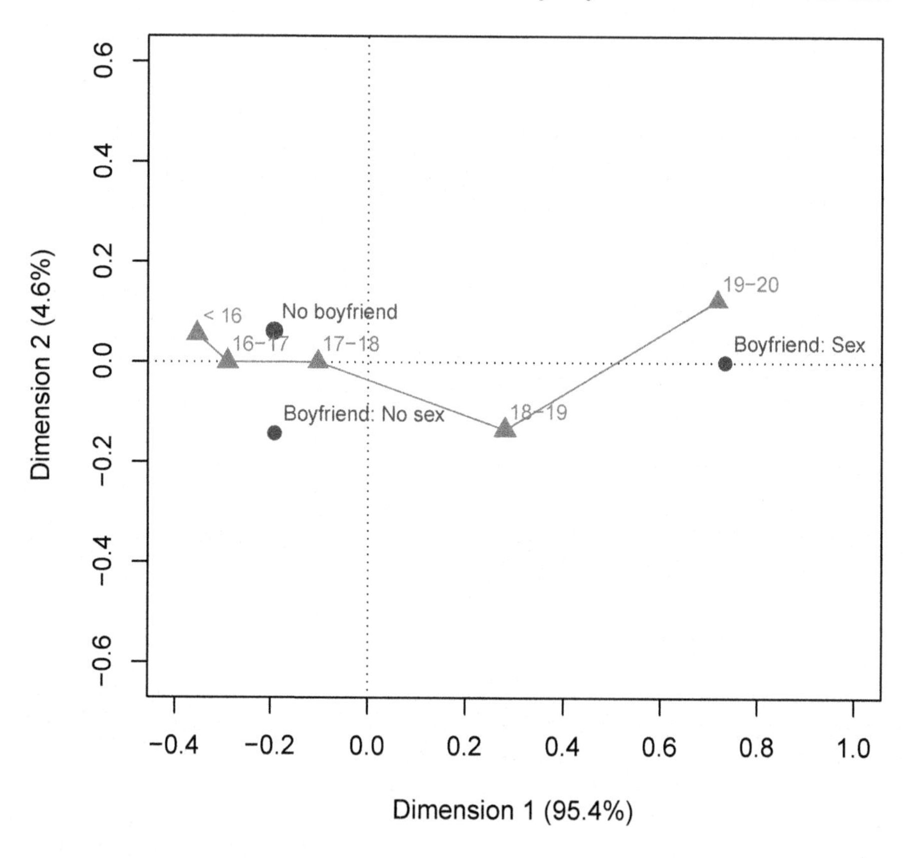

FIGURE 14.8
Resulting map from correspondence analysis of the age and the relationships with boyfriends of 139 girls.

then acquiring a boyfriend, and then having sex with their boyfriend, a story that has broken the hearts of fathers everywhere, at least temporarily, until their wives suggest they reflect back to the time when they themselves were teenagers.

14.3.2 Connections of Work Activities and Job Advantages

Let us look at another example of CA that will be a bit more challenging for the interpretation compared to the simple example in the previous subsection. The main reason is that the following example, taken from Lebart et al. (1984), involves larger dimensions, both in respect to the number of observations and the number of categories in the data. The context of the study is a typical survey research, where people fill up a questionnaire with many of the questions being categorical. In one of the questions, the respondents had to select a type of work activity that best describes their job (e.g., "farming", "construction", "social services", "teaching", "domestic work", "small

TABLE 14.9
Types of Work Activities and Main Advantages of Job from 6933 Survey Respondents (Lebart et al., 1984)

	Main Advantage of Job							
Type of Work Activity	VARIETY	FREEDOM	...	LIKE IT	OTHER	NONE	OUTDOORS	NO ANS
Farming–fishing	4	189	...	11	15	12	8	1
Farm–food industry	1	13	...	9	5	11	0	0
Energy–mines	1	9	...	4	3	6	1	0
Steel	5	5	...	2	3	22	0	0
Chemical–glass–oil	2	7	...	3	0	5	0	1
Wood–paper	2	5	...	1	0	3	0	2
Auto–aviation–shipping	2	3	...	6	1	24	0	1
Textile–leather–shoes	3	18	...	6	2	26	0	2
Pharmaceutical industries	3	7	...	2	1	8	0	0
Manufacturing	0	18	...	10	4	26	0	6
Construction	7	63	...	14	8	35	2	2
Food–grocery	2	43	...	6	1	7	0	3
Small business	8	95	...	13	4	18	1	3
Miscellaneous business	5	32	...	8	3	18	0	3
Administrative services	8	26	...	6	9	16	3	4
Telecommunications	1	7	...	1	3	5	0	2
Social services	4	10	...	3	2	1	0	1
Health services	3	31	...	24	1	5	0	5
Teaching–research	2	33	...	18	3	11	1	3
Transportation	2	19	...	3	3	13	0	1
Insurance–banking	8	12	...	3	1	10	0	3
Domestic workers	0	8	...	2	2	11	1	1
Other services	8	35	...	11	4	14	0	1
Printing–publishing	2	13	...	5	4	11	0	2
Private services	3	26	...	10	3	8	0	2
No answer	0	14	...	5	1	3	0	3

business" etc.). In another question, they had to choose the main advantage in their job, again from a set of given alternatives (e.g., "variety", "freedom", "salary", "being outdoors" etc.) including categories "other" and "none". Missing answers constituted their own categories on both questions. After the usual, time-consuming hard work of data cleanup, recoding, joining of some categories as well as other initial tasks and necessary checkings, the data set, consisting of 6933 observations was ready for the analyses.

Again we begin with a cross table, as both variables of interest are clearly nominal measurements with no real order within their categories. The variables have been cross-classified to form a contingency table of 26 categories for the type of the work activity and 17 categories for the main advantage of job. Part of the table is shown in Table 14.9. The result of the correspondence analysis is the two-dimensional map of the categories shown in Figure 14.9.

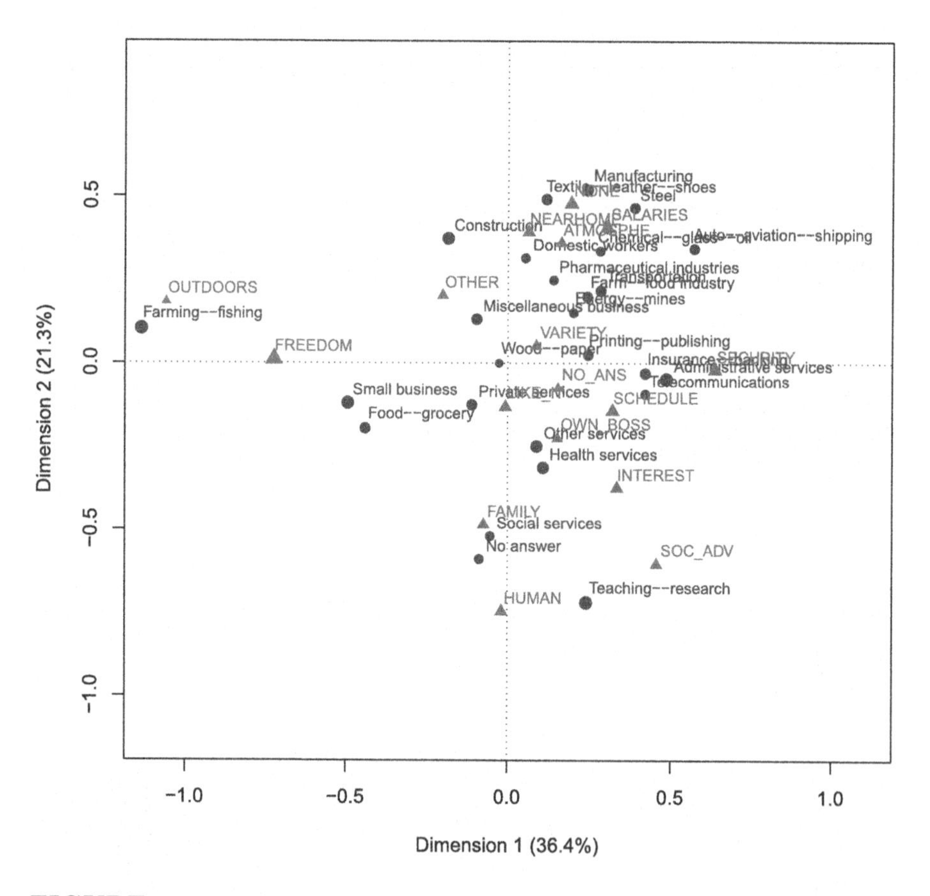

FIGURE 14.9
Resulting map from correspondence analysis of the work activities and main advantages of jobs (symbol sizes proportional to the row and column masses of the points).

A clear difference to the earlier example is that only about 57% of the total inertia of the table is explained. The first dimension accounts for 36.4% and the second dimension for 21.3% of the inertia. The adequacy of the fit would need to be addressed in more detail and perhaps additional dimensions should be analyzed. Another difference is that, because of the size of the data, understanding any associations in the data by looking at the plain numbers is quite impossible. A graph will certainly reveal to the eye much more than $26 \times 17 = 442$ numbers.

We restrict ourselves to a brief interpretation of the map in Figure 14.9. The horizontal dimension seems to be related to the size of the workplace, from smaller ones on the left to the various large companies on the right. The corresponding advantages vary from being outdoors, having freedom, through family and human aspects to salaries, social advantages, schedules, and security. The vertical dimension goes from various service activities on the bottom to factory work, where again the different advantages (including "near home" and a bit depressed sounding "none"!) correspond to the work activities perhaps in a quite expected way. The "no answer" categories of both variables appear about in the middle of their scales.

Correspondence analysis and its variations such as multiple correspondence analysis, joint correspondence analysis, and canonical correspondence analysis are described in detail in Greenacre (2016). Examples of the application of multiple correspondence analysis appear in Greenacre and Blasius (2006). Readers contemplating using correspondence analysis are strongly advised to consult such references using the necessarily brief account given in this chapter only as an initial stepping stone.

14.4 Summary

- Multidimensional scaling and correspondence analysis both aim to help in understanding particular types of data by displaying the data graphically.
- Multidimensional scaling applied to proximity matrices is often useful in uncovering the dimensions on which similarity judgments are made.
- For more complete accounts of multidimensional scaling, including least-squares and other methods, see, for example, Groenen and Borg (2014), Cox and Cox (2001) or Everitt and Rabe-Hesketh (1997).
- Correspondence analysis is often a useful supplement to the routine chi-square test applied to contingency tables.
- For more information of the connections between correspondence analysis and other multivariate methods, together with more general biplot techniques to visualize the results of these methods, see Greenacre (2010).

14.5 Exercises

14.1 Return to the data on road distances in Finland and now use the fit criteria defined in Technical Section 14.1 to determine the number of dimensions required to represent these distances. Construct an appropriate plot (or plots) to visualize the solution with this number of dimensions.

14.2 The elements of Table 14.10 give the Mahalanobis distances (see Chapter 12 for a definition) between ten types of galaxy calculated from raw data on seven variables such as diameter, brightness, and color observed on 273 galaxies. Apply classical scaling to these distances and construct what you consider relevant diagrams to visualize the solution. (For an interesting interpretation of the diagram(s), you should take a look at Barnett, 1981 or Nathanson, 1971.)

TABLE 14.10
Inter-group Mahalanobis Distances for Ten Types of Galaxy

	I	**SBc**	**Sc**	**Sbb**	**Sb**	**Sba**	**Sa**	**SBO**	**SO**	**E**
I(rregulars)	0.00									
SBc	3.29	0.00								
Sc	2.79	1.13	0.00							
Sbb	3.52	1.75	1.45	0.00						
Sb	3.77	2.97	1.71	2.02	0.00					
Sba	3.27	3.01	2.13	1.89	1.27	0.00				
Sa	3.93	3.72	3.00	2.25	1.86	0.68	0.00			
SBO	3.86	5.12	4.11	3.24	3.15	1.59	1.51	0.00		
SO	3.77	5.70	4.85	3.85	3.41	1.74	2.05	0.91	0.00	
E(llipticals)	4.12	6.88	6.02	7.03	5.38	4.09	4.03	2.24	1.87	0.00

14.3 The elements of the matrix in Table 14.11 result from averaging the ratings of 18 students on the degree of similarity between 12 nations on a scale ranging from 1, indicating "very different", to 9 for "very similar". No instructions were given concerning the characteristics on which these similarity judgments were to be made. (Further details of the study are given in Kruskal and Wish, 1978.) Apply both classical scaling and nonmetric scaling to the similarity judgments and compare the results with the help of appropriate plots to represent each solution. Which technique is likely to be most appropriate to use on these data and why?

TABLE 14.11
Proximity Matrix of Similarity Ratings for 12 Nations Averaged over 18 Students

	Brz	Zai	Cub	Egy	Fra	Ind	Isr	Jpn	Chi	USSR	USA	Yug
Brazil	0.00											
Zaire	4.83	0.00										
Cuba	5.28	4.26	0.00									
Egypt	3.44	5.00	5.17	0.00								
France	4.72	4.00	4.11	4.78	0.00							
India	4.50	4.83	4.00	5.83	3.44	0.00						
Israel	3.83	3.33	3.61	4.67	4.00	4.11	0.00					
Japan	3.50	3.39	2.94	9.84	4.11	4.50	4.83	0.00				
China	2.39	4.00	5.50	4.39	3.67	4.11	3.00	4.17	0.00			
USSR	3.06	3.39	5.44	4.39	5.06	4.50	4.17	4.61	5.72	0.00		
USA	5.39	2.39	3.17	3.33	5.94	4.28	5.94	6.06	2.56	5.00	0.00	
Yug	3.17	3.50	5.11	4.28	4.72	4.00	4.44	4.28	5.06	6.67	3.56	0.00

Note: Yug = Yugoslavia

14.4 Table 14.12 gives data on the cross-classification of people in Caithness, Scotland, by eye and hair color (Fisher 1940). The region of the UK is particularly interesting as there is a mixture of people of Nordic, Celtic, and Anglo-Saxon origin. Apply simple correspondence analysis and compute the corresponding chi-square test. Can you notice "correspondances" or connections between the eye and hair colors?

TABLE 14.12
Cross-classification of 5387 People in Caithness, Scotland, by Eye and Hair Color

	Hair Color				
Eye Color	**Fair**	**Red**	**Medium**	**Dark**	**Black**
Blue	326	38	241	110	3
Light	688	116	584	188	4
Medium	343	84	909	412	26
Dark	98	48	403	681	85

15

Exploratory Factor Analysis

15.1 Introduction

In many areas of psychology, and other disciplines in the behavioral sciences, it is often not possible to measure directly the concepts of primary interest. Two obvious examples are intelligence and social class. In such cases, the researcher is forced to examine the concepts indirectly by collecting information on variables that can be measured or observed directly, and which may also realistically be assumed to be indicators, in some sense, of the concepts of real interest. The psychologist who is interested in an individual's "intelligence," for example, may record examination scores in a variety of different subjects in the expectation that these scores are dependent in some way on what is widely regarded as intelligence but are also subject to random errors. Further, a sociologist, say, concerned with people's "social class," might pose questions about a person's occupation, educational background, home ownership, etc., on the assumption that these do reflect the concept in which he or she is really interested.

Both intelligence and social class are what are generally referred to as *latent variables*, that is, concepts that cannot be measured directly but can be assumed to relate to a number of measurable or *manifest variables*. The method of analysis most generally used to help uncover the relationships between the assumed latent variables and the manifest variables is *factor analysis*. The model on which the method is based is essentially that of multiple linear regression, except that now the manifest variables are regressed on the unobservable latent variables (often referred to in this context as *common factors*), so that direct estimation of the corresponding regression coefficients (known now as *factor loadings*) is not possible.

A point to be made at the outset is that factor analysis comes in two distinct varieties; the first is *exploratory factor analysis*, which is used to investigate the relationship between manifest variables and factors without making too strict assumptions about which manifest variables are related to which factors, and the second is *confirmatory factor analysis*, which is used to test whether a specific factor model postulated a priori provides an adequate fit for the covariances or correlations between the manifest variables.

In this chapter, we shall focus on the exploratory approach, while the confirmatory factor analysis, and its extension to what is known as *structural equation modeling*, will be the topics described in the next chapter. Although we discuss these two approaches in separate chapters, it should be noted that factor analysis need not be strictly of either type. In practice, many studies are to some extent both exploratory and confirmatory.

15.2 The Factor Analysis Model

The basis of factor analysis is a regression model, linking the manifest variables to a set of unobserved (and unobservable) latent variables. In essence, the model assumes that the observed relationships between the manifest variables (as measured by their covariances or correlations) are the results of the relationships of these variables to the latent variables. A relatively brief account of the factor analysis model is given in Technical Section 15.1.

Technical Section 15.1: Factor Analysis Model

The q observed or manifest variables are represented as the vector $\mathbf{x}' = [x_1, x_2, \ldots, x_q]$ and are all assumed, for convenience, to have zero mean (it is only information about the relationships between the manifest variables as contained in their covariance or correlation matrix that is of interest in factor analysis so the zero means assumption is of no consequence). The manifest variables are assumed to be related by a regression-type model to a smaller number of unobserved latent variables, the common factors, represented as the vector $\mathbf{f}' = [f_1, f_2, \ldots, f_k]$ where $k < q$. We can write the assumed model as a series of regression-like equations:

$$
\begin{aligned}
x_1 &= \lambda_{11} f_1 + \lambda_{12} f_2 + \cdots + \lambda_{1k} f_k + u_1 \\
x_2 &= \lambda_{21} f_1 + \lambda_{22} f_2 + \cdots + \lambda_{2k} f_k + u_2 \\
&\vdots \\
x_q &= \lambda_{q1} f_1 + \lambda_{q2} f_2 + \cdots + \lambda_{qk} f_k + u_q
\end{aligned}
$$

The λ_{ij} values are essentially regression coefficients showing how each x_i depends on the k common factors; in this context, they are known as factor loadings. When estimated from a sample correlation matrix, the factor loadings are the estimated correlations between factors and manifest variables. The factor loadings are used in the interpretation of the factors, that is, larger values relate a factor to the corresponding observed variables, and from looking at which variables load highly on a factor, we can try to come up with a meaningful description or label for each factor. The u_i values are analogous to the residual terms in the usual multiple linear regression model (see Chapter 4), but as they are specific to each x_i in the factor analysis context, they are more commonly known as *specific variates.* The series of regression equations above may be written more concisely as

$$\mathbf{x} = \mathbf{\Lambda f} + \mathbf{u}$$

where

$$\mathbf{\Lambda} = \begin{bmatrix} \lambda_{11} & \cdots & \lambda_{1k} \\ \vdots & \vdots & \vdots \\ \lambda_{q1} & \cdots & \lambda_{qk} \end{bmatrix}, \quad \mathbf{f} = \begin{bmatrix} f_1 \\ \vdots \\ f_k \end{bmatrix}, \text{ and } \mathbf{u} = \begin{bmatrix} u_1 \\ \vdots \\ u_q \end{bmatrix}$$

We assume that the residual terms $u_1, \dots, u_q$ are uncorrelated with each other and with the common factors $f_1, \dots, f_k$. The two assumptions imply that, given the values of the factors, the manifest variables are independent, that is, the correlations of the observed variables arise from their relationships with the factors.

Since the factors are unobserved, we can fix their location and scale arbitrarily. We will assume that they occur in standardized form with mean 0 and standard deviation 1. We will also assume, initially at least, that the factors are uncorrelated with one another, in which case the factor loadings are the correlations of the manifest variables and the factors. With these additional assumptions about the factors, the factor analysis model implies that σ_i^2, the variance of variable x_i, is given by

$$\sigma_i^2 = \sum_{j=1}^{k} \lambda_{ij}^2 + \psi_i$$

where ψ_i is the variance of u_i. So, the factor analysis model implies that the variance of each observed variable can be split into two parts. The first part, h_i^2, given by $h_i^2 = \sum_{j=1}^{k} \lambda_{ij}^2$, is known as the *communality* of the variable and represents the variance shared with the other variables via the common factors. The second part, ψ_i, is called the *specific or unique variance* and relates to the variability in x_i not shared with the other variables. In addition, the factor model leads to the following expression for σ_{ij}, the covariance of variables x_i and x_j:

$$\sigma_{ij} = \sum_{l=1}^{k} \lambda_{il}\lambda_{jl}$$

The covariances are not dependent on the specific variates in any way; it is the relationships of the manifest variables to the common factors that account for the relationships between the manifest variables. Collecting together the equations above relating observed variances and covariances to the factor loadings and the variances of the specific variates, we find that the factor analysis model implies that the population covariance matrix of the observed variables, $\mathbf{\Sigma}$, has the form

$$\boldsymbol{\Sigma} = \boldsymbol{\Lambda}\boldsymbol{\Lambda}' + \boldsymbol{\Psi}$$

where

$$\boldsymbol{\Psi} = \text{diag}\,(\psi_i)$$

The converse also holds: if $\boldsymbol{\Sigma}$ can be decomposed into the form given here, then the k-factor model holds for $\mathbf{x}$. In practice, of course, $\boldsymbol{\Sigma}$ will need to be estimated by the sample covariance matrix $\mathbf{S}$. Alternatively, the model will be applied to the correlation matrix $\mathbf{R}$, and we will need to obtain estimates of $\boldsymbol{\Lambda}$ and $\boldsymbol{\Psi}$ so that the observed covariance matrix takes the form required by the model. (See the next section for an account of estimation methods.) We will also need to determine the value of k, the number of factors, so that the model provides the most parsimonious but adequate fit for $\mathbf{S}$ or $\mathbf{R}$.

To apply the factor analysis model outlined in Technical Section 15.1 to a sample of multivariate observations, we need to estimate the parameters of the model, factor loadings, and specific variances in some way. The estimation problem in factor analysis is essentially that of finding the estimates $\hat{\boldsymbol{\Lambda}}$ and $\hat{\boldsymbol{\Psi}}$ for which

$$\mathbf{S} \approx \hat{\boldsymbol{\Lambda}}\hat{\boldsymbol{\Lambda}}' + \hat{\boldsymbol{\Psi}}$$

(If the x_i values are standardized, then $\mathbf{S}$ is replaced by $\mathbf{R}$.)

In some very simple cases, an exact solution is possible, that is, one in which the approximately equal sign in this equation becomes an equals sign. Considering such an example may be helpful before moving on to consider estimation in more realistic situations. The example we shall use here is one originally discussed by Spearman (1904), and concerns children's examination marks in three subjects: Classics (x_1), French (x_2), and English (x_3). The sample correlation matrix calculated by Spearman is as follows:

$$\mathbf{R} = \begin{bmatrix} 1.00 & & \\ 0.83 & 1.00 & \\ 0.78 & 0.67 & 1.00 \end{bmatrix}$$

If we assume a single factor, then the appropriate factor analysis model is

$$\begin{aligned} x_1 &= \lambda_1 f + u_1, \\ x_2 &= \lambda_2 f + u_2, \\ x_3 &= \lambda_3 f + u_3 \end{aligned}$$

In this example, the common factor f might be equated with intelligence or general intellectual ability, and the specific variates u_1, u_2, u_3 will have small variances if their associated observed variable is closely related to f. Here, the number of parameters in the model (six) is equal to the number of independent elements in $\mathbf{R}$, and so, by equating elements of the observed correlation matrix to the corresponding values predicted by the single-factor model, we will be able to find estimates of $\lambda_1, \lambda_2, \lambda_3, \psi_1, \psi_2$, and ψ_3 such that the model fits exactly. The six equations derived from the matrix equality implied by the factor analysis model, that is,

$$\mathbf{R} = \begin{bmatrix} \lambda_1 \\ \lambda_2 \\ \lambda_3 \end{bmatrix} \begin{bmatrix} \lambda_1 & \lambda_2 & \lambda_3 \end{bmatrix} + \begin{bmatrix} \psi_1 & 0 & 0 \\ 0 & \psi_2 & 0 \\ 0 & 0 & \psi_3 \end{bmatrix}$$

are

$$\begin{aligned} \hat{\lambda}_1\hat{\lambda}_2 &= 0.83 \\ \hat{\lambda}_1\hat{\lambda}_3 &= 0.78 \\ \hat{\lambda}_1\hat{\lambda}_4 &= 0.67 \\ \hat{\psi}_1 &= 1.0 - \hat{\lambda}_1^2 \\ \hat{\psi}_2 &= 1.0 - \hat{\lambda}_2^2 \\ \hat{\psi}_3 &= 1.0 - \hat{\lambda}_3^2 \end{aligned}$$

The solutions of these equations are

$$\begin{array}{lll} \hat{\lambda}_1 = 0.99 & \hat{\lambda}_2 = 0.84 & \hat{\lambda}_3 = 0.79 \\ \hat{\psi}_1 = 0.02 & \hat{\psi}_2 = 0.30 & \hat{\psi}_3 = 0.38 \end{array}$$

These values, when plugged in to the formula for the correlation matrix implied by the factor model, will reproduce the observed correlation matrix, but here, the factor model is not of any practical use because it uses six parameters to model the same number of independent elements in **R**. The model does not provide a simplified description of the relationships between the three examination scores.

15.3 Estimating the Parameters in the Factor Analysis Model

We now have to consider how to estimate the parameters in the factor analysis model in those situations of practical interest in which the number of parameters in the model is less (and hopefully, considerably less) than the number of independent elements in the covariance or correlation matrix of the manifest variables, so that the factor analysis model provides a genuinely more parsimonious description of the relationships between the manifest variables. The term "factoring" is often used for this approach.

There are two main methods of estimation or factoring leading to what are known as *principal factor analysis* and *maximum likelihood factor analysis*, both of which are briefly described in Technical Section 15.2.

Technical Section 15.2: Estimating the Parameters in the k-Factor Analysis Model

1. Principal Factor Analysis

Principal factor analysis (or principal axis factoring) is similar in many respects to principal components analysis (see Chapter 13); however, it does not operate directly on $\mathbf{S}$, the covariance matrix of the observed variables (or directly on $\mathbf{R}$, the correlation matrix), but on what is known as the *reduced covariance matrix* $\mathbf{S}^*$, defined as

$$\mathbf{S}^* = \mathbf{S} - \hat{\mathbf{\Psi}}$$

where $\hat{\mathbf{\Psi}}$ is a diagonal matrix with entries $\hat{\psi}_i$ that are estimates of the specific variances ψ_i. We remember from Technical Section 15.1 that, given a set of estimated loadings, the variance of variable x_i, s_i^2, implied by the model is

$$s_i^2 = \sum_{j=1}^{k} \hat{\lambda}_{ij}^2 + \hat{\psi}_i$$

So, the diagonal elements of $\mathbf{S}^*$ are given by $\sum_{j=1}^{k} \hat{\lambda}_{ij}^2$for $i = 1, \ldots, q$; these values are the estimated communalities—the parts of the variance of each observed variable that can be explained by the common factors. Unlike principal components analysis, factor analysis does not try to account for all observed variance; only that which is shared through the common factors. A matter of more concern in factor analysis is accounting for the covariances or correlations between the manifest variables rather than their variances.

To calculate the reduced covariance matrix $\mathbf{S}^*$ (or $\mathbf{R}^*$ with $\mathbf{R}$ replacing $\mathbf{S}$), we need values for the estimated communalities that are calculated from estimated factor loadings. But initially, we have no estimates of factor loadings. To get round this seemingly hen-or-egg situation, we need to find a sensible way of calculating initial values for the communalities that does not depend on having estimated factor loadings. When factor analysis is based on the correlation matrix of the manifest variables, there are two frequently used approaches:

- Take the initial communality of a variable x_i as the square of the multiple correlation coefficient of x_i with the other observed variables.
- Take the initial communality of x_i as the largest of the absolute values of the correlation coefficients between x_i and one of the other variables.

Each of these possibilities will lead to higher values for the initial communality when x_i is highly correlated with at least some of the other manifest variables, which is essentially what is required. Given initial communality values, a principal components analysis is performed on

$\mathbf{S}^*$, and the first k eigenvectors are used to provide the estimates of the loadings in the k-factor model. The estimation process can stop here, or the loadings obtained at this stage can provide revised communality estimates calculated as $\sum_{j=1}^{k} \hat{\lambda}_{ij}^2$, where the $\hat{\lambda}_{ij}$ values are the loadings estimated in the previous step. The procedure is then repeated until some convergence criterion is satisfied. Difficulties can sometimes arise with this iterative approach if at any time a communality estimate exceeds the variance of the corresponding manifest variable, resulting in a negative estimate of the variable's specific variance. Such a result is known as a *Heywood case* (Heywood, 1931) and is clearly unacceptable since we cannot have negative variances.

2. Maximum Likelihood Factor Analysis
Maximum likelihood (ML) is regarded, by statisticians at least, as perhaps the most respectable method of estimating the parameters in the factor analysis model. The essence of this approach is to define a fit function

$$F_{ML}(\mathbf{\Lambda}, \mathbf{\Psi}) = \log|\mathbf{\Sigma}| + \text{trace}\left(\mathbf{S}\mathbf{\Sigma}^{-1}\right) - \log|\mathbf{S}| - q$$

between the observed covariance matrix $\mathbf{S}$ and the covariance matrix $\mathbf{\Sigma} = \mathbf{\Lambda}\mathbf{\Lambda}' + \mathbf{\Psi}$ implied by the factor analysis model. Estimates of factor loadings and specific variances are found by minimizing F_{ML} using an iterative procedure that begins with initial estimates of the parameters found from a principal factor analysis; details are given in Lawley and Maxwell (1971), Mardia et al. (1979), and Everitt (1984, 1987). Minimizing F_{ML} is essentially equivalent to maximizing L, the likelihood function for the k-factor model, under the assumption of multivariate normality of data because it can be shown that $L = -\frac{1}{2}nF$ plus a function of the observations. In practice, the assumption of multivariate normality quite seldom holds, but the ML method has been found to be very robust to departures of normality. More details can be found from Jöreskog (2007).

Nowadays, maximum likelihood is the recommended method for most applications of factor analysis.

15.4 Determining the Number of Factors

Determining how many factors, k, are needed to give an adequate representation of the observed covariances or correlations is generally critical when fitting an exploratory factor analysis model. A k and $k+1$ factor solution will often produce quite different factors and factor loadings for all factors, unlike a principal component analysis in which the first k components will be identical in each solution. Further, as pointed out by Jolliffe (2002), with too few factors, there will be too many high loadings, and with too many factors, factors may be fragmented and difficult to interpret convincingly. Choosing k might be done by examining solutions corresponding to different values of k and deciding subjectively which can be given the most convincing interpretation—not an entirely convincing method in many circumstances.

An advantage of the maximum likelihood approach is that it has an associated formal hypothesis-testing procedure for the number of factors. It involves the value of the minimized fit function F_{ML} defined in Technical Section 15.2. The test statistic is

$$U = n' \min(F_{ML})$$

where

$$n' = n - \frac{1}{6}(2q + 5) - \frac{2}{3}(k + 1)$$

If k common factors are adequate to account for the observed covariances or correlations of the q manifest variables, then U has, asymptotically, a chi-squared distribution with v degrees of freedom, where

$$v = \frac{1}{2}(q - k)^2 - \frac{1}{2}(q + k)$$

In most exploratory studies, k cannot be exactly specified in advance, and so, a sequential procedure may be used. Starting with some small value for k (usually $k = 1$), the parameters in the corresponding factor analysis model are estimated by maximum likelihood. If U is not significant, the current value of k is accepted; otherwise, k is increased by 1 and the process repeated. If at any stage the degrees of freedom of the test become 0, then either no non-trivial solution is appropriate or, alternatively, the factor model itself with its assumption of linearity between observed and latent variables is questionable.

15.5 Fitting the Factor Analysis Model: An Example

As an example of fitting the factor analysis model, we will consider the following correlation matrix that arises from the scores of 220 boys in six school subjects:

$$\mathbf{R} = \begin{array}{l} \text{French} \\ \text{English} \\ \text{History} \\ \text{Arithmetic} \\ \text{Algebra} \\ \text{Geometry} \end{array} \begin{bmatrix} 1.00 & & & & & \\ 0.44 & 1.00 & & & & \\ 0.41 & 0.35 & 1.00 & & & \\ 0.29 & 0.35 & 0.16 & 1.00 & & \\ 0.33 & 0.32 & 0.19 & 0.59 & 1.00 & \\ 0.25 & 0.33 & 0.18 & 0.47 & 0.46 & 1.00 \end{bmatrix}$$

Applying the test for the number of factors described in the previous section to one-, two-, and three-factor models gives the following results:

Test of the hypothesis that one factor is sufficient:
 The chi-square statistic is 51.6 on 9 degrees of freedom.
 The *p*-value is 5.37e-08.
Test of the hypothesis that two factors are sufficient:
 The chi-square statistic is 2.18 on 4 degrees of freedom.
 The *p*-value is 0.703.
Test of the hypothesis that three factors are sufficient:
 The degrees of freedom for the model is 0 and the fit was 0.001.

The results indicate that two factors would be sufficient to account for the relationships between the six school subjects. Three factors would probably be too many, because there are no degrees of freedom left for the test. With more factors, the degrees of freedom would tend negative, indicating a clear overfitting of the model.

The estimated parameters of the two-factor model are shown in Table 15.1. The first two columns include the factor loadings that represent the correlations between the measured variables and the unmeasured, hypothesized factors. They are used for the interpretation of the factors. The loadings on the second factor have a mixture of positive and negative signs, which can make interpretation difficult. The loadings do not seem to reveal any clear pattern.

The rightmost column in Table 15.1 gives the communalities of the variables, that is, rowwise sums of squares of the loadings, while the columnwise sums of squares appear on the bottom line. The sum of the either sums of squares is 2.8, which is the total amount of the common variance "explained" by the model. The total variance is 6, as there are six variables with variances equal to 1 (the diagonal entries in the correlation matrix). Although the focus in factor analysis is more on explaining the covariances or correlations between

TABLE 15.1
Maximum Likelihood Two-Factor Solution for Correlations of Six School Subjects

	Factor Loadings		
Subject	**Factor 1**	**Factor 2**	**Communality**
French	0.56	0.42	0.49
English	0.57	0.29	0.41
History	0.39	0.45	0.36
Arithmetic	0.74	−0.28	0.62
Algebra	0.72	−0.21	0.56
Geometry	0.59	−0.13	0.37
Sum of squares	2.20	0.60	2.80

the variables instead of maximizing the variance explained, these numbers are often used in reporting the results. We may, for example, calculate the proportion of the common variance explained to be $2.2/2.8 = 0.79$ for Factor 1 and $0.6/2.8 = 0.21$ for Factor 2.

The columnwise sums of squares are also called the factor variances, and they correspond to so called *eigenvalues* of the correlation matrix. As the variance of a standardized variable is 1, there is perhaps not much point in such factors that would have variance less than 1. This argument is sometimes used in determining the number of factors based on the eigenvalues, but the "eigenvalue rule" is better suited to principal components analysis (see Chapter 13) where the focus is on the total variance.

Interpretation of the two factors will be left until later in the chapter for reasons that will hopefully become clear after the next section. However, we can use the estimated loadings and specific variances to calculate the correlation matrix of the six school subjects implied by the fitted two-factor model using the formula

$$\hat{\mathbf{R}} = \hat{\mathbf{\Lambda}}\hat{\mathbf{\Lambda}}' + \hat{\mathbf{\Psi}}$$

which gives

$$\hat{\mathbf{R}} = \begin{matrix} \text{French} \\ \text{English} \\ \text{History} \\ \text{Arithmetic} \\ \text{Algebra} \\ \text{Geometry} \end{matrix} \begin{bmatrix} 1.00 & & & & & \\ 0.44 & 1.00 & & & & \\ 0.41 & 0.35 & 1.00 & & & \\ 0.29 & 0.34 & 0.16 & 1.00 & & \\ 0.31 & 0.35 & 0.19 & 0.59 & 1.00 & \\ 0.28 & 0.30 & 0.17 & 0.48 & 0.46 & 1.00 \end{bmatrix}$$

We can see that there are only minor differences between the original correlations $\mathbf{R}$ and the reproduced correlations $\hat{\mathbf{R}}$, and hence the two-factor model provides a parsimonious and adequate fit for $\mathbf{R}$.

15.6 Rotation of Factors

Up to now, we have conveniently ignored a problem with the factor analysis model—that the factor loadings are not uniquely determined by this model. What this means is explained in Technical Section 15.3.

Technical Section 15.3: The Lack of Uniqueness of Factor Loadings

As we have seen previously, the k-factor model can be written in terms of a $q \times k$ matrix of factor loadings $\mathbf{\Lambda}$, a vector of k-common factors f, and a vector of q residuals u, as $\mathbf{x} = \mathbf{\Lambda\Lambda}'\mathbf{f} + \mathbf{u}$. Now, let us introduce a $k \times k$ orthogonal matrix $\mathbf{M}$, such that $\mathbf{MM}' = \mathbf{I}$ and rewrite the basic regression equation linking the observed variables to the common factors as $\mathbf{x} = \mathbf{\Lambda MM}'\mathbf{f} + \mathbf{u}$. This satisfies all the requirements of a k-factor model as outlined previously with new factors $\mathbf{f}^* = \mathbf{M}'\mathbf{f}$, and new factor loadings $\mathbf{\Lambda M}$. A model with these factor loadings and factors implies that the population covariance matrix of x-variables is given by

$$\mathbf{\Sigma} = (\mathbf{\Lambda M})(\mathbf{\Lambda M})' + \mathbf{\Psi}$$

because $\mathbf{MM}' = \mathbf{I}$ reduces it to the original form of $\mathbf{\Sigma} = \mathbf{\Lambda\Lambda}' + \mathbf{\Psi}$. This implies that factors $\mathbf{f}$ with loadings $\mathbf{\Lambda}$, and factors $\mathbf{f}^*$ with loadings $\mathbf{\Lambda M}$ are, for any orthogonal matrix $\mathbf{M}$, completely equivalent for explaining the covariance matrix of the observed variables.

The result of the lack of uniqueness of factor loadings is that, essentially, there are an infinite number of solutions to the factor analysis model, as previously formulated. Consequently, to define a unique solution, it becomes necessary to introduce some constraints on the parameters in the original factor analysis model. In general, what is done is to require the first factor to make maximal contribution to the common variance of the observed variables, the second to make maximal contribution to this variance subject to being uncorrelated to the first, etc. (Compare the derivation of principal components in Chapter 13.) Such constraints ensure a unique solution and lead to uncorrelated (orthogonal) factors that are arranged in descending order of "importance." However, these properties are not inherent in the factor analysis model, and merely considering such a solution may hinder interpretation; two consequences of the constraints, for example, are:

- The factorial complexity of variables is likely to be greater than 1 regardless of the underlying true model; consequently, variables may have substantial loadings on more than one factor.

- Except for the first factor, the remaining factors are often bipolar, that is, they have a mixture of positive and negative loadings.

It may be possible that a more interpretable solution can be achieved using the equivalent model with loadings $\mathbf{\Lambda}^* = \mathbf{\Lambda M}$ for a particular orthogonal matrix $\mathbf{M}$. Such a process is generally known as *factor rotation*, but before we consider how to choose $\mathbf{M}$, that is, how to "rotate" the factors, we need to address the question "is factor rotation an acceptable process?" Certainly in the past, factor analysis has been the subject of severe criticism because of the possibility of rotating factors. Critics have suggested that this apparently allows the investigator to impose on the data whatever type of solution he or she is looking for; some have even gone so far as to suggest that factor analysis has become popular in some areas precisely because it does enable users to impose their preconceived ideas of the structure behind the observed correlations (Blackith and Reyment, 1971). But, on the whole, such suspicions are not justified, and factor rotation can be a useful procedure for simplifying an exploratory factor analysis.

Factor rotation merely allows the fitted factor analysis model to be described as simply as possible; rotation does not alter the overall structure of a solution but only how the solution is described. Rotation is a process by which a solution is made more interpretable without changing its underlying mathematical properties. Initial factor solutions with variables loading on several factors and with bipolar factors can be difficult to interpret. Interpretation is more straightforward if each variable is highly loaded on at the most one factor, and if all factor loadings are either large and positive, or near 0, with few intermediate values. The variables are thus split into disjoint sets, each of which is associated with a single factor. This aim is essentially what Thurstone (1931) referred to as *simple structure*. In more detail, such structure has the following properties:

- Each row of the factor-loading matrix should contain at least one zero.
- Each column of the loading matrix should contain at least k zeros.
- Every pair of columns of the loading matrix should contain several variables whose loadings vanish in one column but not in the other.
- If the number of factors is four or more, every pair of columns should contain a large number of variables with zero loadings in both columns.
- Conversely, for every pair of columns of the loading matrix, only a small number of variables should have nonzero loadings in both columns.

When simple structure is achieved, the observed variables will fall into mutually exclusive groups whose loadings are high on single factors, perhaps moderate to low on a few factors, and of negligible size on the remaining factors.

The search for simple structure or something close to it begins after an initial factoring has determined the number of common factors necessary and the communalities of each observed variable. The factor loadings are then transformed by post multiplication by a suitably chosen orthogonal matrix. Such a transformation is equivalent to a rigid rotation of the axes of the originally identified factor space.

15.6.1 A Simple Example of Graphical Rotation

Let us now return to the example of factoring the six school subjects. The factor loadings of the maximum likelihood solution were given in Table 15.1. Figure 15.1, where the loadings are plotted in the two-factor space, reveals that the school subjects form two separate groups: the mathematical subjects are grouped together below the horizontal axis (hence the negative loadings

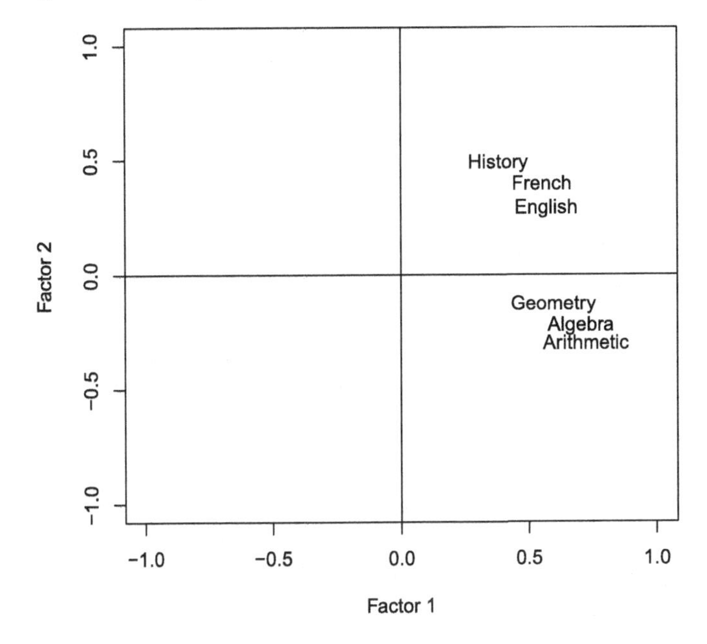

FIGURE 15.1
Plot of factor loadings estimated from the correlation matrix of six school subjects by maximum likelihood method.

on Factor 2), while the verbal subjects appear above the axis. All variables are located on the right side of the origin, as they all have positive loadings on Factor 1, but that does not help much with the interpretation. Perhaps the initial factor solution could be made more interpretable by a suitable rotation of the coordinate axes of the factor space.

In Figure 15.2a, new orthogonal axes shown by bold lines correspond to a rotation of the original axes through about 40°. Figure 15.2b shows the positive area of the factor space after the rotation. By referring each variable to these new axes, a new set of loadings can be obtained. Indeed, the new rotated factor loadings in Table 15.2 (now all positive) are far easier to interpret than the unrotated ones—the first being perhaps a "mathematical" factor, and the second a "verbal" factor. The highest loadings (on which the interpretation is based) are highlighted and the rows are sorted hierarchically according to the loadings on each factor, which helps the interpretation, especially with larger factor models.

Note that the communalities of each observed variable remain the same in the unrotated and the rotated solution, but the factor variances (columnwise sums of squares of the loadings) are redistributed, and hence do not anymore correspond to the eigenvalues of the correlation matrix. Both factors have now variances greater than 1 so there is no doubt about the merits of the second factor, compared to its original variance that was less than 1. The sum of the

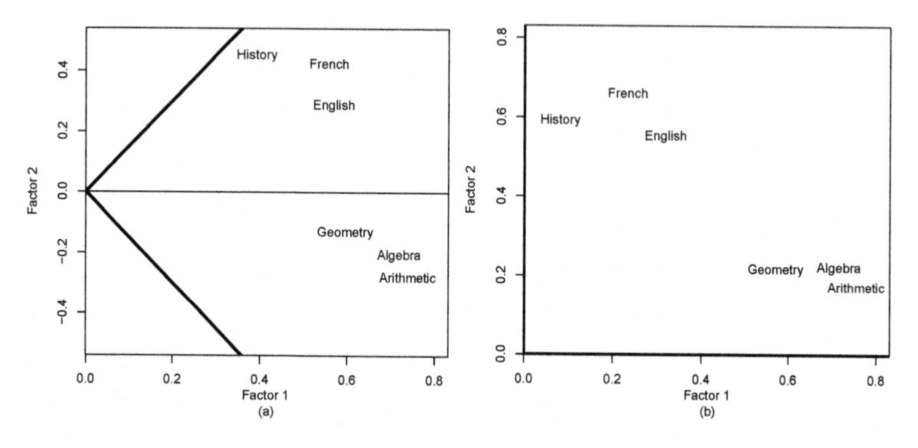

FIGURE 15.2
Plot of factor loadings of six school subjects (a) showing the rotated axes that lead to a simpler interpretation of the two factors, and (b) in the rotated factor space of the two factors.

factor variances (in this case, 2.81), that is the total amount of the common variance "explained" by the model, remains the same, as it is also the sum of the communalities.

The interpretation of the factors gives additional support for the adequacy of the two-factor model. One-factor model (that was rejected by the chi-square test) would represent a "general factor" corresponding to the original idea of factor analysis by Spearman (1904), but here it seems that one factor would not be sufficient. The k-factor model emerged as the result of further development on Spearman's work by Thurstone (1931, 1935, 1947) and others, and it is the factor model most often applied in practice. We can conclude that our

TABLE 15.2
Orthogonal Rotation of the Maximum Likelihood Two-Factor Solution for Correlations of Six School Subjects

	Factor Loadings		
Subject	**Factor 1**	**Factor 2**	**Communality**
Arithmetic	**0.77**	0.17	0.62
Algebra	**0.72**	0.22	0.56
Geometry	**0.57**	0.22	0.37
French	0.23	**0.66**	0.49
History	0.08	**0.59**	0.36
English	0.32	**0.55**	0.41
Sum of squares	1.59	1.22	2.81

two-factor model, rotated to a simpler structure, makes a quite clear distinction between a "mathematical" factor and a "verbal" factor, a setting that might have been the original motivation behind the study of the six school subjects.

15.6.2 Numerical Rotation Methods

When there are more than two factors, the rotation has to be done iteratively for each pair of factors. For a three-factor model, that would involve rotations of factors 1 and 2, then 1 and 3, and finally 2 and 3. It is obvious that graphical rotation—the original approach of Thurstone—would become very tedious, at least with larger factor models. That paved the way for the development of numerical rotation methods.

During the rotation phase, we might also choose to abandon one of the assumptions made previously, namely, that factors are orthogonal, that is, independent (the condition was assumed initially simply for convenience in describing the factor analysis model). Consequently, two types of rotation are possible:

- Orthogonal rotation: Methods restrict the rotated factors to being uncorrelated.
- Oblique rotation: Methods allow the rotated factors to correlate.

So, the first question that needs to be considered when rotating factors is: Should we use an orthogonal or oblique rotation? As for many questions posed in data analysis, there is no single answer to this question. There are advantages and disadvantages to using either type of rotation procedures. As a general rule, if a researcher is primarily concerned with getting results that "best fit" the data, then the researcher should rotate the factors obliquely. If, on the other hand, the researcher is more interested in the generalizability of the results, then orthogonal rotation is probably to be preferred.

One major advantage of an orthogonal rotation is simplicity since the loadings represent correlations between factors and manifest variables. This is not the case with an oblique rotation because of correlations between the factors. With an oblique solution, there are two parts to consider:

- Factor pattern coefficients: Regression coefficients that multiply with factors to produce measured variables according to the common factor model.
- Factor structure coefficients: Correlation coefficients between manifest variables and the factors.

Additionally, there is a matrix of factor correlations to consider. In many cases in which these correlations are relatively small, researchers may prefer to return to an orthogonal solution, where the factor correlations are zeroes and the pattern coefficients are equal to the structural coefficients.

A large variety of rotation algorithms or techniques exist, although only relatively few are in general use. Jennrich (2007) gives an excellent overview of the topic from the work of Thurstone to the present, including the latest steps of development (Jennrich 2004, 2006). See also Mulaik (2010). For orthogonal rotation, the two most commonly used techniques are known as *varimax* and *quartimax*:

- Varimax rotation: Originally proposed by Kaiser (1958), this has as its rationale the aim of factors with a few large loadings and as many near-zero loadings as possible. This is achieved by iterative maximization of a quadratic function of the loadings—details are given in Mardia et al. (1979). It produces factors that have high correlations with one small set of variables and little or no correlation with other sets. There is the tendency for any general factor to disappear because the factor variance is redistributed.
- Quartimax rotation: Originally suggested by Carroll (1953), this approach forces a given variable to correlate highly on one factor and either not at all or very low on other factors. This is far less popular than varimax.

For oblique rotation, the two methods most often used are *oblimin* and *promax*:

- Oblimin rotation: Invented by Jennrich and Sampson (1966), this method attempts to find simple structure with regard to the factor pattern matrix through a parameter that is used to control the degree of correlation between the factors. Fixing a value for this parameter is not straightforward, but Pett et al. (2003) suggest that values between about -0.5 and 0.5 are sensible for many applications.
- Promax rotation: This is a method suggested by Hendrickson and White (1964) that operates by raising the loadings in an orthogonal solution (generally, a varimax rotation) to some power. The goal is to obtain a solution that provides the best structure using the lowest possible power loadings and the lowest correlation between the factors.

Factor rotation has often been regarded as controversial since it apparently allows the investigator to impose on the data whatever type of solution is required. However, this is clearly not the case since, although the axes may be rotated about their origin or allowed to become oblique, the distribution of the points will remain invariant. Rotation is simply a procedure that allows new axes to be chosen so that the positions of the points can be described as simply as possible.

It should be noted that rotation techniques are also often applied to the results from a principal components analysis in the hope that it will aid in their interpretability. Although in some cases this may be acceptable, it does have several disadvantages, which are listed by Jolliffe (1989). The main problem is that the defining property of principal components, namely, that of accounting for maximal proportions of the total variation in the observed variables, is lost after rotation.

TABLE 15.3
Oblique Rotation of the Maximum Likelihood Two-Factor Solution for Correlations of Six School Subjects

	Factor Loadings		
Subject	**Factor 1**	**Factor 2**	**Communality**
Arithmetic	**0.81**	−0.04	0.62
Algebra	**0.73**	0.03	0.56
Geometry	**0.57**	0.07	0.37
French	0.04	**0.68**	0.49
History	−0.11	**0.65**	0.36
English	0.17	**0.53**	0.41
Sum of squares	1.60	1.21	2.81

Note: Factor correlation 0.52.

15.6.3 A Simple Example of Numerical Rotation

We shall return briefly to the six school subjects in order to show the factor loadings after an oblique rotation. The earlier rotation (see Table 15.2) that was shown graphically, was actually achieved with an orthogonal varimax rotation. Now, we proceed with the oblimin rotation method.

The results are shown in Table 15.3. Again, the communalities do not change. And in this case the factor variances are essentially the same as before. However, the loadings look even simpler, as the "mathematical" variables have practically zero loadings on the "verbal" factor and vice versa. But the "price" of this temptating result is that now the factors correlate with each other by 0.52, which must be taken into account when interpreting the solution. (The factor structure coefficients that are not shown, would in this case be quite close to the factor loadings of the orthogonal rotation.)

15.7 Estimating Factor Scores

In most applications, an exploratory factor analysis will consist of the estimation of the parameters in the model and the rotation of the factors, followed by an (often heroic) attempt to interpret the fitted model. There are occasions, however, when the investigator would like to find factor scores for each individual in the sample. Such scores, similar to those derived in a principal components analysis (see Chapter 13), might be useful in a variety of ways. But the calculation of factor scores is not as straightforward as the calculation of principal components scores. In the original equation defining the factor analysis model, the variables are expressed in terms of factors, whereas to calculate scores we require the relationship to be in the opposite direction. Bartholomew et al. (2011) makes the point that to talk about "estimating" factor scores is essentially misleading since they are random variables, and the issue is really one of prediction.

But if we make the assumption of normality, the conditional distribution of $\mathbf{f}$ given $\mathbf{x}$ can be found. It is

$$\mathrm{N}\left[\mathbf{\Lambda}'\mathbf{\Sigma}^{-1}\mathbf{x}, \left(\mathbf{\Lambda}'\mathbf{\Psi}^{-1}\mathbf{\Lambda} + \mathbf{I}\right)^{-1}\right]$$

Consequently, one plausible way of calculating factor scores would be to use the sample version of the mean of this distribution, that is,

$$\hat{\mathbf{f}} = \hat{\mathbf{\Lambda}}'\mathbf{S}^{-1}\mathbf{x}$$

where the vector of scores for an individual, $\mathbf{x}$, is assumed to have zero means, that is, sample means for each variable have already been subtracted. Other possible methods for deriving factor scores are described in Rencher and Christensen (2012). In many respects, the most challenging problem with factor analysis is not the rotational indeterminacy of the loadings but the indeterminacy of the factor scores.

15.7.1 Analyzing the Crime Rates by Factor Analysis

As an illustration of the estimation (prediction) of factor scores, we shall return to the crime rate data introduced in Chapter 13. We shall proceed through all the phases of the exploratory factor analysis that we explained in the previous sections of this chapter.

We begin from the test for the number of factors to one-, two-, and three-factor models. It gives the following results:

Test of the hypothesis that one factor is sufficient:
  The chi-square statistic is 64.63 on 14 degrees of freedom.
  The *p*-value is 1.78e-08.
Test of the hypothesis that two factors are sufficient:
  The chi-square statistic is 20.02 on 8 degrees of freedom.
  The *p*-value is 0.0102.
Test of the hypothesis that three factors are sufficient:
  The chi-square statistic is 4.9 on 3 degrees of freedom.
  The *p*-value is 0.179.

The results indicate that three factors are needed to account for the relationships between the seven crime rates. The maximum likelihood estimated parameters of the three-factor model are shown in Table 15.4.

Following the typical procedure of explorative factor analysis, we will seek an easier solution for the interpretation of the factors. Here, we proceed with varimax rotation on the three-factor solution. The details of the rotated solution are given in Table 15.5, now in the form where the rows are sorted and highest loadings are highlighted.

TABLE 15.4
Estimated Parameters for the Three-Factor Model Fitted to the Crime Rate Data by Maximum Likelihood

Specific Variance Estimates

Murder	Rape	Robbery	Assault	Burglary	Theft	Vehicle
0.030	0.360	0.185	0.253	0.136	0.279	0.005

Estimated Factor Loadings

	Factor 1	Factor 2	Factor 3
Murder	0.654	0.727	−0.115
Rape	0.611	0.307	0.415
Robbery	0.828	0.344	−0.103
Assault	0.697	0.479	0.181
Burglary	0.625	0.330	0.604
Theft	0.422	0.231	0.700
Vehicle	0.992	−0.106	−0.010
Sum of squares	3.523	1.145	1.083

Looking at the loadings on the first factor of the rotated solution in Table 15.5, we see that burglary, theft and, to a slightly lesser extent, rape, are highly correlated with this factor. The second factor has a large correlation with the murder rate, and also with robbery and assault. The third factor is dominated by its correlation with vehicle crime. If we ignore the loading for rape on the first factor, it could perhaps be labeled property crime, and the second factor might be crime against the person. However, this type of exercise needs more knowledge of the subject matter, so we leave it to readers to amuse themselves with devising other labels. The example does demonstrate that even rotated solutions are not always open to easy interpretation.

TABLE 15.5
Varimax-Rotated Three-Factor Solution for Crime Rate Data

	Factor Loadings			
Variable	**Factor 1**	**Factor 2**	**Factor 3**	**Communality**
Theft	**0.831**	0.132	0.120	0.721
Burglary	**0.828**	0.330	0.263	0.864
Rape	**0.645**	0.369	0.297	0.640
Murder	0.259	**0.922**	0.229	0.970
Robbery	0.243	**0.664**	0.561	0.815
Assault	0.492	**0.629**	0.331	0.747
Vehicle	0.285	0.317	**0.902**	0.995
Sum of squares	2.240	2.049	1.463	5.752

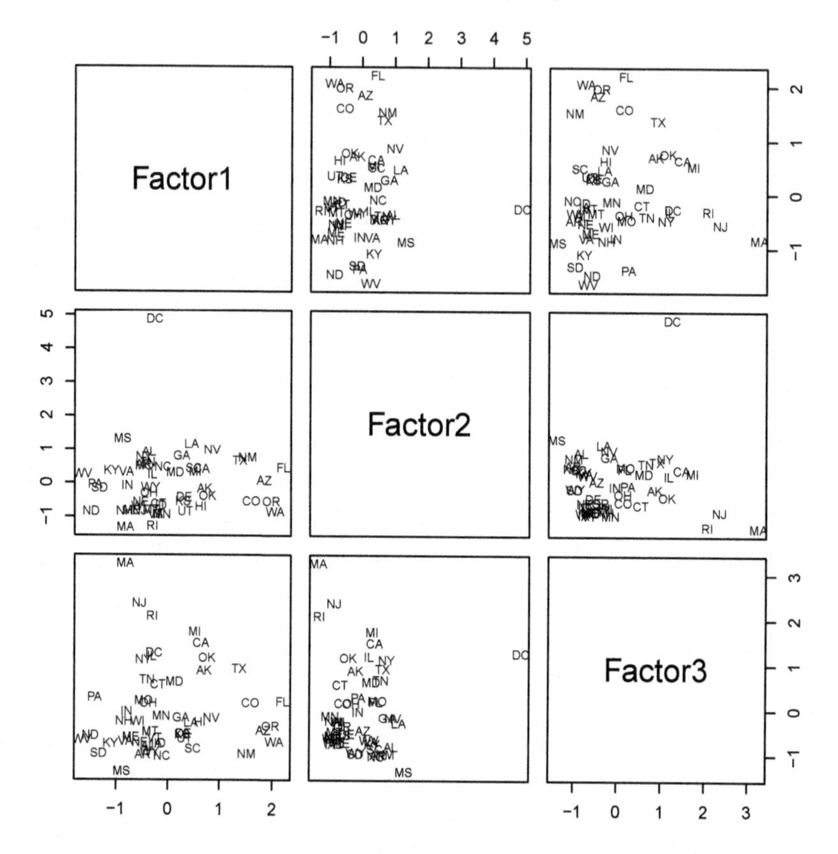

FIGURE 15.3
Scatterplot matrix of factor scores from the three-factor model fitted to crime rate data.

Finally, using the estimation (prediction) method for the factor scores mentioned in the beginning of this section, we can work further with the three-factor solution and illustrate the individual factor scores for the US states in the crime rate data with a scatterplot matrix given in Figure 15.3; points are labeled by state. The most notable feature of this plot is again the position of DC, which has a very high score on Factor 2, crime against the person; perhaps better to steer clear of DC!

15.8 Exploratory Factor Analysis and Principal Component Analysis Compared

Factor analysis, like principal components analysis, is an attempt to describe a set of multivariate data using a smaller number of dimensions than one begins with, but quite different approaches are used by each technique to achieve this goal. Some differences between principal components analysis and exploratory factor analysis are:

- Factor analysis tries to explain the covariances or correlations of the observed variables by postulating a small number of underlying latent variables, the common factors in this context, to which the manifest variables are related. Principal components analysis constructs linear functions of the observed variables that account for decreasing proportions of the variance of these variables, so that the first few components may account for a substantial proportion of the variance and thus provide a useful summary of the data.
- If the number of retained components is increased, say, from m to $m + 1$, the first m components are unchanged. This is not the case in factor analysis, where there can be substantial changes in all factors if the number of factors is changed.
- The calculation of principal component scores is straightforward; the calculation of factor scores is more complex, and a variety of methods have been suggested.
- There is usually no relationship between the principal components extracted from the sample correlation matrix and those based on the sample covariance matrix. For maximum likelihood factor analysis, however, the results of analyzing either matrix are essentially equivalent (this is not true of principal factor analysis).

Despite these differences, the results from both types of analysis are frequently very similar. Certainly, if the specific variances are small, we would expect both forms of analysis to give similar results. However, if the specific variances are large, they will be absorbed into all the principal components, both retained and rejected, whereas factor analysis makes special provision for them. A detailed comparison between these two methods is provided by Widaman (2007).

Lastly, it should be remembered that both principal components analysis and factor analysis are similar in one important respect—they are both pointless if the observed variables are almost uncorrelated. In this case, factor analysis has nothing to explain, and principal components analysis will simply lead to components that are similar to the original variables.

15.9 Summary

- Exploratory factor analysis models attempt to explain the relationships between a set of manifest variables, as measured by the variables covariance or correlation matrix in terms of the relationships of the observed variables to a small number of underlying latent variables—the common factors.

- Initial factor solutions are almost always "rotated" before any attempt is made to interpret the factors. Rotation methods attempt to achieve "simple structure," and rotated solutions may be allowed to be orthogonal (uncorrelated common factors) or oblique (correlated common factors). Orthogonal solutions are more commonly used because they are easier to interpret.

- Principal components analysis and exploratory factor analysis both attempt to simplify multivariate data by reducing their dimensionality. Principal components analysis involves a straightforward mathematical transformation of the observed variables, and has the primary aim of accounting for the variance of these variables. Exploratory factor analysis postulates the existence of underlying latent variables and aims to account for the correlations or covariances of the observed variables with a statistical model.

- Factor analysis has had its fair share of criticisms by statisticians and others primarily due to the lack of uniqueness of the factor loadings, but the editors of the book *Factor Analysis at 100* that collects the papers of a conference that in 2004 commemorated the 100 year anniversary of Spearman's (1904) famous article, have no doubts about the technique as is demonstrated by the following quotation taken from the Preface of the book (Cudeck and MacCallum, 2007):

  > Factor analysis is one of the great success stories of statistics in the social sciences because the primary focus of attention has always been on the relationships among fundamental traits such as intelligence, social class, or health status, that are unmeasureable. Factor analysis provided a way to go beyond empirical variables, such as tests and questionnaires, to the corresponding latent variables that underlie them. The great accomplishment of Spearman was in advancing a method to investigate fundamental factors. It has proven its utility in hundreds of scientific studies, across dozens of disciplines, and over a hundred years.

15.10 Exercises

15.1 Returning to the small exploratory factor analysis example described in Section 15.2, suppose now that the observed correlations had been

$$\mathbf{R} = \begin{array}{l} \text{Classics} \\ \text{French} \\ \text{English} \end{array} \begin{bmatrix} 1.00 & & \\ 0.84 & 1.00 & \\ 0.60 & 0.35 & 1.00 \end{bmatrix}$$

Find the values of the parameters in a one-factor model fitted to these correlations. Are there any problems with the solution?

15.2 Apply principal factor analysis to the crime rate data, and compare the varimax-rotated solution to that given in the text and found using the maximum likelihood estimation. Also compare the predicted correlation matrices from the principal factor analysis and maximum likelihood factor analysis solutions.

15.3 Investigate the use of alternative rotation methods to varimax on the crime rate data.

15.4 The following matrix gives the correlations between ratings on pain made by 123 people suffering from extreme pain. The nine statements are

1. Whether or not I am in pain in the future depends on the skills of the doctors.
2. Whenever I am in pain, it is usually because of something I have done or not done.
3. Whether or not I am in pain depends on what the doctors do for me.
4. I cannot get any help for my pain unless I go to seek medical advice.
5. When I am in pain, I know that it is because I have not been taking proper exercise or eating the right food.
6. People's pain results from their own carelessness.
7. I am directly responsible for my pain.
8. Relief from pain is chiefly controlled by the doctors.
9. People who are never in pain are just plain lucky.

Each statement was scored on a scale from 1 (total agreement) to 6 (total disagreement).

$$
\mathbf{R} = \begin{bmatrix}
1.00 & & & & & & & & \\
-0.04 & 1.00 & & & & & & & \\
0.61 & -0.07 & 1.00 & & & & & & \\
0.45 & -0.12 & 0.59 & 1.00 & & & & & \\
0.03 & 0.49 & 0.03 & -0.08 & 1.00 & & & & \\
-0.29 & 0.43 & -0.13 & -0.21 & 0.47 & 1.00 & & & \\
-0.30 & 0.30 & -0.24 & -0.19 & 0.41 & 0.63 & 1.00 & & \\
0.45 & -0.31 & 0.59 & 0.63 & -0.14 & -0.13 & -0.26 & 1.00 & \\
0.30 & -0.17 & 0.32 & 0.37 & -0.24 & -0.15 & -0.29 & 0.40 & 1.00
\end{bmatrix}
$$

a. Perform a principal components analysis on these data and examine the associated scree plot to decide on the appropriate number of components.

b. Apply maximum likelihood factor analysis and use the test described in the chapter to select the necessary number of common factors.

c. How do the principal components and factor analysis solutions compare?

d. Rotate the factor solution selected, using both an orthogonal and an oblique procedure, and interpret the results.

16

Confirmatory Factor Analysis and Structural Equation Models

16.1 Introduction

An exploratory factor analysis (EFA) as described in Chapter 15 is typically used in the early investigation of a set of multivariate data to determine whether the factor analysis model is useful in providing a parsimonious way of describing and accounting for the relationships between the observed variables. The analysis will determine which observed variables are most highly correlated with the common factors and how many common factors are needed to give an adequate description of the data. Essentially, a researcher uses EFA to determine factor structure. In EFA, no constraints are placed on which observed manifest variables load on which latent factors, but a simple structure is aimed at using a suitable rotation of the factor space so that the factor structure is easily interpretable (this is not always achieved!).

The second type of factor analysis is known as *confirmatory factor analysis* (CFA) which is used to test whether a *specific* factor model postulated a priori on the basis of theory and/or empirical research (often involving phases of EFA), provides an adequate fit for the observed covariation of the manifest variables. Byrne (2005) nicely summarizes the distinction between the two approaches of factor analysis:

> Whereas EFA operates inductively in allowing the observed data to determine the underlying factor structure *a posteriori*, CFA operates deductively in postulating the factor structure *a priori*.

In this chapter, we will give an account of confirmatory factor analysis models, intertwined with a brief account of what is known as *structural equation modeling*, also referred to as *covariance structure modeling*. In these models both response and explanatory latent variables are allowed and these are linked by a series of linear equations. Although more complex than CFA models, the aim of structural equation models (SEM) is essentially the same, namely to explain the covariation of the manifest variables in terms of hypothesized relationships of these variables to the assumed underlying latent variables along with the relationships postulated between the latent variables themselves. This allows the researcher to set and test hypotheses of *causal relations*, which will evidently require a fairly good knowledge of the substance theory in addition to the skills of statistical modeling. For more information on the interesting possibilities of working with causal questions, we recommend *The Book of Why* by Pearl and Mackenzie (2018), where (on p. 5–6) the authors refer to the turn of the 20th century, when statistics came into being through the work of Galton, Pearson, Spearman and others:

> This was a critical moment in the history of science. The opportunity to equip causal questions with a language of their own came very close to being realized but was squandered. In the following years, these questions were declared unscientific and went underground. ... [C]ausal vocabulary was virtually prohibited for more than a half century. ... Because of this prohibition, mathematical tools to manage causal questions were deemed unnecessary, and statistics focused exclusively on how to summarize data, not on how to interpret it. A shining exception was path analysis, invented by geneticist Sewall Wright in the 1920s and a direct ancestor of the methods we will entertain in this book.

The central feature of path analysis is a *path diagram* that is used to illustrate confirmatory factor analysis models and structural equation models visually. Such a diagram shows how manifest variables are related to latent variables, which latent variables are correlated and, in SEM models, how response latent variables are related to explanatory latent variables. Examples of such diagrams will be given in the subsequent sections.

16.2 Estimation, Identification, and Assessing the Fit for Confirmatory Factor Analysis and Structural Equation Models

The topics related to the estimation and identification of the models and assessing their fit is similar with structural equation models and confirmatory factor analysis models. In the following subsections, we shall give a short introduction to each of these issues.

16.2.1 Estimation

Structural equation models including confirmatory factor analysis models will contain a number of terms that are fixed by the researcher and a number of free parameters that need to be estimated from the data. In a confirmatory factor analysis, for example, the loadings for some manifest variables on some of the postulated latent variables will be set a priori to zero; additionally some correlations between the latent variables might also be fixed at zero. Such a model is fitted to a set of data by estimating its free parameters so that the variances and covariances of the manifest variables as implied by the model of interest are as close as possible in some sense to the corresponding observed values. See Technical Section 16.1 for more details of the estimation process.

Technical Section 16.1: Estimation in SEM and CFA models

Estimation in SEM and CFA models involves finding values for the model parameters that minimize a discrepancy function indicating the magnitude of the differences between the elements of $\mathbf{S}$, the observed covariance matrix of the manifest variables and those of $\boldsymbol{\Sigma}(\boldsymbol{\theta})$, the covariance matrix implied by the fitted model (i.e., a matrix the elements of which are functions of the parameters of the model), contained in the vector $\boldsymbol{\theta} = (\theta_1, \dots, \theta_t)'$. There are a number of possibilities for discrepancy functions; for example, the ordinary least squares discrepancy function, F_{LS}, is

$$F_{LS}(\mathbf{S}, \boldsymbol{\Sigma}(\boldsymbol{\theta})) = \sum_{i<j}\sum_{j} \left(s_{ij} - \sigma_{ij}(\boldsymbol{\theta})\right)^2,$$

where s_{ij} and $\sigma_{ij}(\boldsymbol{\theta})$ are the elements of $\mathbf{S}$ and $\boldsymbol{\Sigma}(\boldsymbol{\theta})$. But this criterion has several problems that make it unsuitable for estimation; for example, it is not independent of the scale of the manifest variables, and so different estimates of the model parameters would be produced using the sample covariance matrix and the sample correlation matrix. Other problems with the least squares criterion are detailed in Everitt (1984).

The most commonly used method of estimating the parameters in SEM and CFA models is maximum likelihood under the assumption that the observed data have a multivariate normal distribution. It is easy to show that maximizing the likelihood is now equivalent to minimizing the discrepancy function, F_{ML}, given by

$$F_{ML}(\mathbf{S}, \boldsymbol{\Sigma}(\boldsymbol{\theta})) = \log|\boldsymbol{\Sigma}(\boldsymbol{\theta})| + \mathrm{trace}\left(\mathbf{S}\boldsymbol{\Sigma}(\boldsymbol{\theta})^{-1}\right) - \log|\mathbf{S}| - q$$

(cf. maximum likelihood factor analysis in Chapter 15). We see that by varying the parameters $\theta_1, \dots, \theta_t$ so that $\boldsymbol{\Sigma}(\boldsymbol{\theta})$ becomes more like $\mathbf{S}$, F_{ML} becomes smaller. Iterative numerical algorithms are needed to minimize the function F_{ML} with respect to the parameters, but for details see Everitt (1984) and Everitt and Dunn (2001).

16.2.2 Identification

Consider the following simple example of a model in which there are three manifest variables, x, x', and y, and two latent variables, u and v, with the relationships between the manifest and latent variables being

$$x = u + \delta$$
$$y = v + \epsilon$$
$$x' = u + \delta'$$

If we assume that δ, δ', and ϵ have expected values of zero, that δ and δ' are uncorrelated with each other and with u, and that ϵ is uncorrelated with v, then the covariance matrix of the three manifest variables may be expressed in terms of parameters representing the variances and covariances of the residuals and the latent variables as

$$\boldsymbol{\Sigma}(\boldsymbol{\theta}) = \begin{bmatrix} \theta_1 + \theta_2 & & \\ \theta_3 & \theta_4 + \theta_5 & \\ \theta_3 & \theta_4 & \theta_4 + \theta_6 \end{bmatrix}$$

where $\boldsymbol{\theta}' = (\theta_1, \theta_2, \theta_3, \theta_4, \theta_5, \theta_6)$ and $\theta_1 = \text{Var}(v)$, $\theta_2 = \text{Var}(\epsilon)$, $\theta_3 = \text{Cov}(v, u)$, $\theta_4 = \text{Var}(u)$, $\theta_5 = \text{Var}(\delta)$, and $\theta_6 = \text{Var}(\delta')$. It is immediately apparent that estimation of the parameters in this model poses a problem. The two parameters θ_1 and θ_2 are not uniquely determined because one can be, for example, increased by some amount and the other decreased by the same amount without altering the covariance matrix predicted by the model. In other words, in this example, different sets of parameter values (i.e., different θs) will lead to the same predicted covariance matrix, $\boldsymbol{\Sigma}(\boldsymbol{\theta})$. The model is said to be *unidentifiable.*

Formally, a model is identified if and only if $\boldsymbol{\Sigma}(\boldsymbol{\theta}_1) = \boldsymbol{\Sigma}(\boldsymbol{\theta}_2)$ implies that $\boldsymbol{\theta}_1 = \boldsymbol{\theta}_2$. In SEM and CFA models, identifiability depends on the choice of model and on the specification of fixed, constrained (for example, two parameters constrained to equal one another), and free parameters. If a parameter is not identified, it is not possible to find a consistent estimate of it. Establishing model identification in SEM and CFA models can be difficult because there are no simple, practicable, and universally applicable rules for evaluating whether a model is identified, although there is a simple necessary (but not sufficient) condition for identification, namely that the number of free parameters in a model, t, be less than $q(q+1)/2$. For a more detailed discussion of the identifiability problem, see Bollen and Long (1993) or Skrondal and Rabe-Hesketh (2004).

16.2.3 Assessing the Fit

Once a model has been pronounced identified and its parameters estimated, the next step becomes that of assessing how well the model-predicted covariance matrix fits the covariance matrix of the manifest variables. A global measure of fit of a model is provided by the likelihood ratio statistic given by $X^2 = (N - 1) \min F_{ML}$, where N is the sample size and $\min F_{ML}$ is the minimized value of the maximum likelihood discrepancy function F_{ML} given in Technical Section 16.1. If the sample size is sufficiently large, the X^2 statistic provides a test that the population covariance matrix of the manifest variables is equal to the covariance implied by the fitted model against the alternative hypothesis that the population matrix is unconstrained. Under the equality hypothesis, X^2 has a chi-squared distribution with degrees of freedom ν given by $\frac{1}{2}q(q+1) - t$, where t is the number of free parameters in the model.

The likelihood ratio statistic is often the only measure of fit quoted for a fitted model, but on its own it has limited practical use because in large samples even relatively trivial departures from the equality null hypothesis will lead to its rejection. Consequently, in large samples most models may be rejected as statistically untenable. A more satisfactory way to use the test is for a comparison of a series of nested models where a large difference in the statistic for two models compared with the difference in the degrees of freedom of the models indicates that the additional parameters in one of the models provide a genuine improvement in fit.

Perhaps the best way to assess the fit of a model is to use the X^2 statistic alongside one or more of the following procedures:

- Visual inspection of the residual covariances (i.e., the differences between the covariances of the manifest variables and those predicted by the fitted model). These residuals should be small when compared with the values of the observed covariances or correlations.
- Examination of the standard errors of the parameters and the correlations between these estimates. If the correlations are large, it may indicate that the model being fitted is almost unidentified.
- Estimated parameter values outside their possible range; i.e., negative variances or absolute values of correlations greater than unity are often an indication that the fitted model is fundamentally wrong for the data.

In addition, a number of *fit indices* have been suggested that can be useful in assessing the fit of a model. Developing the fit indices has been an active topic of research, resulting in varying recommendations of their usage. Currently, the following four indices are probably the most frequently applied:

- Comparative fit index (CFI),
- Tucker–Lewis index (TLI),
- Root mean square error of approximation (RMSEA), and
- Standardized root mean squared residual (SRMR).

All these indices are defined explicitly in Hu and Bentler (1999). More recent advice concerning their usage, interpretation, and recommendations are given in Jackson et al. (2009) and Bartholomew et al. (2011). For background and details, see also Bollen and Long (1993).

The first two fit indices in the above list (CFI and TLI) represent *relative* (or incremental) fit indices that compare the fit of a model to the fit of some baseline model, typically the independence model where the observed variables are assumed to be uncorrelated). The CFI can take values between zero (no fit) and one (perfect fit); in practice, only values close to one suggest an acceptable level of fit. For the TLI, a value greater than one might indicate overfitting, while values less than 0.9 indicate a poor fit.

The latter two (RMSEA and SRMR) represent *absolute* (or overall) fit indices that help to determine how well the hypothesized model fits the sample data. They may also be called *misfit* indices, as they range from one to zero, with the reasonable fit indicated by values close to zero.

For the RMSEA, it is possible to calculate a confidence interval, whereas the SRMR is the standardized version of the square root of the mean squared differences between the elements in $\mathbf{S}$ and $\hat{\boldsymbol{\Sigma}}$. In addition to the amount of (mis)fit, SRMR provides useful information of the model, as a value of, say, SRMR $= 0.04$ indicates that the model explains the correlations to within an average error of 0.04.

Also helpful in assessing the fit of the model are the summary statistics for the *normed residuals*, which are essentially the differences between corresponding elements of $\mathbf{S}$ and $\hat{\boldsymbol{\Sigma}}(\boldsymbol{\theta})$ but scaled so that they are unaffected by the differences in the variances of the observed variables. The normed residuals, r_{ij}^*, are defined as

$$r_{ij}^* = \frac{s_{ij} - \hat{\sigma}_{ij}}{\sqrt{(\hat{\sigma}_{ij}\hat{\sigma}_j^2 + \hat{\sigma}_{ij}^2)/n}}$$

Generally, the absolute values of the normed residuals should all be less than 2 to claim that the current model fits the data well. Diagnostic tools (such as residual plots) for detecting outliers and influential observations, for example, are common in regression analysis (see Chapters 3 and 4) but still under development in structural equation models (see Hildreth, 2013).

Although as mentioned above a proposed CFA model might arise on the basis of an EFA, the CFA model must be tested on a *fresh* set of data; models must *not* be generated from and tested on the same data.

16.3 Examples of Confirmatory Factor Analysis

We will now illustrate the application of confirmatory factor analysis with three examples.

16.3.1 Ability and Aspiration

Calsyn and Kenny (1977) recorded the values of the following six variables for 556 eighth-grade students:

1. Self-concept of ability, SCA (x_1)
2. Perceived parental evaluation, PPE (x_2)
3. Perceived teacher evaluation, PTE (x_3)
4. Perceived friend's evaluation, PFE (x_4)
5. Educational aspiration, EA (x_5)
6. College plans, CP (x_6)

Calsyn and Kenny postulated that two underlying latent variables, *ability* and *aspiration* generated the relationships between the observed variables. The first four of the manifest variables were assumed to be indicators of ability, and the last two indicators of aspiration. In addition, ability and aspiration were allowed to be correlated. So, the regression-like equations specifying the postulated model are

$$x_1 = \lambda_1 f_1 + 0 f_2 + u_1$$

$$x_2 = \lambda_2 f_1 + 0 f_2 + u_2$$

$$x_3 = \lambda_3 f_1 + 0 f_2 + u_3$$

$$x_4 = \lambda_4 f_1 + 0 f_2 + u_4$$

$$x_5 = 0 f_1 + \lambda_5 f_2 + u_5$$

$$x_6 = 0 f_1 + \lambda_6 f_2 + u_6$$

where f_1 represents the ability latent variable, and f_2 represents the aspiration latent variable. Note that unlike the exploratory factor analysis, a number of factor loadings are fixed at 0 and play no part in the estimation process. The model has a total of 13 parameters to estimate: six factor loadings (λ_1 to λ_6), six specific variances (ψ_1 to ψ_6), and one correlation between ability and aspiration (ρ). (To be consistent with the nomenclature used earlier in this chapter, all parameters should be suffixed thetas; this could, however, become confusing, so we have changed the nomenclature and use lambdas, etc., in a manner similar to how they are used in Chapter 15.) The observed correlation matrix given in Table 16.1 has six variances and 15 correlations, a total of 21 terms. Consequently, the postulated model has $21 - 13 = 8$ degrees of freedom. A path diagram for the correlated, two-factor model is shown in Figure 16.1.

The results from fitting the ability and aspiration model to the observed correlations are shown in Table 16.2 (note that the two latent variables f_1 and f_2 have their variances fixed at one, although it is the fixing that is important,

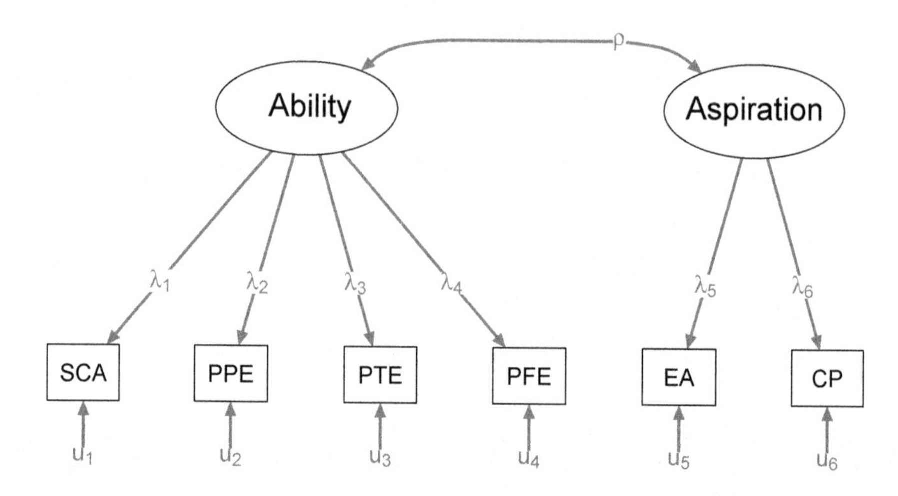

FIGURE 16.1
Path diagram for ability and aspiration model.

not the value at which they are fixed; these variances cannot be free parameters to be estimated). The estimates divided by their standard errors (SE) represent z-values that test whether parameters are significantly different from zero. In this case, all z-values would have very small associated p-values giving support for the importance of the parameters in the model. Note that with CFA models the standard errors of parameters assume importance because they allow the investigator to assess whether parameters might be dropped from the model to find a more parsimonious model that still provides an adequate fit to the data. (In EFA, standard errors of factor loadings can be calculated, but they are hardly ever used; instead, an informal interpretation of factors is made.)

Of particular note amongst the parameter estimates is the correlation between the two latent variables; this estimate (0.66 with a standard error of 0.03) is a *disattenuated correlation*, that is, the correlation between "true" ability and "true" aspiration uncontaminated by measurement error in the

TABLE 16.1
Observed Correlations for the Ability and Aspiration Example

	x_1	x_2	x_3	x_4	x_5	x_6
x_1	1.00					
x_2	0.73	1.00				
x_3	0.70	0.68	1.00			
x_4	0.58	0.61	0.57	1.00		
x_5	0.46	0.43	0.40	0.37	1.00	
x_6	0.56	0.52	0.48	0.41	0.72	1.00

TABLE 16.2
Results of Fitting the Ability and Aspiration, Correlated Two-Factor Model to the Correlations in Table 16.1

Parameter	Estimate	Standard Error (SE)	Estimate/SE
λ_1	0.863	0.035	24.558
λ_2	0.849	0.035	23.961
λ_3	0.805	0.035	22.115
λ_4	0.695	0.039	18.000
λ_5	0.775	0.040	19.206
λ_6	0.929	0.039	23.569
ψ_1	0.255	0.023	19.911
ψ_2	0.279	0.024	11.546
ψ_3	0.352	0.027	13.070
ψ_4	0.516	0.035	14.876
ψ_5	0.399	0.038	10.450
ψ_6	0.137	0.044	3.152
ρ	0.667	0.032	21.521

Note: $X^2 = 9.26$, df $= 8$, $p = 0.321$
CFI $= 0.999$, TLI $= 0.999$, RMSEA $= 0.017$, SRMR $= 0.012$
Range of normalized residuals: $[-0.441, 0.533]$

observed indicators of the two latent variables. An approximate 95% confidence interval for the disattenuated correlation is [0.60, 0.73].

The fit of the model can be partially judged by a chi-square statistic described in previous section, which in this case takes the value 9.26 with eight degrees of freedom and an associated p-value of 0.32, suggesting that the postulated model fits the data very well. (The chi-square test for the null model is simply a test that the population covariance matrix of the observed variables is diagonal; i.e., that the observed variables are independent. In most cases, this null model will be rejected; if it is not, a further model-fitting exercise is a waste of time.)

The various fit indices given in Table 16.2 indicate that the model is a good fit for the data. Also the absolute values of the normed residuals (defined earlier) seem to be all (much) less than 2, claiming that the current model fits the ability and aspiration data well.

16.3.2 Drug Usage among Students

For our second example of fitting a CFA model, we return to the drug usage among students data introduced in Chapter 13. In the original investigation of these data reported by Huba et al. (1981), a CFA model was postulated for the model arising from consideration of previously reported research on student drug usage. The model consisted of the following three latent variables:

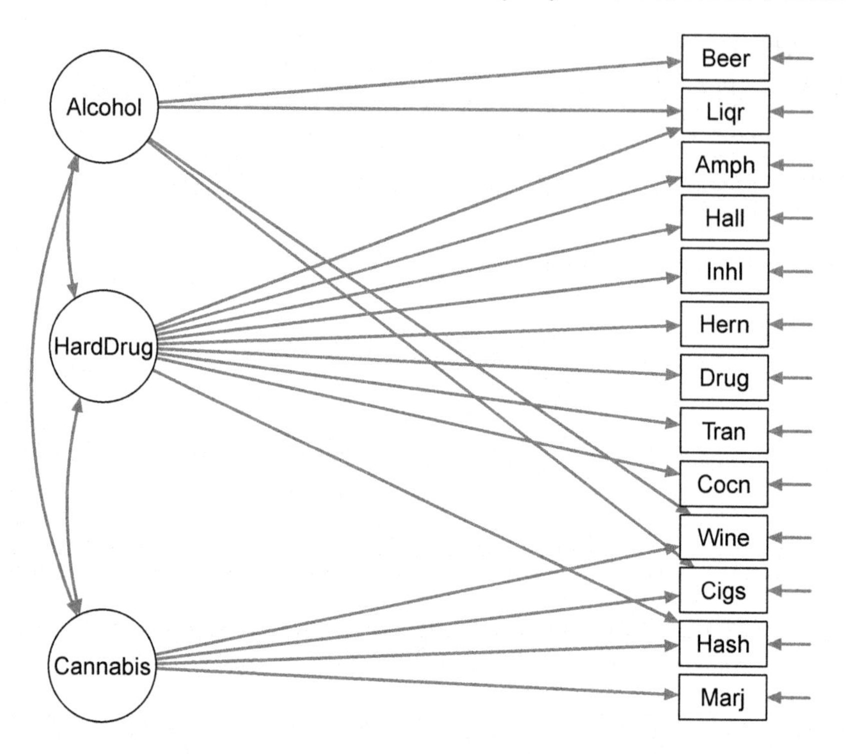

FIGURE 16.2
Path diagram for the drug usage model.

- f_1: Alcohol use—with nonzero loadings on beer, wine, liquor, and cigarettes usage.
- f_2: Cannabis use—with nonzero loadings on marijuana, hashish, cigarettes, and wine usage. The cigarettes variable is assumed to load on both the first and second latent variables because it sometimes occurs with both alcohol and marijuana use and, at other times, does not. The nonzero loading on wine was allowed because of reports that wine is frequently used with marijuana and, consequently, some of the use of wine might be an indicator of tendencies toward cannabis use.
- f_3: Hard drug use—with nonzero loadings on amphetamines, tranquilizers, hallucinogenics, hashish, cocaine, heroin, drugstore medication, inhalants, and liquor. The use of each of these substances was considered to suggest a strong commitment to the notion of psychoactive drug use.

Each of the three latent variables is assumed to have a variance of one.

Correlations between each pair of the three factors are allowed to be free parameters to be estimated. So, the proposed model can be specified by the following series of equations:

$$
\begin{aligned}
\text{cigarettes} &= \lambda_1 f_1 + \lambda_2 f_2 + 0 f_3 + u_1 \\
\text{beer} &= \lambda_3 f_1 + 0 f_2 + 0 f_3 + u_2 \\
\text{wine} &= \lambda_4 f_1 + \lambda_5 f_2 + 0 f_3 + u_3 \\
\text{liquor} &= \lambda_6 f_1 + 0 f_2 + \lambda_7 f_3 + u_4 \\
\text{cocaine} &= 0 f_1 + 0 f_2 + \lambda_8 f_3 + u_5 \\
\text{tranquilizers} &= 0 f_1 + 0 f_2 + \lambda_9 f_3 + u_6 \\
\text{drugstore medication} &= 0 f_1 + 0 f_2 + \lambda_{10} f_3 + u_7 \\
\text{heroin} &= 0 f_1 + 0 f_2 + \lambda_{11} f_3 + u_8 \\
\text{marijuana} &= 0 f_1 + \lambda_{12} f_2 + 0 f_3 + u_9 \\
\text{hashish} &= 0 f_1 + \lambda_{13} f_2 + \lambda_{14} f_3 + u_{10} \\
\text{inhalants} &= 0 f_1 + 0 f_2 + \lambda_{15} f_3 + u_{11} \\
\text{hallucinogenics} &= 0 f_1 + 0 f_2 + \lambda_{16} f_3 + u_{12} \\
\text{amphetamines} &= 0 f_1 + 0 f_2 + \lambda_{17} f_3 + u_{13}
\end{aligned}
$$

The proposed model also allows for nonzero correlations between each pair of latent variables, and so, has a total of 33 parameters to estimate: 17 loadings (λ_1 to λ_{17}), 13 specific variances (ψ_1 to ψ_{13}) and 3 between latent variables correlations (ρ_1 to ρ_3). Consequently, the model has $91 - 33 = 58$ degrees of freedom. A path diagram for the model is shown in Figure 16.2. (For the complete names corresponding to the abbreviated labels of the manifest variables, see the above equations and Chapter 13.)

Note that the model includes several *cross-loadings*, that is, factor loadings where a manifest variable is assumed to load on more than one factor. Cross-loadings are typical in EFA but they should usually be avoided in CFA models, unless there are substantial grounds for including them in the model. In this case, four manifest variables, namely, cigarettes, hashish, liquor, and wine were assumed to measure (and hence load on) two factors (as indicated by the above equations and the arrows of the path diagram in Figure 16.2).

The results of fitting the proposed model are given in Table 16.3. Here, the chi-square test for goodness of fit takes the value 323.35, which with 58 degrees of freedom has an associated p-value that is very small; the model does not appear to fit very well. The fit indices are quite good, but the absolute values of the normed residuals seem to be much larger than 2. Readers are referred to the original Huba et al. (1981) paper for details of how the model was modified to try to achieve a better fit.

TABLE 16.3
Results of Fitting a Correlated Three-Factor Model to Drug Usage Data

Parameter	Estimate	Standard Error(SE)	Estimate/SE
λ_1	0.358	0.035	10.371
λ_2	0.332	0.035	9.401
λ_3	0.792	0.023	35.021
λ_4	0.875	0.038	23.285
λ_5	−0.152	0.037	−4.158
λ_6	0.722	0.024	30.673
λ_7	0.123	0.023	5.439
λ_8	0.465	0.026	18.079
λ_9	0.676	0.024	28.182
λ_{10}	0.359	0.025	13.602
λ_{11}	0.476	0.026	18.571
λ_{12}	0.912	0.030	29.958
λ_{13}	0.396	0.030	13.379
λ_{14}	0.381	0.029	13.050
λ_{15}	0.543	0.025	21.602
λ_{16}	0.618	0.025	25.233
λ_{17}	0.763	0.023	32.980
ψ_1	0.611	0.024	25.823
ψ_2	0.374	0.020	18.743
ψ_3	0.379	0.024	16.052
ψ_4	0.408	0.019	21.337
ψ_5	0.784	0.029	26.845
ψ_6	0.544	0.023	23.222
ψ_7	0.871	0.032	27.653
ψ_8	0.773	0.029	26.735
ψ_9	0.169	0.044	3.846
ψ_{10}	0.547	0.022	24.593
ψ_{11}	0.705	0.027	25.941
ψ_{12}	0.618	0.025	24.655
ψ_{13}	0.418	0.021	19.713
ρ_1	0.634	0.027	23.369
ρ_2	0.313	0.029	10.674
ρ_3	0.499	0.027	18.412

Note: $X^2 = 323.35$, df $= 58$, $p < 0.0001$
CFI = 0.959, TLI = 0.945, RMSEA = 0.053, SRMR = 0.039
Range of normalized residuals: [−3.031, 4.577]

16.4 Eight Factors of Systems Intelligence

Finally, as a more complex example of using CFA, we shall turn to the field of *Systems Intelligence* (SI) which is defined as a person's ability to behave or act intelligently in complex systems (such as organizations, family, and everyday life) that involve interaction and feedback (Saarinen and Hämäläinen, 2004).

Based on a combination of theoretical considerations and empirical research involving EFA, Törmänen et al. (2016) developed a 32 variable inventory of SI. They postulated a factor structure where eight factors of SI would generate the relationships between the 32 observed variables. The eight factors can be described briefly as follows:

- Systemic perception: Ability to see the systems around us;
- Attunement: Capability to feel and tune into people and systems;
- Attitude: Our overall approach to life in systems;
- Spirited discovery: Passionate engagement with new ideas;
- Reflection: Capacity to reflect on our thoughts and think about our thinking;
- Wise action: Ability to behave with understanding and a long time horizon;
- Positive engagement: Character of our communicative interactions; and
- Effective responsiveness: Talent at taking timely, appropriate actions.

For a more verbose description of the SI factors, see Hämäläinen et al. (2014).

In order to maintain good content validity and a balanced inventory, each factor was supposed to be measured by four manifest variables that were short, mostly positively oriented sentences, resulting from an iterative process of selection based on theoretical work and EFA. Taking as an example one variable of each factor, the wordings of the sentences were as follows (with the description of the postulated factor and its abbreviation in the parentheses):

- "I form a rich overall picture of situations" (Systemic perception, PER),
- "I approach people with warmth and acceptance" (Attunement, ATT),
- "I have a positive outlook on the future" (Attitude, ATD),
- "I like to play with new ideas" (Spirited discovery, DIS),
- "I view things from many different perspectives" (Reflection, REF),
- "I am willing to take advice" (Wise action, WIS),

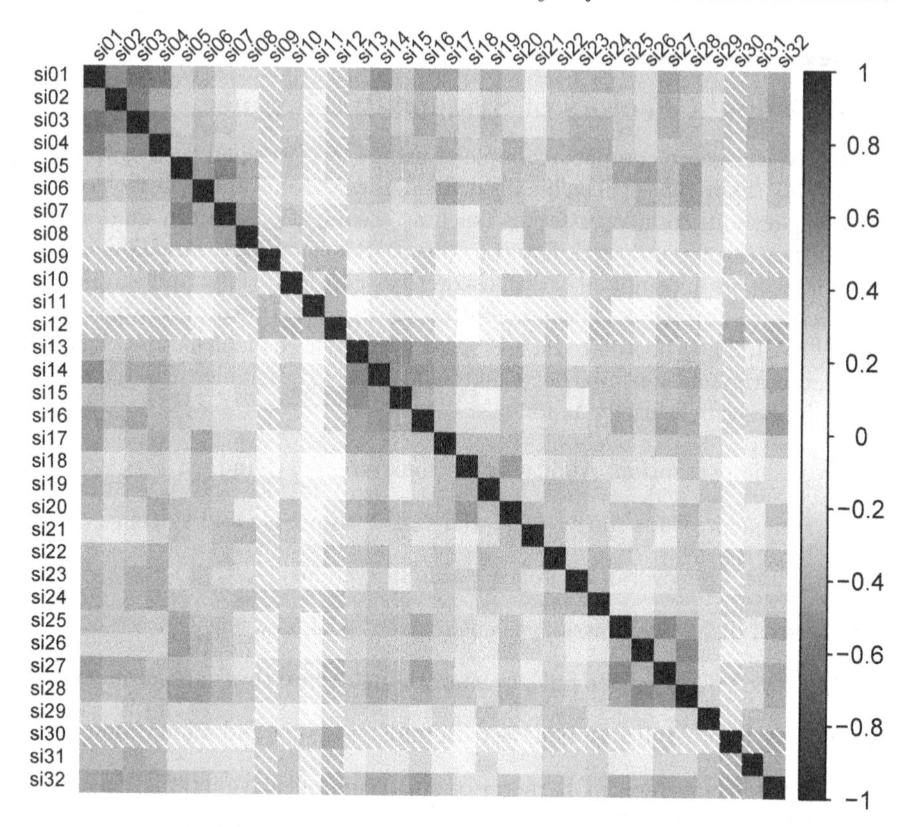

FIGURE 16.3
Corrgram for the Systems Intelligence data.

- "I contribute to the shared atmosphere in group situations" (Positive engagement, ENG), and
- "I prepare myself for situations to make things work" (Effective responsiveness, EFF).

Rather than showing the correlation matrix of the 32 variables in this example, which would be difficult to interpret given that it contains $32 \times 32 = 1024$ numerical values, we prefer to use a *corrgram* (Friendly, 2002; Murdoch and Chow, 1996) which shows the correlation matrix in a condensed graphical form. The plot is shown in Figure 16.3; we see that the correlations are mostly positive (only four variables have negative correlations with the other ones), reflecting the positive orientations of the sentences. As expected, the variables do form several clusters around the diagonal of the matrix, but the structure calls for a more detailed investigation by the means of confirmatory factor analysis.

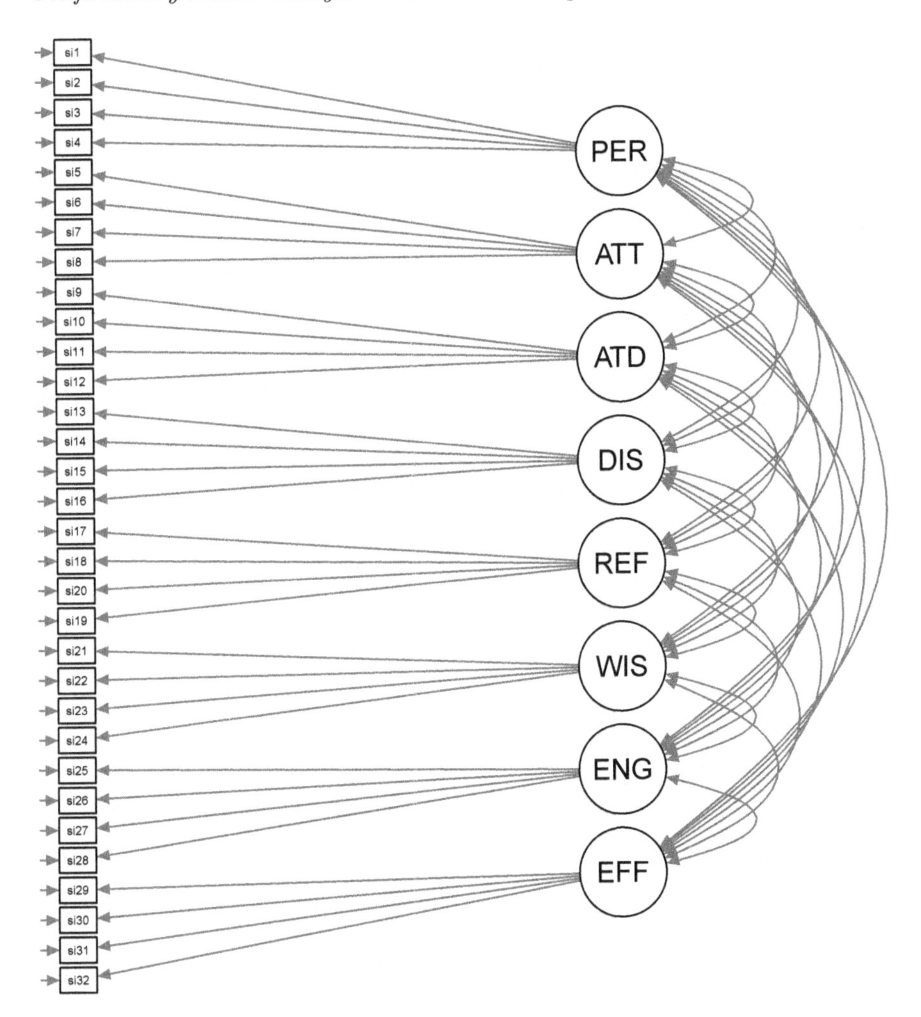

FIGURE 16.4
Path diagram for the hypothesized Systems Intelligence factor model.

16.4.1 Testing the Factorial Validity of the SI Inventory

For testing the factorial validity of the eight-dimensional factor structure of the Systems Intelligence inventory, Törmänen et al. (2016) postulated a specific model, where four loadings per factor were left free for estimation while most loadings on each of the factors were fixed at zero because they were "small" in the simple structure obtained by the EFA. Based on the results of EFA, the factors were allowed to be correlated in the CFA model. We shall skip the detailed equations that are also called the *measurement model*, and instead display the model graphically. Figure 16.4 shows the path diagram of the eight postulated SI factors and the 32 manifest variables with the specific factors, drawn as arrows pointing to the manifest variables.

As mentioned previously, whilst it is perfectly appropriate to arrive at a factor model to submit to a CFA from an EFA, the CFA model *must* be tested on a fresh set of data. This procedure was followed by Törmänen et al. (2016), as they completed their analyses based on an extensive EFA using a set of *learning data*, and then tested the CFA model with *validation data* that comprised of a completely different sample of respondents. We shall summarize the results based on the validation data that consist of 815 complete observations.

Again, it is a good practice to check the degrees of freedom of the postulated CFA model. To begin from the eight factors, there are as many as 28 factor correlations (ρ_1 to ρ_{28}). In addition, there are 32 factor loadings (λ_1 to λ_{32}) and 32 specific variances (ψ_1 to ψ_{32}) to be estimated, so the proposed model has a total of 92 parameters. (The variances of the factors are again fixed at one, which is a quite typical choice made in CFA models.) Consequently, based on the correlation matrix (shown in visual form in Figure 16.3) of 32 manifest variables and hence 528 unique items, the initial SI model will have $528 - 92 = 436$ degrees of freedom.

We summarize the results of fitting the proposed model briefly. Here, the chi-square test takes the value 1256.67, which with 436 degrees of freedom has an associated p-value of almost zero; the model does not fit very well according to the goodness of fit test. (This is also quite typical in practice.) The fit indices, however, are fairly good, as CFI = 0.951, TLI = 0.944, RMSEA = 0.048, and SRMR = 0.068, but the normed residuals that range from -4 to 5, might perhaps reveal some problems in the model. Probably further research is required to find the best possible indicators for the eight factors of systems intelligence. We also note that the correlations between the latent factors are fairly high, ranging between -0.87 and 0.89. Obviously, there is some overlap with those eight dimensions that is reflected in the factor correlations.

Finally, one area that we would like to mention here, as it deserves careful attention in developing and validating a measurement instrument, is the *measurement quality* that includes both *validity* and *reliability* of the measurements. The validity is much related to the substance itself, while the reliability is more of a statistical issue. The reliability of measurements can be examined in the context of CFA and SEM models, and in a recent study, McNeish et al. (2018) discuss interesting connections between the fit indices and measurement quality in the structural equation modeling. But for details, we recommend the references given in the summary of this chapter. For an introduction to the topics of reliability (and validity) of measurements and their effects in statistical modeling, see Tarkkonen and Vehkalahti (2005), Vehkalahti et al. (2007) and Alwin (2007).

16.5 Structural Equation Models

Structural equation models represent the convergence of relatively independent research traditions in psychiatry, psychology, econometrics, and biometrics. The idea of latent variables in psychometrics arises from Spearman's (1904) early work on general intelligence. The concept of simultaneous directional influences of some variables on others has been part of economics since Haavelmo (1943), and the resulting *simultaneous equation models* have been used extensively by economists but essentially only with observed variables. *Path analysis* was introduced very early by Wright (1921, 1934) in a biometrics context as a method for studying the direct and indirect effects of variables. The quintessential feature of path analysis is a path diagram showing how a set of explanatory variables influence a dependent variable under consideration. How the paths are drawn determines whether the explanatory variables are correlated causes, mediated causes, or independent causes.

Later, path analysis was taken up by sociologists such as Blalock (1961, 1963) and then by Duncan (1966), who demonstrated the value of combining path-analytic representation with simultaneous equation models. And, finally, in the 1970s, several workers most prominent of whom were Jöreskog (1970, 1973, 1978), Bentler (1980), and Browne (1974), combined all these various approaches into a general method that could in principle deal with extremely complex models in a routine manner.

Structural equation models have been intensively studied for many decades, and in recent years, supported by the development of more and more efficient computing facilities, the models have spread to several new fields of studies with applications and data sets of increasing complexity. For a wide coverage of the state-of-the-art of the structural equation modeling, see Kline (2016), for example.

To finish this chapter, we shall take a brief look at a basic example of a structural equation model.

16.5.1 Example of a Structural Equation Model

In the following, we shall shortly consider a typical SEM example related to job satisfaction. Houghton and Jinkerson (2007) studied the relationships among four latent factors: constructive thought strategies, dysfunctional thought processes, subjective well-being, and job satisfaction. Three composite indicators for each of the factors were formed using summed items and subscales from various measurement instruments, such as EBA (Evaluating Beliefs and Assumptions), ST (Self Talk), MI (Mental Imagery), and DAS (Dysfunctional Attitude Survey), UF (Underwood and Froming), FOR (Fordyce), and JDI-W (Job Descriptive Index Work Subscale).

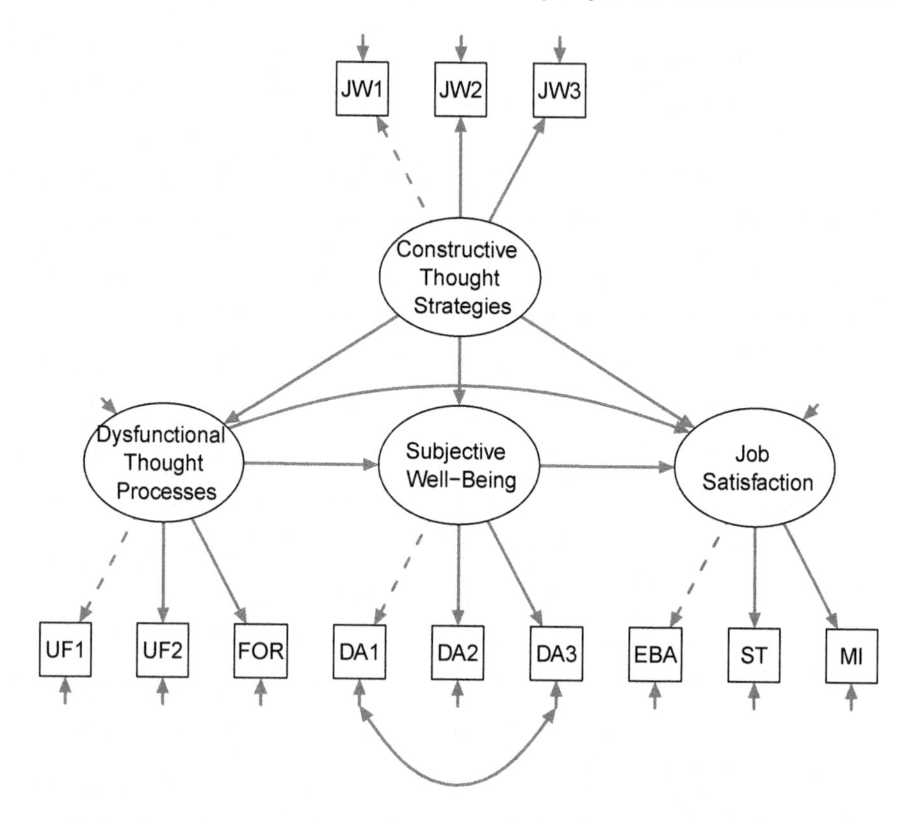

FIGURE 16.5
Path diagram for the hypothesized structural equation model for job satisfaction data.

The path diagram of the hypothesized structural equation model appears in Figure 16.5, where the job satisfaction is the response latent variable of primary interest. The other factors are explanatory latent variables, but simultaneously, two of them (dysfunctional thought processes and subjective well-being) act also as response latent variables. Constructive thought processes is the only factor that is not in the role of a response variable in the hypothesized model. In addition, the model includes one residual covariance term specified between two manifest variables (DA1 and DA3). For details, see the paper by Houghton and Jinkerson (2007). Unlike in the earlier examples of this chapter, the variances of each of the latent factors are not fixed at one. Instead, using another typical alternative, the first factor loading of each set of the manifest variables is fixed at one, which is indicated in the diagram with dashed lines.

The fit of the model is fairly good, with the chi-square test having the value 56.7 with 47 degrees of freedom, and hence having an associated (clearly non-significant) p-value of 0.16. The four fit indices (see the previous sections) would clearly support the good model fit, as CFI = 0.991, TLI = 0.987, RMSEA = 0.028, and SRMR = 0.037. There were in total six hypotheses specifying various relations and mediation effects between the latent variables. For example, the first hypothesis was that the subjective well-being would be positively related to job satisfaction, and the second one that dysfunctional thought processes would be negatively related to subjective well-being. According to the results, both hypotheses were fully supported, as the corresponding regression coefficients were statistically significant and had a proper sign. For more details about the results, including the mediation effects, see the original article by Houghton and Jinkerson (2007).

16.6 Summary

- The fitting of confirmatory factor analysis models becomes possible when the investigator has a specific factor model in mind. This model may have arisen from an earlier exploratory analysis or from theoretical considerations. In the former case, the derived model must be tested on new data.
- Structural equation modeling extends the confirmatory factor model by allowing for both dependent and explanatory latent variables and paths between them. In addition, the various contributions on CFA and SEM in the *Encyclopedia of Behavioral Science* edited by Everitt and Howell (2005) are all worth reading.
- For detailed accounts of confirmatory factor analysis and structural equation models, readers are referred to Kline (2016), Mulaik (2009), and Bollen (1989).

16.7 Exercises

16.1 The following matrix gives the correlations between ratings on pain made by 123 people suffering from extreme pain. The nine statements are

1. Whether or not I am in pain in the future depends on the skills of the doctors.
2. Whenever I am in pain, it is usually because of something I have done or not done.

3. Whether or not I am in pain depends on what the doctors do for me.
4. I cannot get any help for my pain unless I go to seek medical advice.
5. When I am in pain, I know that it is because I have not been taking proper exercise or eating the right food.
6. People's pain results from their own carelessness.
7. I am directly responsible for my pain.
8. Relief from pain is chiefly controlled by the doctors.
9. People who are never in pain are just plain lucky.

Each statement was scored on a scale from 1 (total agreement) to 6 (total disagreement).

$$\mathbf{R} = \begin{bmatrix} 1.00 & & & & & & & & \\ -0.04 & 1.00 & & & & & & & \\ 0.61 & -0.07 & 1.00 & & & & & & \\ 0.45 & -0.12 & 0.59 & 1.00 & & & & & \\ 0.03 & 0.49 & 0.03 & -0.08 & 1.00 & & & & \\ -0.29 & 0.43 & -0.13 & -0.21 & 0.47 & 1.00 & & & \\ -0.30 & 0.30 & -0.24 & -0.19 & 0.41 & 0.63 & 1.00 & & \\ 0.45 & -0.31 & 0.59 & 0.63 & -0.14 & -0.13 & -0.26 & 1.00 & \\ 0.30 & -0.17 & 0.32 & 0.37 & -0.24 & -0.15 & -0.29 & 0.40 & 1.00 \end{bmatrix}$$

Using the correlations between the first eight statements of the exercise, fit a correlated two-factor model in which questions 1, 3, 4, and 8 are assumed to be indicators of the latent variable *Doctor's Responsibility* and questions 2, 5, 6, and 7 are assumed to be indicators of the latent variable *Patient's Responsibility*. Find a 95% confidence interval for the correlation between the two latent variables.

* 16.2 Yule et al. (1969) collected the scores on each of ten cognitive tests from the Wechsler series on 150 children. By means of their scores on a test of reading ability, the children were divided into "good" readers and "poor" readers. The covariance matrices of the two groups are shown in Table 16.4.

1. Calculate the correlation matrix for each group.
2. Prior information suggests that the first five tests are "verbal" and the last five "performance" tests. Fit a correlated two-factor model to the correlation matrix of both the "good" readers and the "poor" readers. Construct the relevant path diagrams and assess the goodness of fit of the model in each group.
3. Does the two-factor model fit in both or either groups?

TABLE 16.4
Covariance Matrices of the "Good" and "Poor" Readers' Groups

Good readers ($n = 75$)

	Variable									
	1	**2**	**3**	**4**	**5**	**6**	**7**	**8**	**9**	**10**
1	6.92									
2	2.75	6.55								
3	2.23	1.86	6.50							
4	1.62	1.55	1.88	5.20						
5	2.45	2.23	1.77	1.14	3.72					
6	−0.28	0.78	1.24	1.31	0.85	4.84				
7	0.63	1.36	1.24	0.99	1.06	2.27	7.02			
8	−0.64	−0.34	0.59	0.38	0.78	1.70	2.41	6.00		
9	1.07	0.20	1.67	1.50	1.34	0.23	1.00	2.55	8.76	
10	0.63	0.97	2.36	1.96	1.09	1.32	2.81	2.38	2.20	5.06

Poor readers ($n = 75$)

	Variable									
	1	**2**	**3**	**4**	**5**	**6**	**7**	**8**	**9**	**10**
1	9.06									
2	6.12	10.05								
3	4.76	4.43	5.71							
4	3.90	4.11	2.42	5.62						
5	5.36	6.10	3.88	3.06	7.95					
6	3.05	2.01	2.12	2.45	1.27	6.97				
7	4.07	3.86	3.28	2.40	3.18	2.53	5.43			
8	4.08	3.28	2.42	1.59	3.52	1.61	3.86	8.70		
9	3.54	2.45	2.96	1.69	3.08	0.82	1.64	3.69	9.55	
10	3.43	4.29	3.13	2.05	2.83	3.06	3.17	4.70	2.97	5.95

4. How would you proceed to test that the two-factor model is identical in each group? Does such a model fit the data?
5. Summarize what you have found out about the covariance structure in the two groups of readers.

* 16.3 Smith and Patterson (1984) describe a study in which samples of people in residential neighborhoods were interviewed regarding such issues as victimization experiences, neighborhood safety and evaluation of police performance. Observations were made on seven variables:

1.–3. How likely do you think you will be a victim of robbery (burglary, vandalism) during the next year?

4. Number of self-reported prior victimizations in the last 12 months,
5. Respondent's age,

6. Respondent's sex (a binary variable coded 1 for women and 0 for men),
7. The rate of personal and property victimization per 100 households in the respondent's neighborhood.

The correlations between the seven variables from observations of 1500 people living alone are given in Table 16.5. (We shall conveniently ignore the objections that might be raised about the binary variable.)

TABLE 16.5
Correlation Matrix between Seven Variables (n = 1500)

	Variable						
	1	**2**	**3**	**4**	**5**	**6**	**7**
1	1.00						
2	0.58	1.00					
3	0.54	0.60	1.00				
4	0.17	0.24	0.25	1.00			
5	−0.01	−0.14	−0.13	−0.18	1.00		
6	−0.02	−0.09	−0.09	−0.15	0.24	1.00	
7	0.22	0.22	0.18	0.17	−0.03	−0.10	1.00

(a) Fit a model where variables one to three act as indicators of a latent variable, perceived risk of victimization which is "caused" by the last four variables; note that here the latent variable appears as a response (dependent) variable rather than an explanatory variable. (The model is known as a *multiple indicator multiple cause model*—MIMIC model for short.)

(b) Construct the path diagram for the fitted model showing parameter estimates.

(c) Assess the fit of the model.

(d) Summarize your findings.

17

Cluster Analysis

17.1 Introduction

An intelligent being cannot treat every object it sees as a unique entity unlike anything else in the universe. It has to put objects in categories so that it may apply its hard-won knowledge about similar objects encountered in the past, to the object at hand.

Steven Pinker, *How the Mind Works*, 1997

One of the most basic abilities of living creatures involves the grouping of similar objects to produce a *classification*. The idea of sorting similar objects into categories is clearly a primitive one because early humans, for example, must have been able to realize that many individual objects shared certain properties such as being edible, or poisonous, or ferocious, and so on. Further, classification in its widest sense is needed for the development of language, which consists of words that help us recognize and discuss the different types of events, objects, and people we encounter. Each noun in a language, for example, is essentially a label used to describe a class of objects that have striking features in common; thus, animals are called cats, dogs, horses, etc., and each name collects individuals into groups. Naming and classifying are essentially synonymous.

As well as being a basic human conceptual activity, classification of the phenomena being studied is an important component of virtually all scientific research. In the behavioral sciences, for example, these "phenomena" may be individuals or societies, or even patterns of behavior or perception. The investigator is usually interested in finding a classification in which the items of interest are sorted into a small number of *homogeneous groups* or *clusters*, the terms being synonymous. Most commonly, the required classification is one in which the groups are *mutually exclusive* (an item belongs to a single group) rather than overlapping (items can be members of more than one group). At the very least, any derived classification scheme should provide a convenient method of organizing a large, complex set of multivariate data with the class labels providing a parsimonious way of describing the patterns of similarities and differences in the data. In market research, for example, it might be useful to group a large number of potential customers according to their needs in a particular product area. Advertising campaigns might then be tailored to the different types of consumers as represented by the different groups.

But often, a classification may seek to serve a more fundamental purpose. In psychiatry, for example, the classification of psychiatric patients with different symptom profiles into clusters might help in the search for the causes of mental illnesses and even perhaps to improved therapeutic methods. These twin aims of prediction (separating diseases that require different treatments) and etiology (searching for the causes of disease) for classifications will be the same in other branches of medicine.

Clearly, a variety of classifications will always be possible for whatever is being classified. Human beings could, for example, be classified with respect to economic status into groups labeled lower class, middle class, and upper class, or they might be classified by the annual consumption of alcohol into low, medium, and high. Clearly, different classifications may not collect the same set of individuals into groups, but some classifications will be more useful than others—a point made clearly by the following extract from Needham (1967), in which he considers the classification of human beings into men and women:

> The usefulness of this classification does not begin and end with all that can, in one sense, be strictly inferred from it, namely, a statement about sexual organs. It is a very useful classification because classifying a person as man or woman conveys a great deal more information about probable relative size, strength, certain types of dexterity, and so on. When we say that persons in class man are more suitable than persons in class woman for certain tasks, and conversely, we are only incidentally making a remark about sex, our primary concern being strength, endurance, etc. The point is that we have been able to use a classification of persons that conveys information on many properties. On the contrary, a classification of persons into those with hairs on their forearms between $^3/_{16}$ and $^1/_4$ in. long and those without, though it may serve some particular use, is certainly of no general use, for imputing membership in the former class to a person conveys information on this property alone. Put another way, there are no known properties that divide up a set of people in a similar way.

In a similar vein, a classification of books based on subject matter into classes such as dictionaries, novels, biographies, and so on is likely to be far more useful than one based on, say, the color of the book's binding. Such examples illustrate that any classification of a set of multivariate data is likely to be judged on its usefulness.

It should be noted here that this chapter is concerned only with the problems of classifying previously unclassified material, and so begins with both the number of groups and their composition as unknowns. The situation when groups are known a priori, and the aims are to assess whether these groups differ on a set of variables or to derive a rule for classifying new individuals on the basis of their scores on these variables, is taken up in Chapter 18.

17.2 Cluster Analysis

Cluster analysis is a generic term for a wide range of numerical methods with the common goal of uncovering or discovering groups or clusters of observations that are homogeneous and separated from other groups. Clustering techniques essentially try to formalize what human observers do so well in two or three dimensions. Consider, for example, the scatterplot shown in Figure 17.1. The conclusion that there are three natural groups or clusters of dots is reached with no conscious effort or thought. Clusters are identified by the assessment of the relative distances between points, and, in this example, the relative homogeneity of each cluster and the degree of separation between the clusters makes the task very simple. The examination of scatterplots based either on the original data or perhaps on the first few principal component scores of the data is often a very helpful initial phase when intending to apply some form of cluster analysis to a set of multivariate data.

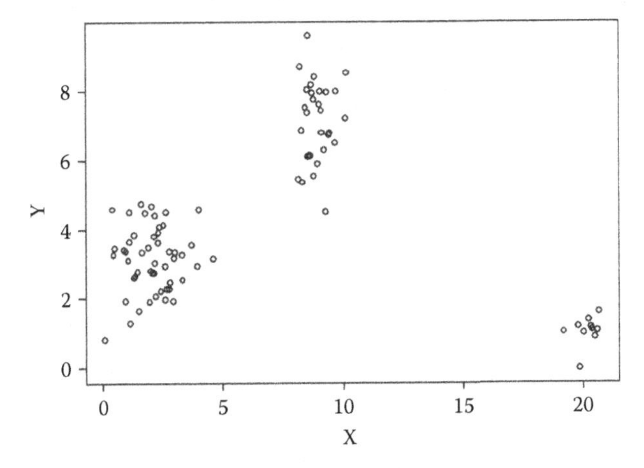

FIGURE 17.1
Bivariate data clearly showing the presence of three clusters.

Cluster analysis techniques are described in detail in Gordon (1987, 1999) and Everitt et al. (2011). In this chapter, we give a relatively brief account of three types of clustering methods: agglomerative hierarchical techniques, k-means clustering, and model-based clustering.

17.3 Agglomerative Hierarchical Clustering

In a *hierarchical classification*, the data are not partitioned into a particular number of classes or groups at a single step. Instead, the classification consists of a series of partitions that may run from a single "cluster" containing all individuals, to n clusters, each containing a single individual. *Agglomerative hierarchical clustering* techniques produce partitions by a series of successive fusions of the n individuals into groups. With such methods, fusions, once made, are irreversible, so that when an agglomerative algorithm has placed two individuals in the same group, they cannot subsequently appear in different groups. Since all agglomerative hierarchical techniques ultimately reduce the data to a single cluster containing all the individuals, the investigator seeking the solution with the "best" fitting number of clusters will need to decide which division to choose. The problem of deciding on the "correct" number of clusters will be taken up later.

An agglomerative hierarchical clustering procedure produces a series of partitions of the data, $P_n, P_{n-1}, \ldots, P_1$. The first, P_n, consists of n single-member clusters, and the last, P_1, consists of a single group containing all n individuals. The basic operation of all methods is similar:

Start: Clusters $C_1, C_2, \ldots, C_n$, each containing a single individual.

Step 1: Find the nearest pair of distinct clusters, say, C_i and C_j, merge C_i and C_j, delete C_j, and decrease the number of clusters by one.

Step 2: If number of clusters equals one, then stop; else return to Step 1.

However, before the process can begin, an interindividual *distance matrix* or *similarity matrix* needs to be calculated (similarly as in multidimensional scaling, see Chapter 14). There are many ways to calculate distances or similarities between pairs of individuals, but here we only deal with a commonly used distance measure, namely, Euclidean distance, defined as follows:

$$d_{ij} = \sqrt{\sum_{k=1}^{q} (x_{ik} - x_{jk})^2}$$

where d_{ij} is the Euclidean distance between individual i with variable values $x_{i1}, x_{i2}, \ldots, x_{iq}$, and individual j with variable values $x_{j1}, x_{j2}, \ldots, x_{jq}$. (Details of other possible distance measures and similarity measures are given in Everitt et al., 2011.) The Euclidean distances between each pair of individuals can be arranged in a matrix that is symmetric because $d_{ij} = d_{ji}$ and has zeros on the main diagonal. Such a matrix is the starting point of many clustering examples, although the calculation of Euclidean distances from the raw data may not be sensible when the variables are on very different scales. In such cases, the variables can be standardized in the usual way before calculating the distance matrix, although this can be unsatisfactory in some cases (see Everitt et al., 2011, for details).

Given an interindividual distance matrix, the hierarchical clustering can begin and, at each stage in the process, the methods fuse individuals or groups of individuals formed earlier who are closest (or most similar). So, as groups are formed, the distance between an individual and a group containing several individuals, and the distance between two groups of individuals will need to be calculated. How such distances are defined leads to a variety of different techniques. Two simple intergroup measures are

$$d_{\text{min}}(A, B) = \min_{i \in A, j \in B} d_{ij} \quad \text{and} \quad d_{\text{max}}(A, B) = \max_{i \in A, j \in B} d_{ij}$$

where $d(A, B)$ is the distance between two clusters A and B, and d_{ij} is the distance between individuals i and j found from the initial interindividual distance matrix.

The intergroup distance measure $d_{\text{min}}(A, B)$ is the basis of *single linkage clustering*, $d_{\text{max}}(A, B)$ that of *complete linkage clustering*. Both these techniques have the desirable property that they are invariant under monotone transformations of the original interindividual distances, that is, they only depend on the ranking on these distances, not their actual values.

A further possibility for measuring intercluster distance or dissimilarity is

$$d_{\text{mean}}(A, B) = \frac{1}{n_A n_B} \sum_{i \in A} \sum_{j \in B} d_{ij}$$

where n_A and n_B are the number of individuals in clusters A and B. The measure $d_{\text{mean}}(A, B)$ is the basis of a commonly used procedure known as *average linkage clustering*. All three intergroup measures described here are illustrated in Figure 17.2.

Hierarchic classifications may be represented by a two-dimensional diagram known as a *dendrogram*, which illustrates the fusions made at each stage of the analysis. An example of such a diagram is given in Figure 17.3. The structure of Figure 17.3 resembles an evolutionary tree, a concept introduced by Darwin under the term "Tree of Life" in his book *On the Origin of Species by Natural Selection* in 1859 (see Figure 17.4), and it is in biological applications that hierarchical classifications are most relevant and most justified (although this type of clustering has also been used in many other areas).

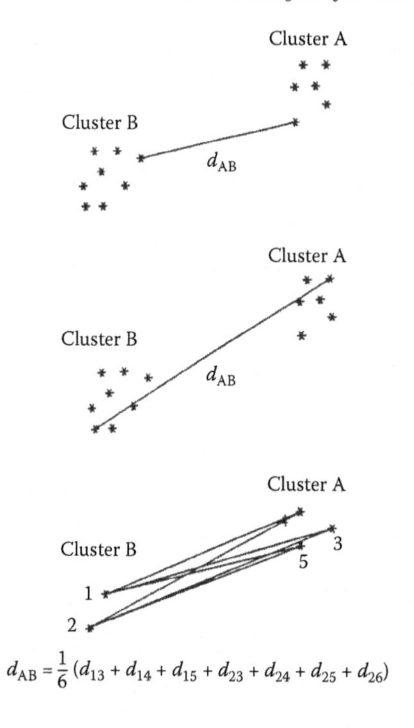

FIGURE 17.2
Intercluster distance measures based on minimum, maximum, and mean of distances between clusters A and B.

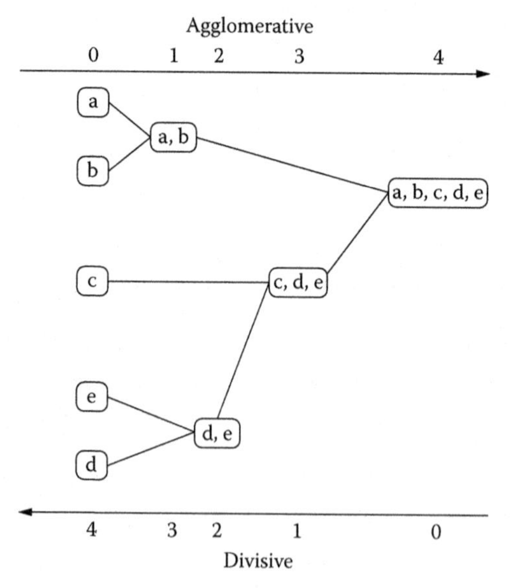

FIGURE 17.3
Example of a dendrogram.

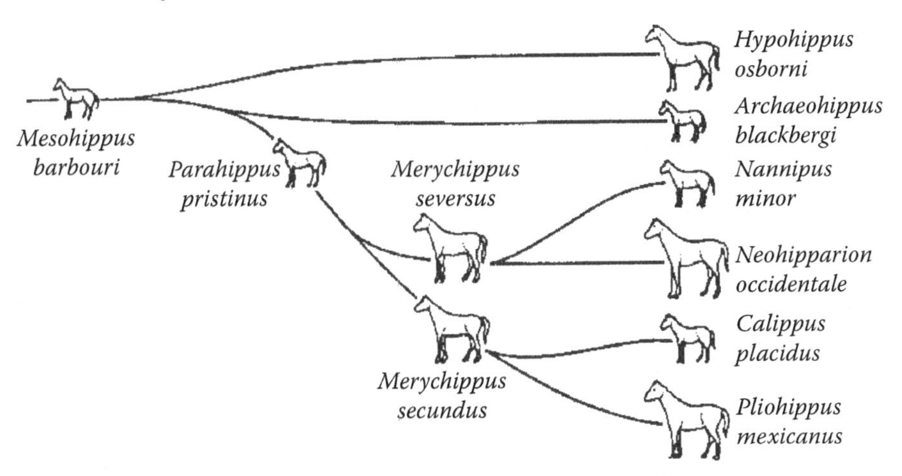

FIGURE 17.4
Evolutionary tree. (From Kaufman, L. and Rousseeuw, P. J. (1990). *Finding Groups in Data: An Introduction to Cluster Analysis*, Wiley, New York. Used with permission of John Wiley & Sons.)

17.3.1 Clustering Individuals Based on Body Measurements

As a first example of the application of the three clustering methods (single linkage, complete linkage, and average linkage), each will be applied to the chest, waist, and hip measurements of 20 individuals given in Chapter 12, Table 12.1.

First Euclidean distances are calculated on the unstandardized measurements; application of each of the three methods to this distance matrix gives the three dendrograms shown in Figure 17.5. How do we select specific partitions of the data from the complete dendrogams? The answer is that we "cut" the dendrogram at some height, and this will give a partition with a particular number of groups. How do we choose where to cut or, in other words, how do we decide on a particular number of groups that is, in some sense, optimal for the data? One informal approach is to examine the sizes of the changes in height in the dendrogram and take a "large" change to indicate the appropriate number of clusters for the data. (More formal approaches are described in Everitt et al., 2011.)

Even using this informal approach on the dendrograms in Figure 17.5, it is not easy to decide where to "cut." So, instead, because we know that these data consist of measurements on 10 men and 10 women, we will look at the two-group solutions from each method that are obtained by cutting the dendrograms at suitable heights. We can display and compare the three solutions graphically by plotting the first two principal component scores of the data, labeling the points to identify the cluster solution of one of the methods. Such plots are also shown in Figure 17.5. The plot associated with the single linkage

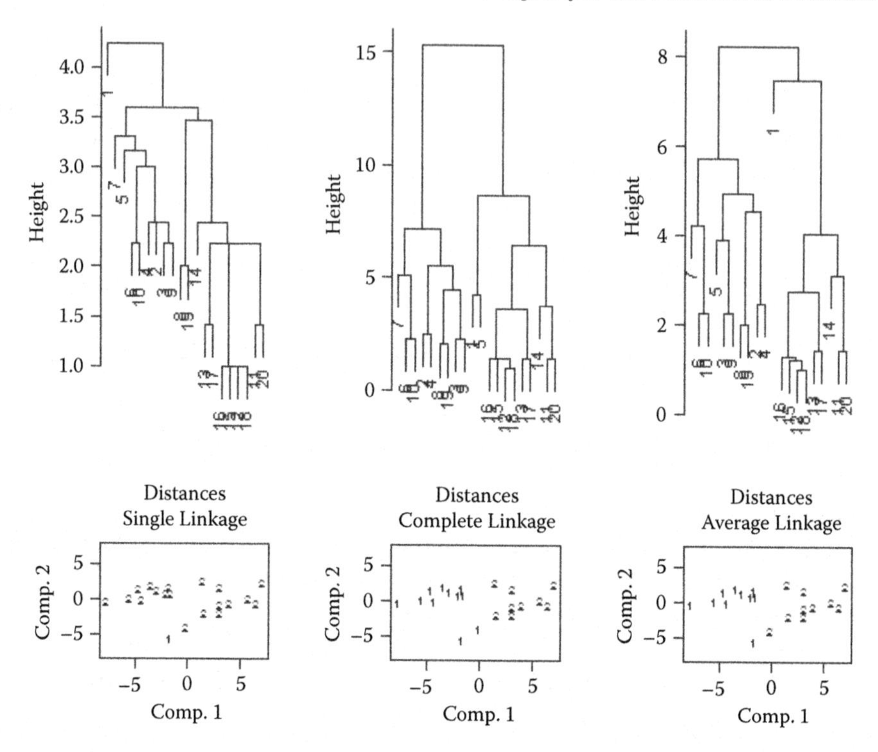

FIGURE 17.5
Dendrograms for single linkage, complete linkage, and average linkage applied to body measurements data.

solution immediately demonstrates one of the problems with using this method in practice, and that is a phenomenon known as *chaining*, which refers to the tendency to incorporate intermediate points between clusters into an existing cluster rather than initiating a new one. As a result, single linkage solutions often contain long "straggly" clusters that do not give a useful description of the data. The two-group solutions from complete linkage and average linkage, also shown in Figure 17.5, are similar and, in essence, place the men (observations 1 to 10) together in one cluster and women (observations 11 to 20) in the other.

17.3.2 Clustering Countries on the Basis of Life Expectancy

The yearbook of the United Nations (2016) is full of population statistics related to all countries and special regions of the world. To be more specific about the countries, we refer to *sovereign states.* Updating (or even compiling) a list of such entities around the world is far from trivial, as some areas might have a *de facto* status of a sovereign state, although they might not be official

TABLE 17.1
Life Expectancies at Different Ages for Men in Seven Countries

	Birth	Aged 25	Aged 50	Aged 75	Aged 100
Japan	80.8	56.3	32.4	12.0	2.2
Italy	80.3	55.9	31.9	11.6	1.8
Spain	80.0	55.5	31.5	11.7	3.4
United Kingdom	79.0	54.7	31.1	11.2	2.2
Finland	78.5	54.1	30.5	11.1	1.7
Cuba	76.5	52.5	29.2	11.0	2.0
United States	76.4	52.6	29.8	11.2	2.1

states etc. Ignoring however such complications, the data to be used in this subsection have been extracted from the UN databases listing 46 variable values related to life expectancy by age and gender in six regions: Africa, America (North), America (South), Asia, Europe, and Oceania. From the 134 sovereign states in these data here we shall concentrate on the analysis of a subset relating to men in 50 countries where life expectancy data are available for ages from birth to 100 years, the latter being an ever increasing possibility for many "lucky" people! (Repeating the analysis described here for the corresponding data for women is left as an exercise for readers—see Exercise 17.6.)

Table 17.1 shows life expectancies for seven countries that are precisely those whose birth and death rates were shown based on their Euclidean distance matrix in Chapter 14. Here, we shall have a somewhat broader view, although we focus on the life expectancies, and select just five crucial ages (0, 25, 50, 75, 100) to represent the potential lifespan of a human-being. We quickly check the variances of the life expectancy of men at those ages in the 50 countries:

Birth variance = 21.6
Aged 25 variance = 18.5
Aged 50 variance = 11.5
Aged 75 variance = 2.11
Aged 100 variance = 0.41

As might have been predicted, the variances are quite different, so calculating the initial intercountry Euclidean distance matrix on the life expectancies as they are in Table 17.1 would not necessarily appear to be very sensible, and thus we could standardize each of the five life expectancies to have the variance of one and then calculate the required Euclidean distances on the standardized data. However, we shall first work with the raw data.

Here, we shall apply only complete linkage to the data, and the resulting dendrogram is shown in Figure 17.6. Again, there is no completely "best" place to cut the dendrogram, but the four-group cluster solution produced by cutting at a height of 10 is shown in Table 17.2. The countries grouped together in each cluster are perhaps different from what might have been expected from intuition. The 17 countries in group 1 have the highest life expectancies at each age, while the 16 countries in group 2 have the second highest figures. Group 3 (10 countries) and group 4 (7 countries) follow this trend with their patterns for mean profiles. The only exception is that the life expentancy at age 100 is somewhat higher in group 4 than in group 3.

We can illustrate the cluster solution graphically in a number of ways. Here, we shall first plot the four-group cluster solution on the scatterplot matrix of the five life expectancies to give Figure 17.7. Most of the plots in this scatterplot matrix are not very interesting, as they merely show the

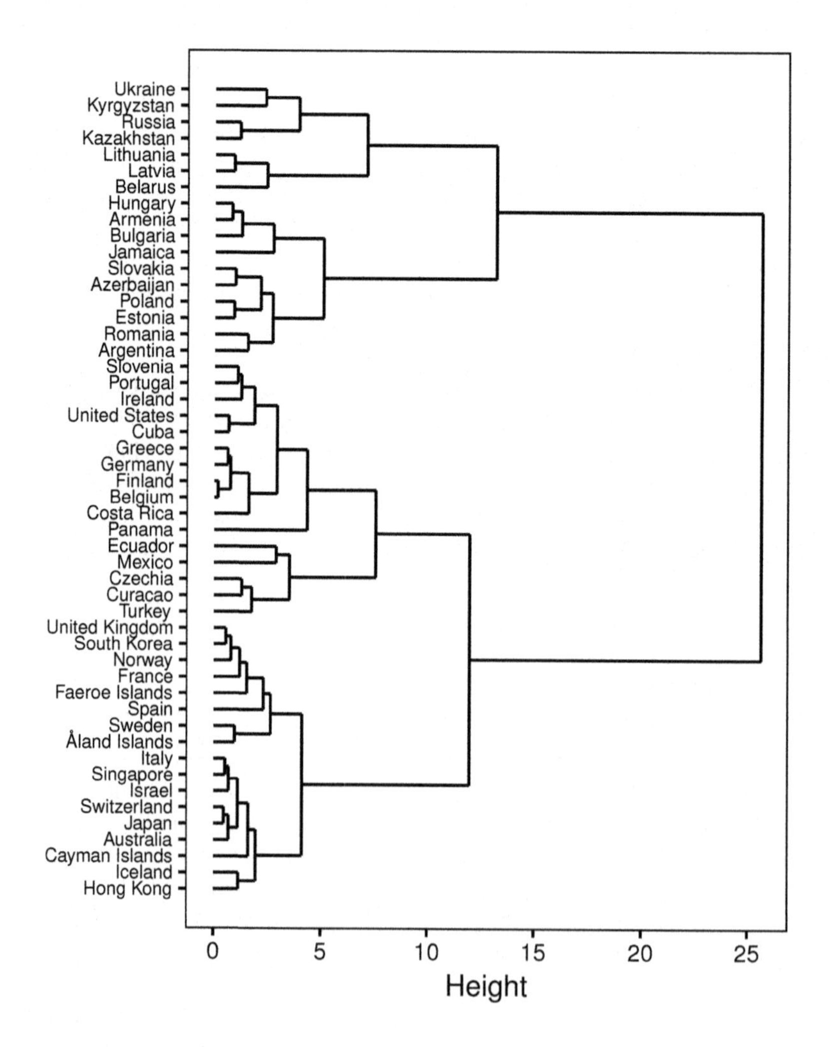

FIGURE 17.6
Complete-linkage dendrogram for life expectancy data.

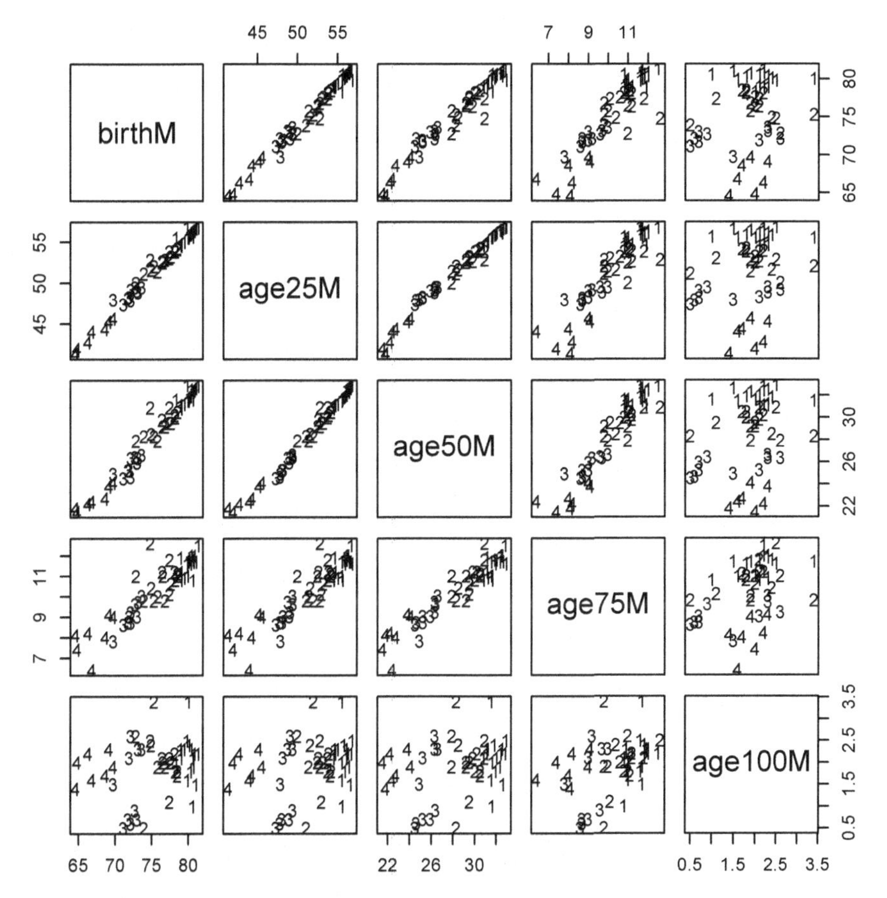

FIGURE 17.7
Scatterplot matrix of life expectancy data showing the four-cluster solution from complete linkage.

order of the groups based on their mean profiles. The most interesting lots in this figure are those with age 100 as the other axis, as they all display the spread of the data. The groups seem to have more variance related to the life expectancies of the countries at age 100.

A further graphic of the cluster solution is shown in Figure 17.8, where the data are plotted in the space of the first two principal components and labeled by cluster membership and the country in question. Now, the data have been standardized, as the principal components have been calculated on the basis of correlations. The first component is essentially the average of the five life expectancies, reflecting the same order and trend that was visible from the mean profiles given in Table 17.2. The second component is mostly displaying the spread of the life expectancy at 100 years of age across the countries in all groups. The longest life expectancies at 100 are in Spain and Turkey while

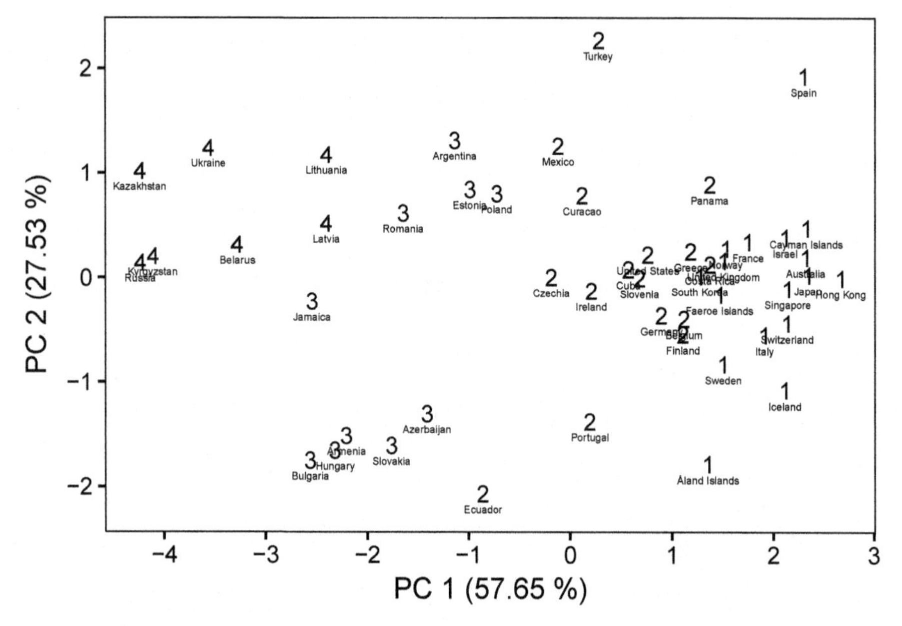

FIGURE 17.8
Four-cluster solution from complete linkage applied to life expectancy data plotted in the space of the first two principal components.

the shortest is in Ecuador. The first two components account for 85% of the variance of the five life expectancies. Figure 17.8 demonstrates that the four clusters differ largely on their average life expectancy over the five ages.

(As an aside, there are many questions of interest that relate to population dynamics, for example, "Why is the world's population increasing?", or "How many young women go to school?", or "How many of us live in poverty?" Answering such questions is not simple and Rosling et al. (2018) in their quite brilliant book, *Factfulness*, demonstrate that a chimpanzee choosing answers at random will consistently outguess "informed" answers by journalists, Nobel laureates and investment bankers—no surprise about the latter!)

17.4 k-Means Clustering

The k-means clustering technique seeks to partition the n individuals in a set of multivariate data into k groups or clusters, $(G_1, G_2, \ldots, G_k)$, where G_i denotes the set of n_i individuals in the ith group, and k is given (or a possible range is specified by the researcher; the problem of choosing the "true" value of k will be taken up later) by minimizing some numerical criterion, low values of which are considered indicative of a "good" solution. The most commonly used

TABLE 17.2
Four Group Solution Produced by Complete Linkage Clustering Applied to the Life Expectancy Data

Countries and Mean Life Expectancies of Males in Each Group

Group 1

Australia, Åland Islands, Cayman Islands, Faeroe Islands, France, Hong Kong, Iceland, Israel, Italy, Japan, Norway, Singapore, South Korea, Spain, Sweden, Switzerland, United Kingdom

Mean

birthM	age25M	age50M	age75M	age100M
80.0	55.7	31.9	11.5	2.1

Group 2

Belgium, Costa Rica, Cuba, Curaçao, Czechia, Ecuador, Finland, Germany, Greece, Ireland, Mexico, Panama, Portugal, Slovenia, Turkey, United States

Mean

birthM	age25M	age50M	age75M	age100M
76.4	52.7	29.4	10.8	2.0

Group 3

Argentina, Armenia, Azerbaijan, Bulgaria, Estonia, Hungary, Jamaica, Poland, Romania, Slovakia

Mean

birthM	age25M	age50M	age75M	age100M
72.1	48.5	25.6	9.0	1.4

Group 4

Belarus, Kazakhstan, Kyrgyzstan, Latvia, Lithuania, Russia, Ukraine

Mean

birthM	age25M	age50M	age75M	age100M
67.1	43.6	22.6	8.0	1.9

implementation of k-means clustering is one that tries to find the partition of the n individuals into k groups that minimizes the within-group sum of squares (WGSS) over all variables; explicitly this criterion is

$$\text{WGSS} = \sum_{j=1}^{q} \sum_{l=1}^{k} \sum_{i \in G_l} \left(x_{ij} - \bar{x}_j^{(l)} \right)^2$$

where

$$\bar{x}_j^{(l)} = \frac{1}{n_l} \sum_{i \in G_l} x_{ij}$$

is the mean of the individuals in group G_l on variable j.

The problem then appears relatively simple; consider every possible partition of the n individuals into k groups, and select the one with the lowest within-group sum of squares. Unfortunately, the problem in practice is not so straightforward. The numbers involved are so vast that complete enumeration of every possible partition remains impossible even with the fastest computer. The scale of the problem is illustrated in the following table:

n	k	Number of Possible Partitions
15	3	2,375,101
20	4	45,232,115,901
25	8	690,223,721,118,368,580
100	5	10^{68}

The impracticability of examining every possible partition has led to the development of algorithms designed to search for the minimum values of the clustering criterion by rearranging existing partitions and keeping the new one only if it provides an improvement. Such algorithms do not, of course, guarantee finding the global minimum of the criterion. The essential steps in these algorithms are as follows:

1. Find some initial partition of the individuals into the required number of groups. Such an initial partition could be provided by a solution from one of the hierarchical clustering techniques described in the previous section.
2. Calculate the change in the clustering criterion produced by "moving" each individual from its own to another cluster.
3. Make the change that leads to the greatest improvement in the value of the clustering criterion.
4. Repeat steps 2 and 3 until no move of an individual causes the clustering criterion to improve.

For a more detailed account of the k-means algorithm, see Steinley (2008).

The k-means approach to clustering using the minimization of the WGSS over all the variables is widely used, but it suffers from two problems:

1. k-means is not scale invariant, that is, different solutions may result from clustering the raw data and the data standardized in some way.
2. k-means imposes a spherical structure on the data, that is, k-means will find clusters shaped like hyper-footballs even if the "true" clusters in the data are of some other shape (see Everitt et al., 2011, for some examples of this phenomenon).

Nevertheless, the k-means method remains very popular. With k-means clustering, the investigator can choose to partition the data into a specified number of groups. In practice, solutions for a range of values for number of groups are found and, in some way, the optimal or "true" number of groups for the data must be chosen. Several suggestions have been made as to how to answer the number of groups question, but none is completely satisfactory. The method we shall use in the forthcoming example is to plot the WGSS associated with the k-means solution for each number of groups. As the number of groups increases, the sum of squares will necessarily decrease, but an obvious "elbow" in the plot may be indicative of the most useful solution for the investigator to look at in detail. (Compare the scree plot described in Chapter 13.)

17.4.1 Clustering Crime Rates

We shall illustrate the application of k-means clustering using the crime rate data introduced in Chapter 13 after removing the outlier, DC, identified in Chapter 13. If we first calculate the variances of the crime rates for the different types of crime we find the following:

	Murder	**Rape**	**Robbery**	**Assault**	**Burglary**	**Theft**
Variance	11.93	209.76	1889.53	19373.54	175895.00	565276.59

The variances are very different, and using k-means on the raw data would not be sensible; we must standardize the data in some way. Here, we standardize each variable by its range. After the standardization, the variances become:

	Murder	**Rape**	**Robbery**	**Assault**	**Burglary**	**Theft**	**Vehicle**
Variance	0.076	0.056	0.046	0.059	0.052	0.062	0.068

The variances of the standardized data are very similar, and we can now progress with clustering the data. First, we plot the WGSS for one- to six-group solutions to see if we can get any indication of number of groups. The plot is shown in Figure 17.9. The only "elbow" in the plot occurs for two groups, and so we will now look at the two-group solution. In Table 17.3, the group membership and means are given. Everything is worse in group 1! A plot of the two-group solution in the space of the first two principal components of the correlation matrix of the data is shown in Figure 17.10. The two groups are created essentially on the basis of the first principal component score, which is, as we have seen in Chapter 13, a weighted average of the crime rates. Perhaps all that cluster analysis is doing here is dividing into two parts a homogenous set of data? This is always a possibility as discussed in some detail in Everitt et al. (2011).

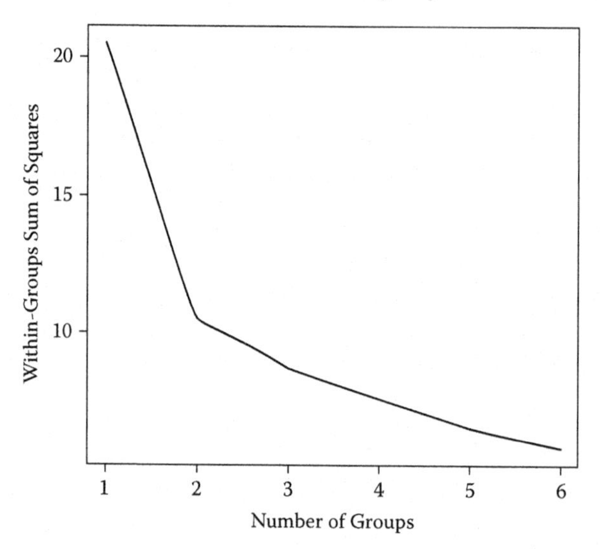

FIGURE 17.9
Plot of within-groups sum of squares against number of clusters.

17.5 Model-Based Clustering

The agglomerative hierarchical and k-means clustering methods described in the two previous sections are based largely on heuristic but intuitively reasonable procedures. However, they are not based on formal models for cluster

TABLE 17.3
Group Membership and Means for the Two-Group Solution from k-Means Applied to Crime Rate Data

Group 1						
MA, NY, NJ, IL, MI, MO, MD, NC, SC, GA, KY, AR, LA, OK, WY, CO, NM, UT, NV, WA, OR, CA						
Mean Crime Rates						
Murder	**Rape**	**Robbery**	**Assault**	**Burglary**	**Theft**	**Vehicle**
9.37	45.37	229.00	394.77	1543.41	3368.05	554.27
Group 2						
ME, NH, VT, RI, CT, PA, OH, IN, WI, MN, IA, ND, SD, NE, KS, DE, DC, VA, WV, FL, TN, AL, MS, TX, MT, ID, AZ, AK, HI						
Mean Crime Rates						
Murder	**Rape**	**Robbery**	**Assault**	**Burglary**	**Theft**	**Vehicle**
4.74	24.80	73.82	182.07	924.21	2564.71	247.04

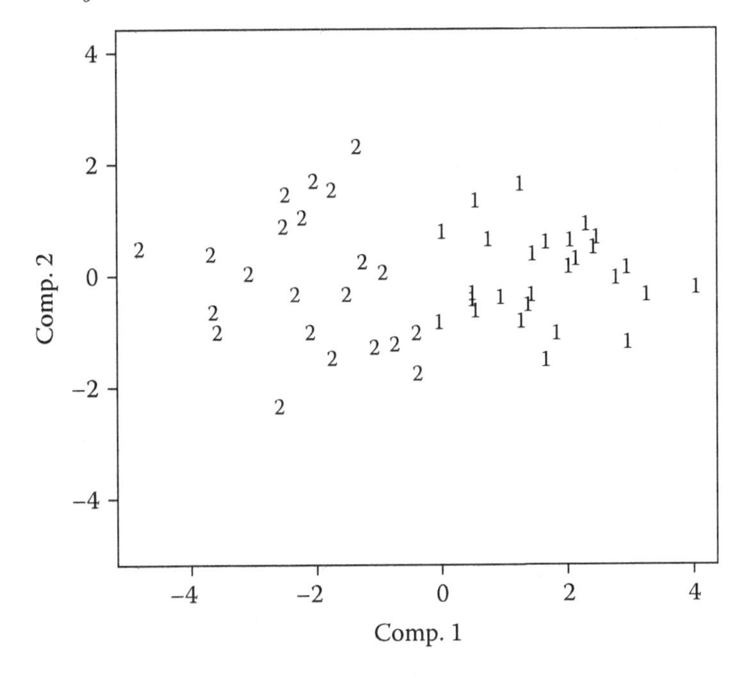

FIGURE 17.10
Plot of k-means two-group solution for standardized crime rate data.

structure in the data, making problems such as deciding between methods, estimating the number of clusters, etc., particularly difficult, and, of course, without a reasonable model, formal inference is precluded. In practice, these may not be insurmountable objections to the use of either the agglomerative methods or k-means clustering because cluster analysis is most often used as an "exploratory" tool for data analysis.

However, if an acceptable model for cluster structure could be found, then cluster analysis based on the model might give more persuasive solutions (more persuasive to statisticians at least). A variety of possibilities have been proposed, but perhaps the most successful approach is that proposed by Scott and Symons (1971) and extended by Banfield and Raftery (1993) and Fraley and Raftery (2002). The technique, which is known as the *classification maximum likelihood* procedure, is described briefly in Technical Section 17.1.

Technical Section 17.1: Classification Maximum Likelihood Clustering

Assume that the population consists of c subpopulations, each corresponding to a cluster of observations, and that the probability density function of a q-dimensional observation $\mathbf{x}' = (x_1, x_2, \ldots, x_q)$ from the jth subpopulation is $f_j(\mathbf{x}, \boldsymbol{\theta}_j)$ for some unknown vector of parameters,

$\boldsymbol{\theta}_j$. Also, assume that $\boldsymbol{\gamma}' = [\gamma_1, \ldots, \gamma_n]$ gives the labels of the subpopulation to which each observation belongs, so $\gamma_i = j$ if $\mathbf{x}_i$ is from the jth subpopulation.

The clustering problem becomes that of choosing $\boldsymbol{\theta}' = [\boldsymbol{\theta}_1, \boldsymbol{\theta}_2, \ldots, \boldsymbol{\theta}_c]$ and $\boldsymbol{\gamma}$ to maximize the likelihood

$$L(\boldsymbol{\theta}, \boldsymbol{\gamma}) = \prod_{i=1}^{n} f_{\gamma_i}\left(\mathbf{x}_i, \boldsymbol{\theta}_{\gamma_i}\right)$$

If $f_j(\mathbf{x}, \boldsymbol{\theta}_j)$ is assumed to be a multivariate normal density with mean vector $\boldsymbol{\mu}_j$ and covariance matrix $\boldsymbol{\Sigma}_j$, this likelihood has the form

$$L(\boldsymbol{\theta}, \boldsymbol{\gamma}) = \prod_{j=1}^{c} \prod_{i:\gamma_i = j} |\boldsymbol{\Sigma}_j|^{-1/2} \exp\left\{-\frac{1}{2}\left(\mathbf{x}_i - \boldsymbol{\mu}_j\right)' \boldsymbol{\Sigma}_j^{-1}\left(\mathbf{x}_i - \boldsymbol{\mu}_j\right)\right\}$$

The maximum likelihood estimator of $\boldsymbol{\mu}_j$ is $\hat{\mathbf{x}}_j = n_j^{-1} \sum_{i:\gamma_i = j} \mathbf{x}_i$. Replacing $\boldsymbol{\mu}_j$ in the likelihood with its maximum likelihood estimator yields the following log- likelihood:

$$\log L(\boldsymbol{\theta}, \boldsymbol{\gamma}) = -\frac{1}{2} \sum_{j=1}^{c} \operatorname{trace}\left(\mathbf{W}_j \boldsymbol{\Sigma}_j^{-1}\right) + n \log |\boldsymbol{\Sigma}_j|$$

where $\mathbf{W}_j$ is the $q \times q$ matrix of sums of squares and cross products of the variables for subpopulation j. Banfield and Raftery (1993) demonstrate the following:

- If the covariance matrix $\boldsymbol{\Sigma}_j$ is σ^2 times the identity matrix for all subpopulations $j = 1, 2, \ldots, c$, that is, the clusters are spherical, then the likelihood is maximized by choosing $\boldsymbol{\gamma}$ to minimize trace($\mathbf{W}$), where $\mathbf{W} = \sum_{j=1}^{c} \mathbf{W}_j$, that is, minimization of the group sum of squares, essentially equivalent to k-means clustering. Use of this criterion in a cluster analysis will tend to produce spherical clusters of largely equal sizes, which may or may not match the "real" clusters in the population. If, of course, the population clusters are not of this type, then a cluster structure may be imposed on the data and the real clusters in the data not found.

- If $\mathbf{S}_j = \mathbf{S}$ for $j = 1, 2, \ldots, c$, that is, if all the clusters in the population have the same "shape" but not necessarily spherical, then the likelihood is maximized by choosing $\boldsymbol{\gamma}$ to minimize $|\mathbf{W}|$ (the determinant of $\mathbf{W}$), a clustering criterion discussed by Friedman and Rubin (1967) and Marriott (1982). The use of this criterion in a cluster analysis will tend to produce clusters with the same elliptical shape, again matching those in the population. Once again, if the population clusters are not of this type, then a cluster structure may be imposed on the data and the real clusters in the data not found.

- If the clusters in the population have different covariance matrices, the likelihood is maximized by choosing γ to minimize $\sum_{j=1}^{c} n_j \log |\mathbf{W}_j|/n_j$, a criterion that allows for different shaped clusters.

Banfield and Raftery (1993) also consider criteria that allow the shape of clusters to be less constrained than with the minimization of trace($\mathbf{W}$) and $|\mathbf{W}|$ criteria, but that are more parsimonious than the model in which all the population clusters are allowed to have different shapes. For example, constraining clusters to be spherical but not to have the same volume, or constraining clusters to have diagonal covariance matrices but allowing their shapes, sizes, and orientations to vary.

The EM algorithm (see Dempster et al., 1977) is used for maximum likelihood estimation—details are given in Fraley and Raftery (2002). Model selection is a combination of choosing both the appropriate clustering model for the population from which the n observations have been taken, that is, are all clusters spherical, all elliptical, all different shapes, or somewhere in between, and the optimal number of clusters. A Bayesian approach is adopted (see Fraley and Raftery, 1998, 1999, 2002), using what is known as the Bayesian information criterion (BIC). The result is a cluster solution that "fits" the observed data as well as possible, and this can include a solution that has only one "cluster," implying that cluster analysis is not really a useful technique for the data.

17.5.1 Clustering European Countries

To illustrate the use of the classification likelihood method, we will apply it to the data shown in Table 17.4. These data arise from a study of what gastroenterologists in Europe tell their cancer patients (Thomsen et al., 1993). A questionnaire was sent to about 600 gastroenterologists in 27 European countries (the study took place before the changes in the political map of the continent) asking what they would tell a patient with newly diagnosed cancer of the colon, and his or her spouse, about the diagnosis. The respondent gastroenterologists were asked to read a brief case history and then to answer six questions with a yes or no answer. The questions were as follows:

Q1: Would you tell this patient that he or she has cancer if he or she asks no questions?

Q2: Would you tell the wife or husband that the patient has cancer? (In the patient's absence.)

Q3: Would you tell the patient that he or she has a cancer if he or she directly asks you to disclose the diagnosis?

(During surgery, the surgeon notices several small metastases in the liver.)

Q4: Would you tell the patient about the metastases (supposing the patient asks to be told the results of the operation)?

Q5: Would you tell the patient that the condition is incurable?

Q6: Would you tell the wife or husband that the operation revealed metastases?

The data in Table 17.4 give the proportion of respondents that answered each question "yes." Applying the classification likelihood clustering approach to these data (see Fraley and Raftery, 1999), we can first examine the resulting plot of BIC values shown in Figure 17.11. In this diagram, the numbers refer to different model assumptions about the shape of clusters:

1. Spherical, equal volume
2. Spherical, unequal volume

TABLE 17.4
Proportion of Respondents Answering Yes to Each of the Questions in the Survey of Gastroenterologists

	Q1	Q2	Q3	Q4	Q5	Q6
Iceland	1.000	1.000	1.000	1.000	1.000	1.000
Norway	0.857	0.833	1.000	1.000	1.000	0.800
Sweden	1.000	0.636	1.000	1.000	0.500	0.667
Finland	1.000	0.667	1.000	1.000	0.833	0.667
Denmark	0.923	0.692	1.000	0.750	0.364	0.538
U.K.	0.633	0.889	1.000	0.950	0.526	1.000
Ireland	1.000	0.667	1.000	0.000	0.000	1.000
Germany	1.000	1.000	1.000	0.857	0.154	0.929
Netherlands	1.000	1.000	1.000	0.875	0.714	0.875
Belgium	0.000	1.000	1.000	0.500	0.000	1.000
Switzerland	1.000	1.000	1.000	0.500	0.000	1.000
France	0.300	0.875	0.625	0.200	0.000	0.875
Spain	0.083	1.000	0.800	0.545	0.000	1.000
Portugal	0.167	1.000	0.667	0.500	0.000	1.000
Italy	0.467	1.000	0.929	0.400	0.133	1.000
Greece	0.125	1.000	0.625	0.125	0.000	1.000
Yugoslavia	0.267	1.000	0.533	0.267	0.000	1.000
Albania	0.400	0.600	0.400	0.400	0.600	0.600
Bulgaria	0.000	1.000	0.333	0.000	0.000	1.000
Romania	0.000	1.000	0.143	0.143	0.143	1.000
Hungary	0.200	1.000	0.800	0.000	0.000	1.000
Czechoslovakia	0.061	0.971	0.088	0.000	0.000	0.571
Poland	0.000	1.000	0.263	0.105	0.000	0.947
Russia	0.000	0.857	0.286	0.000	0.000	0.857
Lithuania	0.000	1.000	0.000	0.000	0.000	1.000
Latvia	0.000	1.000	0.000	0.000	0.000	1.000
Estonia	0.667	1.000	1.000	0.000	0.000	1.000

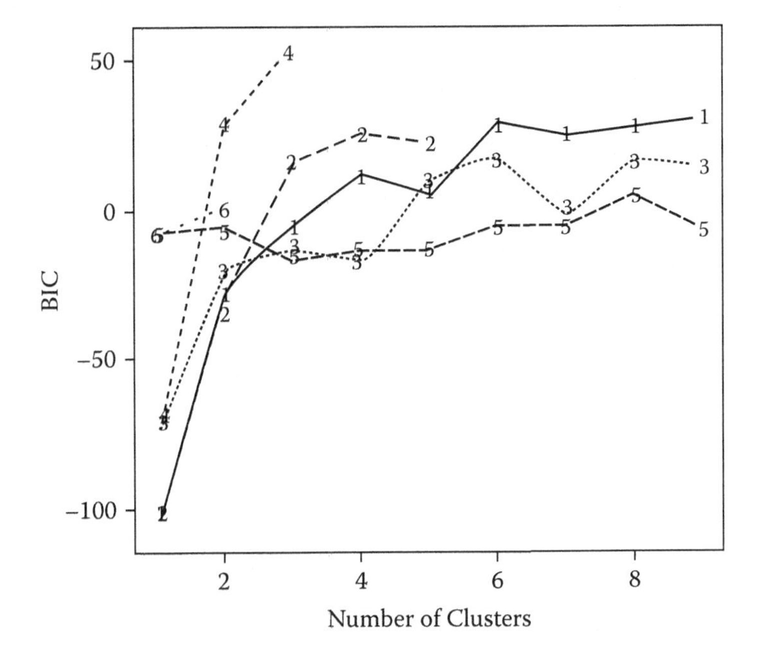

FIGURE 17.11
Plot of BIC values for a variety of cluster structures assumed by the classification likelihood clustering approach and a range of number of clusters.

3. Diagonal, equal volume and shape
4. Diagonal, varying volume and shape
5. Ellipsoidal, equal volume, shape, and orientation
6. Ellipsoidal, varying volume, shape, and orientation

The BIC criterion selects model 4 and three clusters as the optimal solution. Details of this solution are given in Table 17.5. The first cluster consists of countries in which the large majority of respondents gave yes answers to questions 1, 2, 3, 4, and 6, and about half also gave a yes answer to question 5. This cluster includes all the Scandinavian countries, the United Kingdom, Ireland, Germany, the Netherlands, and Albania. Ireland and Albania do not perhaps seem to be "natural" members of this group, but in both countries the number of respondents was small.

In the second cluster, the majority of respondents answered "no" to questions 1, 4, 5 and "yes" to questions 2, 3, and 6; in these countries, it appears that the clinicians do not mind giving bad news to the spouses of patients but not to the patients themselves unless they are directly asked by the patient about his or her condition. This cluster contains Catholic countries such as Spain, Portugal, and Italy.

TABLE 17.5
Results of the Classification-Likelihood Clustering Approach Applied to Cancer Questionnaire Data

	Cluster Means					
	Q1	**Q2**	**Q3**	**Q4**	**Q5**	**Q6**
C1	0.881	0.799	0.940	0.782	0.568	0.808
C2	0.327	0.988	0.798	0.304	0.013	0.988
C3	0.009	0.975	0.159	0.036	0.020	0.911

Cluster Membership

Cluster 1
Iceland, Norway, Sweden, Finland, Denmark, U.K., Ireland, Germany, Netherlands, Albania
Cluster 2
Belgium, Switzerland, France, Spain, Portugal, Italy, Greece, Yugoslavia, Hungary, Estonia
Cluster 3
Bulgaria, Romania, Czechoslovakia, Poland, Russia, Lithuania, Latvia

In cluster three, the large majority of respondents answered "no" to questions 1, 3, 4, and 5, and again, a large majority answered "yes" to questions 2 and 6. In these countries very few clinicians appear to be willing to give the patient bad news even if asked directly by the patient about his or her condition.

17.6 Summary

- Cluster analysis techniques are used to search for clusters/groups in a priori unclassified multivariate data.
- Although clustering techniques are potentially very useful for the exploration of multivariate data, they require care in their application if misleading solutions are to be avoided.
- Many methods of cluster analysis have been developed, and most studies have shown that no one method is best for all types of data. However, the more statistical techniques covered briefly in Section 17.5 and in more detail in Everitt et al. (2011) have definite statistical advantages because the clustering is based on sensible models for the data.
- Cluster analysis is a large area and has been covered only briefly in this chapter. The many problems that need to be considered when using

clustering in practice have barely been touched upon. For a detailed discussion of these problems, see Everitt et al. (2011).

17.7 Exercises

17.1 Reanalyze the data on life expectancies standardizing the variables. How do the results compare with those given in the text for the unstandardized data?

17.2 Apply k-means to the crime rate data after standardizing each variable by its standard deviation. Compare the results with those given in the text found by standardizing by a variable's range.

17.3 The data give the lowest temperatures in degrees Fahrenheit recorded in various months for cities in the United States. Plot the data in any way that you think might be helpful and explore whether there may be clusters of cities using some method of cluster analysis. Display graphically whatever cluster solutions you produce.

17.4 Apply the model-based clustering approach to the data on life expectancies and compare the results with those from the k-means clustering given in the text. Do the same for the crime rate data.

17.5 The data give the protein consumption in 25 European countries for 9 food groups. Is there any evidence that the countries cluster in some way?

17.6 Repeat the analysis of the life expectancy data for the corresponding data for women.

18

Grouped Multivariate Data

18.1 Introduction

The importance of classification in science in general and behavioral science in particular has already been remarked upon in Chapter 17, in which techniques were described for examining multivariate data to discover whether the data consisted of a number of relatively distinct groups or clusters of observations. In this chapter, a further aspect of classification will be discussed, namely that when the groups are known a priori. Such data arises when investigators collect samples of multivariate observations from several different populations, for example, observations on a number of symptoms for patients from different diagnostic categories.

A variety of questions might be asked about grouped multivariate data, and so, there are a variety of (overlapping) approaches to their analysis. In some cases, the investigator will simply be interested in testing whether the groups differ on the variables that have been recorded. When there are two groups, the multivariate analog of Student's t-test, Hotelling's T^2 test, can be used, and when there are more than two groups, *multivariate analysis of variance* (MANOVA) is available. Both methods will be described later in the chapter.

A further question that is often of interest for grouped multivariate data is whether or not it is possible to use the measurements made to construct a classification rule derived from the original observations (the training set) that will allow new individuals having the same set of measurements (the test sample), but no group label, to be allocated to a group in such a way that misclassifications are minimized. The relevant technique is now some form of *discriminant function analysis*, which is the subject of Section 18.2.2. A question that might be posed about constructing such an allocation rule is, "if group labels can be allocated a priori in some definitive fashion, why would we want to use the recorded variables for classification?" The answer might simply be "convenience" if definitive group labeling is costly or lengthy, or it might be "necessity," for example, in medicine if definitive group labeling can only be made by postmortem examination.

We begin by looking at the simplest case of grouped multivariate data, namely, when there are only two groups.

18.2 Two-Group Multivariate Data

18.2.1 Hotelling's T^2 Test

Willerman et al. (1991) collected data on 20 male and 20 female right-handed Anglo psychology students at a large university in the United States. The subjects took three subtests of the Wechsler Adult Intelligence Scale-Revised test. The scores recorded were full-scale IQ (FSIQ), verbal IQ (VIQ), and performance IQ (PIQ). The data for the first five men and the first five women are given in Table 18.1.

TABLE 18.1
Wechsler Adult Intelligence IQ Scores for Five Men and Five Women

Subject	FSIQ	VIQ	PIQ
1	140	150	124
2	139	123	150
3	133	129	128
4	89	93	84
5	133	114	147
6	133	132	124
7	137	132	134
8	99	90	110
9	138	136	131
10	92	90	98

Note: FSIQ = Full-scale IQ; VIQ = verbal IQ; PIQ = performance IQ.

In this case interest lies in testing the hypothesis that the three-dimensional mean vectors of IQ scores are the same for men and women. The appropriate test is Hotelling's T^2 test, the multivariate analog of the independent samples t-test. The test and the assumptions on which it is based are described in Technical Section 18.1.

Technical Section 18.1: Hotelling's T^2

If there are q variables, the null hypothesis is that the means of the variables in the first population equal the means of the variables in the second population.

If $\boldsymbol{\mu}_1$ and $\boldsymbol{\mu}_2$ are the mean vectors of the two populations, the null hypothesis can be written as

$$H_0 : \boldsymbol{\mu}_1 = \boldsymbol{\mu}_2$$

The test statistic T^2 is defined as

$$T^2 = \frac{n_1 n_2}{n_1 + n_2} D^2$$

where n_1 and n_2 are the sample sizes in each group, and D^2 is the generalized distance introduced in Chapter 12, namely,

$$D^2 = (\bar{\mathbf{x}}_1 - \bar{\mathbf{x}}_2)' \, \mathbf{S}^{-1} \, (\bar{\mathbf{x}}_1 - \bar{\mathbf{x}}_2)$$

where $\bar{\mathbf{x}}_1$ and $\bar{\mathbf{x}}_2$ are the two sample mean vectors, and $\mathbf{S}$ is the estimate of the assumed common covariance matrix of the two populations, calculated from the two sample covariance matrices $\mathbf{S}_1$ and $\mathbf{S}_2$ as

$$\mathbf{S} = \frac{(n_1 - 1)\mathbf{S}_1 + (n_2 - 1)\mathbf{S}_2}{n_1 + n_2 - 2}$$

Note that the form of the test statistic in the multivariate case is very similar to that for the univariate independent samples t-test, involving a difference between "means" (here, mean vectors), and an assumed common "variance" (here, a covariance matrix). Under H_0 (and when the assumptions given below hold), the statistic F given by

$$F = \frac{(n_1 + n_2 - q - 1)T^2}{(n_1 + n_2 - 2)q}$$

has a Fisher's F-distribution with q and $n_1 + n_2 - q - 1$ degrees of freedom.

The T^2 test is based on the following assumptions:

- In each population, the variables have a multivariate normal distribution.
- The two populations have the same covariance matrix.
- The observations are independent.

Here, Hotelling's T^2 takes the value 0.27, with the corresponding F-statistic being 0.09, having 3 and 36 degrees of freedom; the associated p-value is 0.97. There is no evidence of a gender difference on the three measures of IQ.

It might be thought that the results of Hotelling's T^2 test would simply reflect those that would be obtained using a series of univariate t-tests, in the sense that if no significant differences are found by the separate t-tests, then the T^2 test will inevitably lead to acceptance of the null hypothesis that the population mean vectors are equal. On the other hand, if any significant difference is found when using the t-tests on the individual variables, then the T^2 statistic must also lead to a significant result. But, these speculations are not correct (if they were, the whole T^2 test would be a waste of time). It is entirely possible to find no significant difference for each separate t-test but a significant result for the T^2 test, and vice versa. An explanation of how this can happen in the case of two variables is provided in Technical Section 18.2.

Technical Section 18.2: Univariate and Multivariate Tests for Equality of Means of Two Variables

Suppose we have a sample of n observations on two variables x_1 and x_2, and we wish to test whether the population means of the two variables μ_1 and μ_2 are both 0. Assume that the mean and standard deviation of the x_1 observations are $\bar{x}_1$ and s_1, respectively, and of the x_2 observations, $\bar{x}_2$ and s_2. If we test separately whether each mean takes the value 0, then we would use two t-tests. For example, to test $\mu_1 = 0$ against $\mu_1 \neq 0$, the appropriate test statistic is

$$t = \frac{\bar{x}_1 - 0}{s_1/\sqrt{n}}$$

The hypothesis $\mu_1 = 0$ would be rejected at the α percent level of significance, if

$$t < -t_{100(1-\alpha/2)} \quad \text{or} \quad t > t_{100(1-\alpha/2)}$$

that is, if $\bar{x}_1$ fell outside the interval

$$\left[-\frac{s_1 t_{100(1-\alpha/2)}}{\sqrt{n}}, +\frac{s_1 t_{100(1-\alpha/2)}}{\sqrt{n}}\right]$$

where $t_{100(1-\alpha/2)}$ is the $100(1-\alpha/2)$ percent point of the t distribution with $n-1$ degrees of freedom. Thus, the hypothesis would not be rejected if $\bar{x}_1$ fell within this interval. Similarly, the hypothesis $\mu_2 = 0$ for the variable x_2 would not be rejected if the mean $\bar{x}_2$ of the x_2 observations fell within a corresponding interval with s_2 substituted for s_1.

The multivariate hypothesis $[\mu_1, \mu_2] = [0, 0]$ would therefore not be rejected if both these conditions were satisfied. If we were to plot the point $(\bar{x}_1, \bar{x}_2)$ against rectangular axes, the area within which the point could lie and the multivariate hypothesis not rejected is given by the rectangle ABCD of Figure 18.1, where AB and DC are of length $2s_1 t_{100(1-\alpha/2)}/\sqrt{n}$, while AD and BC are of length $2s_2 t_{100(1-\alpha/2)}/\sqrt{n}$.

Thus, a sample that gave the means $(\bar{x}_1, \bar{x}_2)$ represented by the point P (see Figure 18.1) would lead to acceptance of the multivariate hypothesis. Suppose, however, that the variables x_1 and x_2 are moderately highly correlated. Then all points (x_1, x_2), and hence, $(\bar{x}_1, \bar{x}_2)$ should lie reasonably close to the straight line MN through the origin marked on the diagram. Therefore, samples consistent with the multivariate hypothesis should be represented by points $(\bar{x}_1, \bar{x}_2)$ that lie within a region encompassing the line MN. When we take account of the nature of the variation of bivariate normal samples that include correlation, this region can be shown to be an ellipse such as that in Figure 18.1. The point P is not consistent with this region and, in fact, *should be rejected* for this sample. Thus, the inference drawn from the two separate univariate

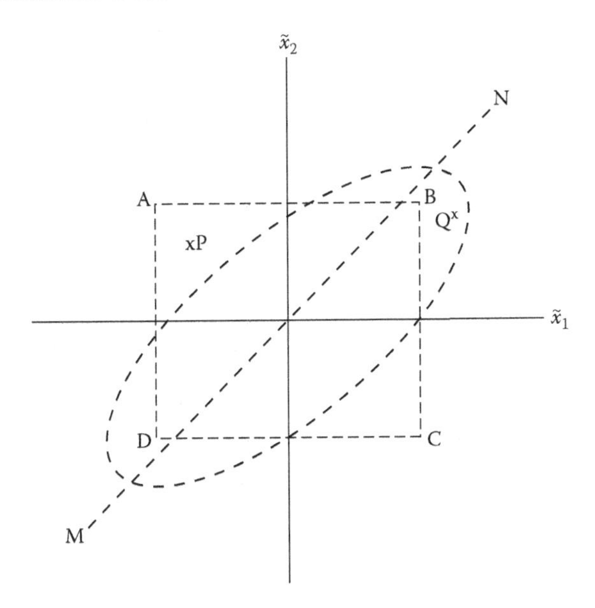

FIGURE 18.1
Why the results of univariate and multivariate tests can differ.

tests conflicts with the one drawn from a single multivariate test, and it is the wrong inference.

A sample giving the $(\bar{x}_1, \bar{x}_2)$ values represented by point Q (again, see Figure 18.1) would give the other type of mistake, where the application of two separate univariate tests leads to the rejection of the null hypothesis, but the correct multivariate inference is that the hypothesis *should not be rejected.* This explanation is taken with permission from Krzanowski (2000).

18.2.2 Fisher's Linear Discriminant Function

Spicer et al. (1987), in an investigation of sudden infant death (SID) syndrome, recorded four variables for each of 16 babies who were victims of SID and for 49 control babies. The babies who died and the control babies all had a gestational age of 37 weeks or more. Part of the data is shown in Table 18.2. The factor 68 variable arises from a particular aspect of 24h recordings or electrocardiograms and respiratory movements made for each child; the SID victims and the controls were matched for age at which these recordings were made. Here, interest lies in deriving a classification rule that could use measurements of the four variables on babies to be able to identify children at risk of SID and, if possible, take appropriate action to prevent the death of the baby.

TABLE 18.2
Part of the SIDs Data

Group	HR	BW	F68	GA
1	108.2	3000	0.321	37
1	131.1	4310	0.450	40
1	129.7	3975	0.244	40
1	142.0	3000	0.173	40
1	145.5	3940	0.304	41
2	139.7	3740	0.409	40
2	121.3	3005	0.626	38
2	131.4	4790	0.383	40
2	152.8	1890	0.432	38
2	125.6	2920	0.347	40

Note: Group = 1 for controls and 2 for SID victims; HR = heart rate (bpm); BW = birth weight (g); F68 = factor 68; GA = gestational age (weeks).

The required classification rule can be constructed using Fisher's *linear discriminant function*, and this is described in Technical Section 18.3.

Technical Section 18.3: Fisher's Linear Discriminant Function

The aim is to find a way of classifying observations into one of two known groups using a set of variables, $x_1, x_2, \ldots, x_q$. Fisher's idea was to find a linear function of the variables $z = a_1x_1 + a_2x_2 + \cdots + a_qx_q$ such that the ratio of the between-group variance of z to its within-group variance is maximized. Therefore, the coefficients $\mathbf{a}' = [a_1, a_2, \ldots, a_q]$ have to be chosen so that V, given by

$$V = \frac{\mathbf{a}'\mathbf{B}\mathbf{a}}{\mathbf{a}'\mathbf{S}\mathbf{a}}$$

is maximized, where **S** is the pooled within-group covariance matrix and **B** the covariance matrix of group means defined as follows:

$$\mathbf{S} = \frac{1}{n-2}\sum_{i=1}^{2}\sum_{j=1}^{n_i}(\mathbf{x}_{ij} - \bar{\mathbf{x}}_j)(\mathbf{x}_{ij} - \bar{\mathbf{x}}_j)'$$

$$\mathbf{B} = \sum_{i=1}^{2} n_i (\bar{\mathbf{x}}_i - \bar{\mathbf{x}})(\bar{\mathbf{x}}_i - \bar{\mathbf{x}})'$$

where $\mathbf{x}_{ij}' = [x_{ij1}, x_{ij2}, \ldots, x_{ijq}]$ represents the set of q variable values for the jth individual in group i, $\bar{x}_j$ is the mean vector of the jth group, and $\bar{\mathbf{x}}$ is the mean vector of all observations. The number of observations in group 1 is n_1, and in group 2 is n_2, with $n = n_1 + n_2$. The vector $\mathbf{a}$ that maximizes V is given by the solution of the following equation:

$$(\mathbf{B} - \lambda\mathbf{S})\mathbf{a} = \mathbf{0}$$

In the two-group situation, the single solution can be shown to be

$$\mathbf{a} = \mathbf{S}^{-1}(\bar{\mathbf{x}}_1 - \bar{\mathbf{x}}_2)$$

The allocation rule is now to allocate an individual with discriminate score z to group 1 if

$$z > \frac{\bar{z}_1 + \bar{z}_2}{2}$$

where $\bar{z}_1$ and $\bar{z}_2$ are the mean discriminant scores in each group. (We are assuming that the groups are labeled such that $\bar{z}_1 > \bar{z}_2$.)

Fisher's discriminant function also arises from assuming that, in the population, the observations in group 1 have a multivariate normal distribution with mean vector $\boldsymbol{\mu}_1$ and covariance matrix $\boldsymbol{\Sigma}$, and those in group 2 have a multivariate normal distribution with mean vector $\boldsymbol{\mu}_2$ and, again, covariance matrix $\boldsymbol{\Sigma}$. Misclassifications are minimized if an individual with vector of scores $\mathbf{x}$ is allocated to group 1 if

$$MVN(\mathbf{x}, \boldsymbol{\mu}_1, \boldsymbol{\Sigma}) > MVN(\mathbf{x}, \boldsymbol{\mu}_2, \boldsymbol{\Sigma})$$

where MVN is shorthand for the multivariate normal density function. Substituting sample mean vectors for $\boldsymbol{\mu}_1$ and $\boldsymbol{\mu}_2$ and the matrix $\mathbf{S}$ defined earlier for $\boldsymbol{\Sigma}$, we are led to the same allocation rule as that given previously. But, the derived classification rule is only valid if the prior probabilities of being in each group are assumed to be the same. If the prior probability of group 1 is π_1 and that of group 2 is π_2, then the new allocation rule becomes allocated to group 1 if

$$z > \frac{\bar{z}_1 + \bar{z}_2}{2} + \log\frac{\pi_2}{\pi_1}$$

To begin, we will use only the factor 68 variable and birth weight in applying Fisher's linear discriminant function to the SID data. We can first construct a scatterplot of the data (see Figure 18.2). The numerical details of the calculations involved are shown in Table 18.3. The discriminant function is

$$z = 0.00195 \times \text{birth weight} \ - 16.077 \times \text{factor 68}$$

Assuming equal prior probabilities (unrealistic, but a point we shall return to later), the allocation rule for a new infant becomes "allocate to group 1 (little SID risk) if $z > 0.506$ (the 'cutoff' value halfway between the discriminant function means of each group), otherwise allocate to group 2 (SID risk)." The

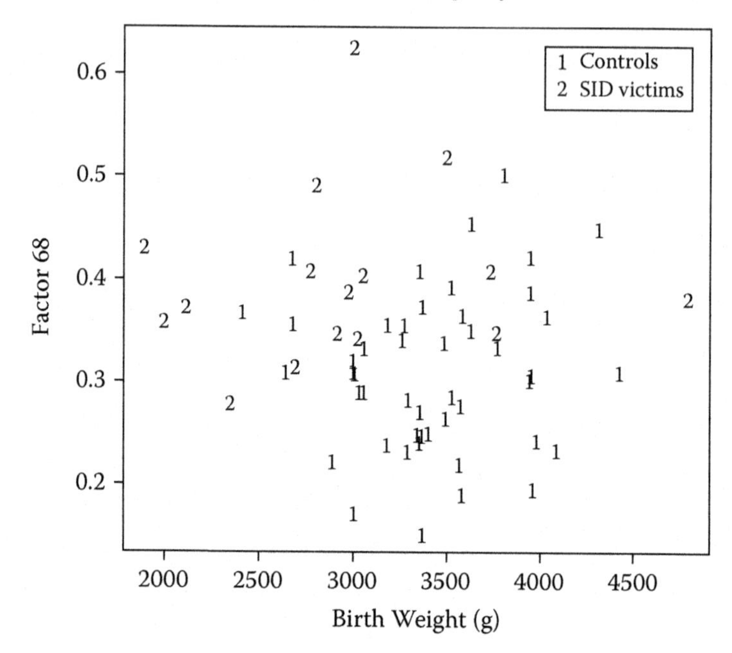

FIGURE 18.2
Scatterplot of the variables birth weight and factor 68 from the SID data.

discriminant function can be shown on the scatterplot of the birth weight and factor 68 variables simply by plotting the line $z - 0.506 = 0$, that is, a line with intercept given by $0.506/a_2$ and a slope of $-a_1/a_2$, where a_1 and a_2 are the discriminant function coefficients, to give Figure 18.3. In terms of this plot, a new infant with values of the two variables leading to a position on the plot above the line would be allocated to the SID risk group, and an infant with a position below the line to the little SID risk group.

A deficiency of the derived allocation rule is that it takes no account of the prior probabilities of class membership in the population under study. Therefore, if used as a screening device for babies at risk of SID in the simple form suggested here many more infants would be considered at risk than is genuinely merited because, fortunately, SID is known to be a relatively rare condition.

A question of some importance about a discriminant function allocation rule is "how well does it perform?" One way this question could be answered is to see how many of the original sample of observations (the training set) it misclassifies. In the case of the discriminant function for the SID data derived

TABLE 18.3
Calculating Fisher's Linear Discriminant Function on SID Data Using Birth Weight and Factor 68 Variables

Group		Controls		SID	
Means		BW	F68	BW	F68
		3437.88	0.31	2964.69	0.41
Covariance		BW	F68	BW	F68
Matrix	BW	1.95e+05	3.24	5.45e+05	7.76
	F68	3.24	0.006	7.76	0.007

Pooled Covariance Matrix

	BW	F68
BW	278612.28	4.32
F68	4.32	0.006

Coefficients of Discriminant Function

BW	0.00195
F68	−16.07705

Discriminant function mean in controls: 1.6984
Discriminant function mean in SID victims: −0.6860, Cutoff value: 0.5062

Note: BW = Birth weight; F68 = factor 68.

here based on birth weight and factor 68, the results of applying it to the data from which it was calculated are

Actual Group	Allocation Rule Group	
	Controls	SID
Controls	41	8
SID	3	13

So, the percentage of misclassifications is 16.9. This method of estimating the misclassification rate is known to be optimistic in many cases. Other more realistic methods for estimating the misclassification rate are described in Everitt and Dunn (2001). (Finding Fisher's linear discriminant function based on all four variables recorded in the SID data is left as an exercise for the reader.)

Fisher's linear discriminant function is optimal when the data arise from populations having multivariate normal distributions with the same covariance matrices. When the distributions are clearly nonnormal, an alternative approach is *logistic discrimination* (see, for example, Anderson, 1972), although the results of both this and Fisher's method are likely to be very similar in most cases. When the two covariance matrices are thought to be unequal, then

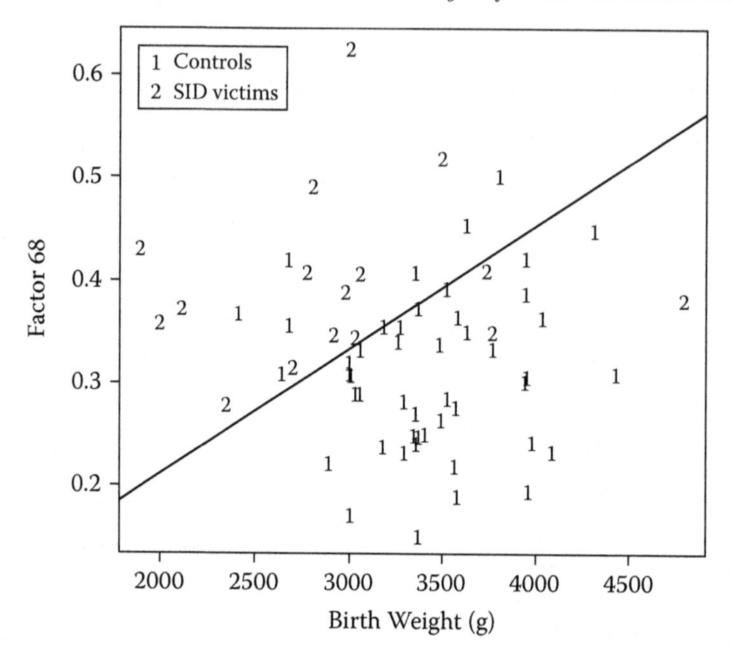

FIGURE 18.3
Scatterplot of factor 68 against birth weight for SID data, showing Fisher's linear discriminant function based on the two variables.

the linear discriminant function is no longer optimal, and a quadratic version may be needed. Details are given in Everitt and Dunn (2001). The quadratic discriminant function has the advantage of increased flexibility compared to the linear version. There is, however, a penalty involved in the form of potential overfitting, making the derived function poor at classifying new observations. Friedman (1989) attempts to find a compromise between the data variability of quadratic discrimination and the possible bias of linear discrimination by adopting a weighted sum of the two, called regularized discriminant analysis.

18.3 More Than Two Groups

18.3.1 Multivariate Analysis of Variance (MANOVA)

Timm (2002) reports the data collected in a large study by Dr. Stanley Jacobs and Mr. Ronald Hritz at the University of Pittsburgh to investigate risk-taking behavior. Students were randomly assigned to three different direction treatments known as Arnold and Arnold (AA), Coombs (C), and Coombs with no penalty (NC) in the direction. Using the three treatment conditions, students were administered two parallel forms of a test given under low and high penalty. Part of the data is shown in Table 18.4. The question of interest here

TABLE 18.4
Part of Data from Investigation of Risk Taking

AA		C		NC	
Low	High	Low	High	Low	High
8	28	46	13	50	55
18	28	26	10	57	51
8	23	47	22	62	52

Note: AA = Arnold and Arnold; C = Coombs; NC = Coombs with no penalty.

is whether the two-dimensional population mean vectors for the three groups are the same. The technique to be used is MANOVA, which is an extension of univariate analysis of variance to multivariate observations. A short account of one-way MANOVA is given in Technical Section 18.4, but MANOVA can, of course, be used with more complex designs when the response is multidimensional.

Technical Section 18.4: One-Way MANOVA

We assume that we have multivariate observations of a sample of individuals from m different populations, where $m \geq 2$, and there are n_i observations sampled from population i. The linear model for observation x_{ijk}, the jth observation on variable k in group i ($k = 1, \ldots, q$, $j = 1, \ldots, n_i$, $i = 1, \ldots, m$) is

$$x_{ijk} = \mu_k + \alpha_{ik} + \varepsilon_{ijk}$$

where μ_k is a general effect for the kth variable, α_{ik} is the effect of group i on the kth variable, and ε_{ijk} is a random disturbance term. The vector $\boldsymbol{\varepsilon}'_{ij} = [\varepsilon_{ij1}, \ldots, \varepsilon_{ijq}]$ is assumed to have a multivariate normal distribution with null mean vector and covariance matrix $\boldsymbol{\Sigma}$, assumed to be the same in all m populations. The error terms of different individuals are assumed independent of one another.

The hypothesis of equal mean vectors in the m populations can be written as

$$H_0 : \alpha_{ik} = 0, \quad i = 1, \ldots, m, \ k = 1, \ldots, q$$

MANOVA is based on two matrices $\mathbf{H}$ and $\mathbf{E}$, the elements of which are defined as follows:

$$h_{rs} = \sum_{i=1}^{k} n_i(\bar{x}_{ir} - \bar{x}_r)(\bar{x}_{is} - \bar{x}_s), \quad r, s = 1, \ldots, q$$

$$e_{rs} = \sum_{i=1}^{k} \sum_{j=1}^{n_i} (\bar{x}_{ijr} - \bar{x}_{ir})(\bar{x}_{ijs} - \bar{x}_{is}), \quad r, s = 1, \ldots, q$$

where $\bar{x}_{ir}$ is the mean of variable r in group i, and $\bar{x}_r$ is the grand mean of variable r. The diagonal elements of $\mathbf{H}$ and $\mathbf{E}$ are, respectively, the between-groups sum of squares for each variable and the within-group sum of squares for the variable. The off-diagonal elements of $\mathbf{H}$ and $\mathbf{E}$ are the corresponding sums of cross products for pairs of variables. In the multivariate situation when $m > 2$, there is no single test statistic that is always the most powerful one for detecting all types of departures from the null hypothesis of the mean vectors of the populations. A number of different test statistics have been proposed that may lead to different conclusions when used in the same data set, although on most occasions they will not. The following are the principal test statistics for MANOVA:

a. Wilks' determinantal ratio $\Lambda = \dfrac{|\mathbf{E}|}{|\mathbf{H}+\mathbf{E}|}$
b. Roy's greatest root; the criterion is the largest eigenvalue of $\mathbf{E}^{-1}\mathbf{H}$
c. Lawley–Hotelling trace $t = \text{trace}\left(\mathbf{E}^{-1}\mathbf{H}\right)$
d. Pillai trace $v = \text{trace}\left[\mathbf{H}\left(\mathbf{H}+\mathbf{E}\right)^{-1}\right]$

Each test statistic can be converted into an approximate F-statistic that allows associated p-values to be calculated. For details, see Morrison (2005).

When there are only two groups, all four test criteria are equivalent and lead to the same F-value as Hotelling's T^2 described in Technical Section 18.1.

Prior to any formal analysis of the data from the risk-taking investigation, it is useful to look at some boxplots, and these are given in Figure 18.4. The "Low" scores appear to increase across the three groups, with the group differences on the "High" score being rather smaller. Applying MANOVA to the data from the investigation of risk taking, we get the results shown in Table 18.5.

TABLE 18.5
MANOVA on Data from the Risk-Taking Investigation

	DF	Pillai	approx. F	Num DF	Den DF	Pr(>F)
Group	1	0.866	268.323	2	83	<2.2e-16
		Wilks				
Group	1	0.134	268.323	2	83	<2.2e-16
		Hotelling-Lawley				
Group	1	6.466	268.323	2	83	<2.2e-16
		Roy				
Group	1	6.466	268.323	2	83	<2.2e-16

Note: Num = Numerator, Den = Denominator.

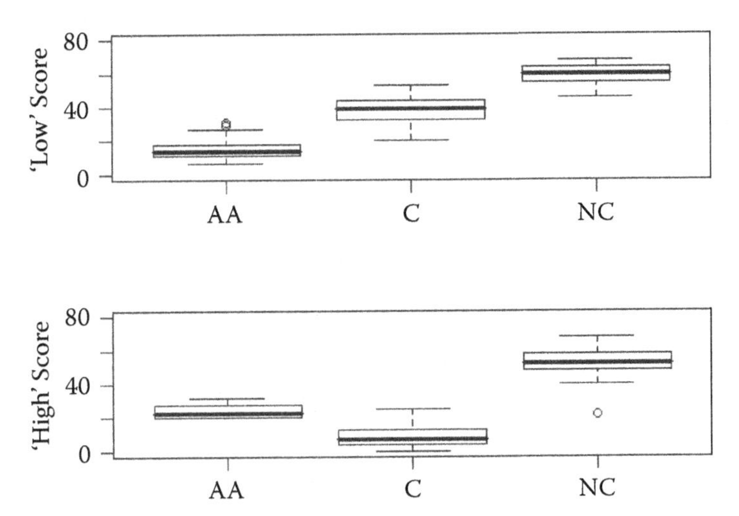

FIGURE 18.4
Boxplots for data from risk-taking experiment.

Clearly, the two-dimensional mean vectors of low and high scores differ in the three groups.

The tests applied in MANOVA assume multivariate normality for the error terms in the corresponding model. An informal assessment of this assumption can be made using the chi-square plot described in Chapter 12, applied to the residuals from fitting the one-way MANOVA model; note that the residuals in this case are each two-dimensional vectors. The plot is shown in Figure 18.5. There is some evidence of departure from the multivariate normal, but the p-values in Table 18.5 are so small that minor departures from the distributional assumption are unlikely to change the conclusions.

About three to four decades ago, MANOVA was a widely used and popular technique particularly amongst psychologists and in particular for analysing repeated measures or longitudinal data sets. But this latter use of the technique has been largely superseded by the methods described in Chapters 8 to 10. And MANOVA is perhaps now largely only of historical interest.

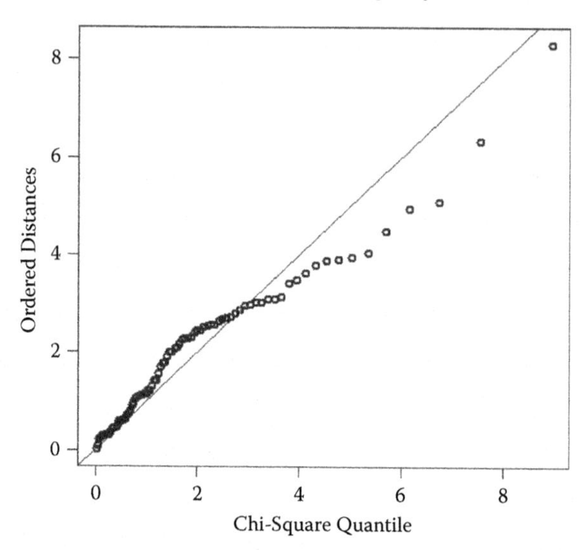

FIGURE 18.5
Chi-square plot of residuals from fitting one-way MANOVA to data from risk-taking experiment.

18.3.2 Classification Functions

In this section we will use data generated during a functional magnetic resonance imaging (fMRI) investigation. Two measures of intensity of each voxel in an image were recorded—PD and T_2. Part of the data is shown in Table 18.6. One aim of the investigation was to derive a rule for allocating each voxel in an image into one of three classes: grey matter, white matter, or cerebrospinal fluid (CSF). When more than two groups are involved, we can again derive classification functions by comparing the assumed multivariate normal densities for each group. Technical Section 18.5 explains how.

TABLE 18.6
Part of fMRI Data

Class	PD	T_2
Grey	124	58
Grey	107	44
White	142	122
White	144	148
CSF	98	45
CSF	87	34

Note: CSF = Cerebrospinal fluid.

Technical Section 18.5: Discriminant Analysis for Three Groups

Assuming that the observations in the three groups have multivariate normal densities with different means, $\boldsymbol{\mu}_1$, $\boldsymbol{\mu}_2$ and $\boldsymbol{\mu}_3$ but a common covariance matrix, $\mathbf{S}$, the allocation rule for an individual with vector of scores $\mathbf{x}$ becomes:

Allocate to group 1 if

$$MVN(\mathbf{x}, \boldsymbol{\mu}_1, \mathbf{S}) > MVN(\mathbf{x}, \boldsymbol{\mu}_2, \mathbf{S})$$

and

$$MVN(\mathbf{x}, \boldsymbol{\mu}_1, \mathbf{S}) > MVN(\mathbf{x}, \boldsymbol{\mu}_3, \mathbf{S})$$

Allocate to group 2 if

$$MVN(\mathbf{x}, \boldsymbol{\mu}_2, \mathbf{S}) > MVN(\mathbf{x}, \boldsymbol{\mu}_1, \mathbf{S})$$

and

$$MVN(\mathbf{x}, \boldsymbol{\mu}_2, \mathbf{S}) > MVN(\mathbf{x}, \boldsymbol{\mu}_3, \mathbf{S})$$

Allocate to group 3 if

$$MVN(\mathbf{x}, \boldsymbol{\mu}_3, \mathbf{S}) > MVN(\mathbf{x}, \boldsymbol{\mu}_1, \mathbf{S})$$

and

$$MVN(\mathbf{x}, \boldsymbol{\mu}_3, \mathbf{S}) > MVN(\mathbf{x}, \boldsymbol{\mu}_2, \mathbf{S})$$

This leads to sample-based allocation rules as follows:

Allocate to group 1 if $h_{12}(\mathbf{x}) > 0$ and $h_{13}(\mathbf{x}) > 0$
Allocate to group 2 if $h_{12}(\mathbf{x}) < 0$ and $h_{23}(\mathbf{x}) > 0$
Allocate to group 3 if $h_{13}(\mathbf{x}) < 0$ and $h_{23}(\mathbf{x}) < 0$

where

$$h_{ij}(\mathbf{x}) = (\bar{\mathbf{x}}_i - \bar{\mathbf{x}}_j)'\mathbf{S}^{-1}\left[\mathbf{x} - \frac{1}{2}(\bar{\mathbf{x}}_i + \bar{\mathbf{x}}_j)\right]$$

and $\bar{\mathbf{x}}_i$ and $\bar{\mathbf{x}}_j$ are group mean vectors, $\mathbf{S}$ is the sample estimate of the assumed common covariance matrix of the three groups and is given by

$$\mathbf{S} = \frac{(n_1 - 1)\mathbf{S}_1 + (n_2 - 1)\mathbf{S}_2 + (n_3 - 1)\mathbf{S}_3}{n_1 + n_2 + n_3 - 3}$$

and $\mathbf{S}_1$, $\mathbf{S}_2$, and $\mathbf{S}_3$ are the estimates of the covariance matrices of each group.

To begin, we can plot the data labeling the three classes. The plot is shown in Figure 18.6. To find the three discriminant functions, we first need to find the mean vectors and covariance matrices of each class and then the pooled

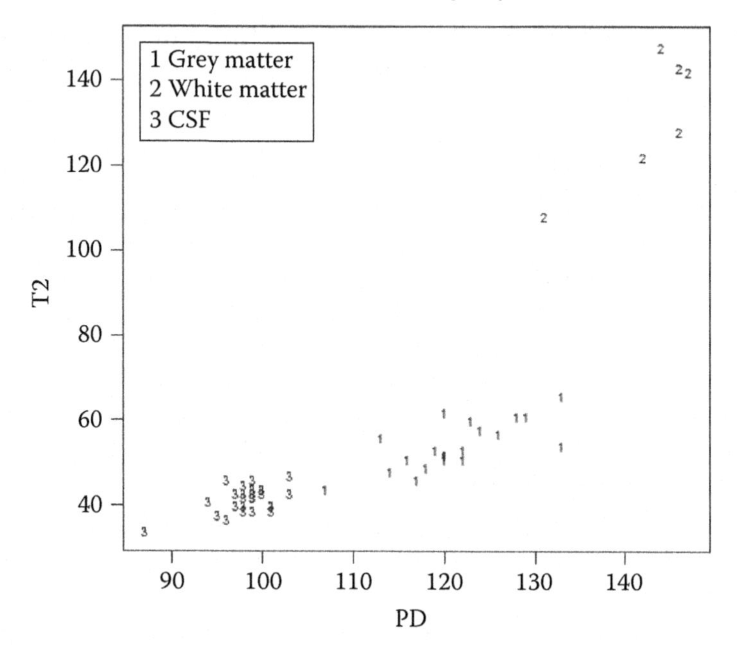

FIGURE 18.6
Scatterplot of imaging data with three classes labeled.

covariance matrix; all are shown in Table 18.7. The coefficients of each linear discriminant function and the thresholds calculated from the information in Table 18.7 are shown in Table 18.8. Each of the discriminant functions can

TABLE 18.7
Means and Covariance Matrices of Each Class in Imaging Data

	Grey Matter ($n = 20$)		**White Matter ($n = 6$)**		**CSF ($n = 24$)**	
	PD	T_2	PD	T_2	PD	T_2
Mean	121.20	54.25	142.67	131.83	98.08	41.67
			Covariance Matrix			
	PD	T_2	PD	T_2	PD	T_2
PD	42.48	26.95	35.87	74.93	10.34	5.77
T_2	26.95	33.25	74.93	233.77	5.77	9.62
			Pooled Covariance Matrix			
		PD		T_2		
PD		26.05		21.69		
T_2		21.69		43.02		

TABLE 18.8
Discriminant Functions and Thresholds for Imaging Data

	Grey Matter		White Matter		CSF	
	PD	T_2	PD	T_2	PD	T_2
Discriminant coefficients	1.17	−2.39	1.11	−0.27	−0.06	2.13
			Thresholds			
	Grey v White		Grey v CSF		White v CSF	
	−68.54		108.86		177.39	

now be shown on the scatterplot of the data using the same approach as that used for a discriminant function for two groups.

The scatterplot showing the three discriminant functions is shown in Figure 18.7. This plot would allow an investigator to classify new, unlabeled voxels, although in practice the three discriminant functions would need to be calculated from a much larger sample of previously labeled voxels.

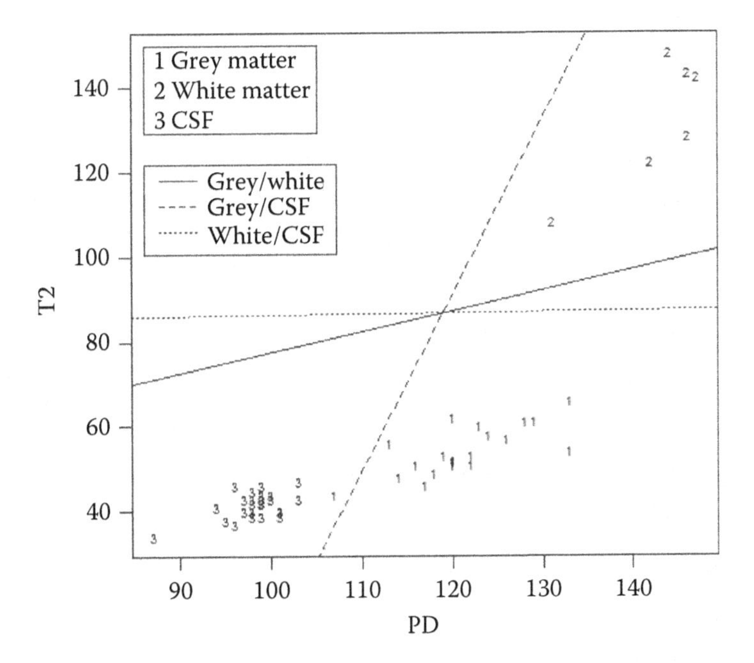

FIGURE 18.7
Scatterplot of imaging data showing the three discriminant functions.

18.4 Summary

- Grouped multivariate data frequently occur in practice.
- The appropriate method of analysis depends on the question of most interest to the investigator.
- Hotelling's T^2 test and MANOVA can be used to assess hypotheses about population mean vectors.
- Fisher's linear discriminant function can be used to construct formal classification rules for allocating new individuals into one of two a priori known groups. And similar rules can be found when there are more than two groups.
- Full details of discriminant function methods are given in Hand (2005).

18.5 Exercises

18.1 Return to the body measurement data introduced in Chapter 12, and find Fisher's linear discriminant function for allocating individuals to be men and women. (We are aware that there is a foolproof method. The first 10 observations in the data set are men, the remaining 10 are women.) Construct a scatterplot of the data showing the group membership and the derived linear discriminant function.

18.2 For the data from the risk-taking investigation, produce a scatterplot of "High" versus "Low" scores showing the three groups. As an exercise, construct discriminant functions for each pair of groups, and show these on the scatterplot.

18.3 In the SID data by coding a variable, group, as controls = 1 and SID victims = −1, show the equivalence of a multiple regression model for group as the response variable and explanatory variables HR, BW, F68, and GA with the discriminant function for the data derived in the text.

18.4 The data show the chemical composition for nine oxides of 48 specimens of Romano-British pottery determined by atomic absorption spectra (for more information on the analysis of *compositional data*, see Greenacre, 2018). Also, given in the data is the label of the kiln site at which the pot was found. Use MANOVA to test whether the pots found at different kiln sites differ in their chemical compositions. The five kiln sites are actually from three different regions with kiln 1 from region 1, kilns 2 and 3 from region 2, and kilns 4 and 5 from region 3. Find the allocation rule for allocating a new pot to one of the three regions, and use the rule on a pot with the following chemical composition:

Al_2O_3	Fe_2O_3	MgO	CaO	Na_2O	K_2O	TiO_2	MnO	BaO
15.5	5.71	2.07	0.98	0.65	3.01	0.76	0.09	0.012

References

Agresti, A. (1996). *An Introduction to Categorical Data Analysis.* John Wiley & Sons, New York.

Aitkin, M. (1978). The analysis of unbalanced cross-classification. *Journal of the Royal Statistical Society, Series A*, 141, 195–223.

Alwin, D. F. (2007). *Margins of Error: A Study of Reliability in Survey Measurement.* John Wiley & Sons, New York.

Anderson, J. A. (1972). Separate sample logistic discrimination. *Biometrika*, 59, 19–35.

Anscombe, F. (1973). Graphs in statistical analysis. *The American Statistician*, 27, 17–21.

Banfield, J. D. and Raftery, A. E. (1993). Model based Gaussian and non-Gaussian clustering. *Biometrics*, 49, 803–821.

Barlow, R. E., Bartholomew, D. J., Bremner, J. M. and Brunk, H. D. (1972). *Statistical Inference under Order Restrictions.* John Wiley & Sons, New York.

Barnard, J. and Rubin, D. B. (1999). Small sample degrees of freedom with multiple imputation. *Biometrika*, 86, 948–955.

Barnard, J., Rubin, D. B. and Schenker, N. (2005). Multiple imputation methods. In *Encyclopedia of Biostatistics*, 2nd edition (eds. P. Armitage and T. Colton). John Wiley & Sons, Chichester, U.K.

Barnett, V., ed. (1981). *Interpreting Multivariate Data.* John Wiley & Sons, Chichester, U.K.

Bartholomew, D. J., Knott, M. and Moustaki, I. (2011). *Latent Variable Models and Factor Analysis: A Unified Approach*, 3rd edition. John Wiley & Sons, Chichester, U.K.

Beck, A. T., Steer, A. and Brown, G. K. (1996). *Beck Depression Inventory Manual.* The Psychological Corporation, San Antonio, Texas.

Bennett, A. (2009). "Heritage rock": Rock music, representation and heritage discourse. *Poetics*, 37, 474–489.

Bentler, P. M. (1980). Multivariate analysis with latent variables: Causal modeling. *Annual Review of Psychology*, 31, 419–456.

Benzécri, J.–P. et al. (1973). *Analyse des Données. Tôme 1: La Classification. Tôme 2: L'Analyse des Correspondances.* [*Data Analysis. Volume 1: Classification. Volume 2: Correspondence Analysis*, in French.] Dunod, Paris.

Benzécri, J.–P. et al. (1992). *Correspondence Analysis Handbook.* Marcel Dekker, New York.

Bertin, J. (1981). *Semiology of Graphics.* University of Wisconsin Press, Wisconsin.

Bickel, P. J., Hammel, E. A. and O'Connell, J. W. (1975). Sex bias in graduate admissions: Data from Berkeley. *Science*, 187, 398–404.

Blackith, R. E. and Reyment, R. A. (1971). *Multivariate Morphometrics.* Academic Press, London.

Blalock, H. M. (1961). Correlation and causality: The multivariate case. *Social Forces*, 39, 246–251.

Blalock, H. M. (1963). Making causal inferences for unmeasured variables from correlations among indicators. *American Journal of Sociology*, 69, 53–62.

Blasius, J. and Greenacre, M., eds. (2014). *Visualization and Verbalization of Data.* Chapman and Hall/CRC, Boca Raton, Florida.

Bollen, K. A. (1989). *Structural Equations with Latent Variables.* John Wiley & Sons, New York.

Bollen, K. A. and Long, J. S., eds. (1993). *Testing Structural Equation Models.* Sage, London.

Borg, I. and Groenen, P. J. F. (2005). *Modern Multidimensional Scaling*, 2nd edition. Springer, New York.

Bradburn, N. M., Sudman, S. and Wansink, B. (2004). *Asking Questions: The Definitive Guide to Questionnaire Design—For Market Research, Political Polls, and Social and Health Questionnaires*, revised edition. Jossey-Bass, San Francisco, California.

Browne, M. W. (1974). Generalized least squares estimators in the analysis of covariance structures. *South African Statistical Journal*, 8, 1–24.

Byrne, B. M. (2005). Factor analytic models: Viewing the structure of an assessment instrument from three perspectives. *Journal of Personality Assessment*, 85, 17–32.

Calsyn, R. J. and Kenny, D. A. (1977). Self-concept of ability and perceived evaluation of others: Cause or effect of academic achievement? *Journal of Educational Psychology*, 69, 136–145.

Carpenter, J., Pocock, S. and Lamm, C. J. (2002). Coping with missing data in clinical trials: A model-based approach applied to asthma trials. *Statistics in Medicine*, 21, 1043–1066.

Carroll, J. B. (1953). An analytical solution for approximating simple structure in factor analysis. *Psychometrika*, 18, 23–38.

Cattell, R. B. (1965). Factor analysis: An introduction to essentials. I: The purpose and underlying models. *Biometrics*, 21, 190–215.

Chambers, J. M., Cleveland, W. S., Kleiner, B. and Tukey, P. A. (1983). *Graphical Methods for Data Analysis.* Wadsworth, Belmont, California.

Chatterjee, S. and Hadi, A. S. (2012). *Regression Analysis by Example*, 5th edition. John Wiley & Sons, New York.

Cleveland, W. S. (1979). Robust locally weighted regression and smoothing scatterplots. *Journal of the American Statistical Association*, 74, 829–836.

Cleveland, W. S. (1993). *Visualizing Data.* Hobart Press, Summit, New Jersey.

Cleveland, W. S. (1994). *The Elements of Graphing Data.* Hobart Press, Summit, New Jersey.

Cleveland, W. S. and McGill, M. E., eds. (1988). *Dynamic Graphics for Statistics.* Wadsworth, Belmont, California.

Cochran, W. G. (1965). The planning of observational studies of human populations. *Journal of the Royal Statistical Society,* Series A, 128, 134–155.

Collett, D. (2003). *Modelling Binary Data,* 2nd edition. Chapman and Hall/CRC, London.

Collett, D. (2015). *Modelling Survival Data in Medical Research,* 3rd edition. Chapman and Hall/CRC, London.

Colman, A. M., ed. (1994). *The Companion Encyclopedia of Psychology.* Routledge, London.

Cook, R. D. and Weisberg, S. (1999). *Applied Regression Including Computing and Graphics.* John Wiley & Sons, New York.

Cook, R. J. (2005). Generalized linear model. In *Encyclopedia of Biostatistics,* 2nd edition (eds. P. Armitage and T. Colton). John Wiley & Sons, Chichester, U.K.

Cox, D. R. (1972). Regression models and life-tables (with discussion). *Journal of the Royal Statistical Society, Series B,* 34, 187–220.

Cox, T. F. and Cox, M. A. A. (2001). *Multidimensional Scaling,* 2nd edition. Chapman and Hall/CRC, London.

Crowder, M. J. and Hand, D. J. (1990). *Analysis of Repeated Measurements.* Chapman and Hall, London.

Cudeck, R. and MacCallum, R. C., eds. (2007). *Factor Analysis at 100: Historical Developments and Future Directions.* Lawrence Elbaum, Mahwah, New Jersey.

Davis, C. S. (1991). Semi-parametric and non-parametric methods for the analysis of repeated measurements with applications to clinical trials. *Statistics in Medicine,* 10, 1959–1980.

Davis, C. S. (2002). *Statistical Methods for the Analysis of Repeated Measurements.* Springer, New York.

De Leeuw, E. D., Hox, J. and Dillman, D., eds. (2008). *International Handbook of Survey Methodology.* Routledge, London.

Dempster, A. P., Laird, N. M. and Rubin, D. B. (1977). Maximum likelihood from incomplete data via the EM algorithm (with discussion). *Journal of the Royal Statistical Society, Series B,* 39, 1–38.

Diggle, P. J. and Kenward, M. G. (1994). Informative drop-out in longitudinal data analysis (with discussion). *Journal of the Royal Statistical Society, Series C (Applied Statistics),* 43, 49–93.

Diggle, P. J., Heagerty, P., Liang, K. and Zeger, S. L. (2002). *Analysis of Longitudinal Data,* 2nd edition. Oxford University Press, Oxford.

Dizney, H. and Gromen, L. (1967). Predictive validity and differential achievement on three MLA comparative foreign language tests. *Educational and Psychological Measurement*, 27, 1127–1130.

Dobson, A. J. and Barnett, A. G. (2018). *An Introduction to Generalized Linear Models*, 4th edition. Chapman and Hall/CRC, London.

Donati, M. A., Chiesi, F. and Primi, C. (2013). A model to explain at-risk/problem gambling among male and female adolescents: Gender similarities and differences. *Journal of Adolescence*, 36, 129–137.

Duncan, O. D. (1966). Path analysis: Sociological examples. *American Journal of Sociology*, 72, 1–16.

Eerola, T. (2011). Are the emotions expressed in music genre-specific? An audio-based evaluation of datasets spanning classical, film, pop and mixed genres. *Journal of New Music Research*, 40, 349–366.

Efron, B. (1998). Foreword in special issue on analyzing non-compliance in clinical trials. *Statistics in Medicine*, 17, 249–250.

Everitt, B. S. (1984). *An Introduction to Latent Variable Models.* Chapman and Hall, London.

Everitt, B. S. (1987). *An Introduction to Optimization Methods and Their Applications in Statistics.* Chapman and Hall, London.

Everitt, B. S. and Dunn, G. (2001). *Applied Multivariate Data Analysis.* Edward Arnold, London.

Everitt, B. S. and Hay, D. F. (1992). *Talking about Statistics: A Psychologist's Guide to Design & Analysis.* Edward Arnold, London.

Everitt, B. S. and Hothorn, T. (2009). *A Handbook of Statistical Analyses Using R*, 2nd edition. Chapman and Hall/CRC, Boca Raton, Florida.

Everitt, B. S. and Howell, D. C., eds. (2005). *Encyclopedia of Statistics in Behavioral Science.* John Wiley & Sons, Chichester, U.K.

Everitt, B. S. and Pickles, A. (2004). *Statistical Aspects of the Design and Analysis of Clinical Trials*, 2nd edition. Imperial College Press, London.

Everitt, B. S. and Rabe-Hesketh, S. (1997). *The Analysis of Proximity Data.* Edward Arnold, London.

Everitt, B. S. and Rabe-Hesketh, S. (2001). *Analysing Medical Data Using S-PLUS.* Springer, New York.

Everitt, B. S. and Wessely, S. (2008). *Clinical Trials in Psychiatry*, 2nd edition. John Wiley & Sons, Chichester, U.K.

Everitt, B. S., Landau, S., Leese, M. and Stahl, D. (2011). *Cluster Analysis*, 5th edition. John Wiley & Sons, Chichester, U.K.

Fisher, R. A. (1940). The precision of discriminant functions. *Annals of Eugenics*, 10, 422–429.

Fitzmaurice, G. M., Laird, N. M. and Ware, J. H. (2011). *Applied Longitudinal Analysis*, 2nd edition. John Wiley & Sons, Hoboken, New Jersey.

Fleiss, J. L. (1986). *The Design and Analysis of Clinical Experiments.* John Wiley & Sons, New York.

Fraley, C. and Raftery, A. E. (1998). How many clusters? Which cluster method? Answers via model-based cluster analysis. *Computer Journal*, 41, 578–588.

Fraley, C. and Raftery, A. E. (1999). MCLUS: Software for the model-based cluster analysis. *Journal of Classification*, 16, 297–306.

Fraley, C. and Raftery, A. E. (2002). Model based clustering, discriminant analysis and density estimation. *Journal of the American Statistical Association*, 97, 611–631.

Friedman, H. P. and Rubin, J. (1967). On some invariant criteria for grouping data. *Journal of the American Statistical Association*, 62, 1159–1178.

Friedman, J. H. (1989). Regularized discriminant analysis. *Journal of the American Statistical Association*, 84, 165–175.

Friendly, M. (2002). Corrgrams. *The American Statistician*, 56, 316–324.

Frison, L. and Pocock, S. J. (1992). Repeated measures in clinical trials: Analysis using mean summary statistics and its implications for design. *Statistics in Medicine*, 11, 1685–1704.

Gabriel, K. R. (1971). The biplot graphic display of matrices with application to principal component analysis. *Biometrika*, 58, 453–467.

Gabriel, K. R. (2002). Goodness of fit of biplots and correspondence analysis. *Biometrika*, 89, 423–436.

Gabriel, K. R. and Odoroff, C. L. (1990). Biplots in biomedical research. *Statistics in Medicine*, 9, 469–485.

Gardner, M. J. and Altman, D. G. (1986). Confidence intervals rather than P values: Estimation rather than hypothesis testing. *British Medical Journal*, 292, 746–750.

Gelman, A. and Unwin, A. (2013). Infovis and statistical graphics: Different goals, different looks. *Journal of Computational and Graphical Statistics*, 22, 2–28.

Georges, P. (2017). Western classical music development: a statistical analysis of composers similarity, differentiation and evolution. *Scientometrics*, 112, 21–53.

Giardiello, F. M., Hamilton, S. R., Krush, A. J. et al. (1993). Treatment of colonic and rectal adenomas with sulindac in familial adenomatous polyposis. *The New England Journal of Medicine*, 328, 1313–1316.

Goldberg, B. P. (1972). *The Detection of Psychiatric Illness by Questionnaire*. Oxford University Press, Oxford.

Goldberg, K. M. and Iglewicz, B. (1992). Bivariate extensions of the boxplot. *Technometrics*, 34, 307–320.

Gordon, A. D. (1987). A review of hierarchical classification. *Journal of the Royal Statistical Society, Series A*, 150, 119–137.

Gordon, A. D. (1999). *Classification*, 2nd edition. Chapman and Hall, London.

Gower, J. C. (1966). Some distance properties of latent root and vector methods used in multivariate analysis. *Biometrika*, 53, 325–338.

Grana, C., Chinol, M., Robertson, C. et al. (2002). Pretargeted adjuvant radioimmunotherapy with Yttrium-90-biotin in malignant glioma patients: A pilot study. *British Journal of Cancer*, 86, 207–212.

Greenacre, M. (1984). *Theory and Applications of Correspondence Analysis*. Academic Press, London.

Greenacre, M. (2010). *Biplots in Practice*. BBVA Foundation, Bilbao.

Greenacre, M. (2016). *Correspondence Analysis in Practice*, 3rd edition. Chapman and Hall/CRC, Boca Raton, Florida.

Greenacre, M. (2018). *Compositional Data Analysis in Practice*. Chapman and Hall/CRC, Boca Raton, Florida.

Greenacre, M. and Blasius, J., eds. (2006). *Multiple Correspondence Analysis and Related Methods*. Chapman and Hall/CRC, Boca Raton, Florida.

Greenwood, M. and Yule, G. U. (1920). An inquiry into the nature of frequency distributions representative of multiple happenings with particular reference to the occurrence of multiple attacks of disease or of repeated accidents. *Journal of the Royal Statistical Society*, 83, 255–279.

Groenen, P. J. F. and van de Velden, M. (2005). Multidimensional scaling. In *Encyclopedia of Statistics in Behavioral Science* (eds. B. S. Everitt and D. C. Howell). John Wiley & Sons, Chichester, U.K.

Groenen, P. J. F. and Borg, I. (2014). Past, present, and future of multidimensional scaling. In *Visualization and Verbalization of Data* (eds. J. Blasius and M. Greenacre). Chapman and Hall/CRC, Boca Raton, Florida.

Groves, R. M., Fowler Jr., F. J., Couper, M. P., Lepkowski, J. M., Singer, E. and Tourangeau, R. (2009). *Survey Methodology*, 2nd edition. John Wiley & Sons, Hoboken, New Jersey.

Haavelmo, T. (1943). The statistical implications of a system of simultaneous equations. *Econometrica*, 11, 1–12.

Hämäläinen, R. P., Jones, R. and Saarinen, E. (2014). *Being Better Better: Living with Systems Intelligence*. Aalto University Publications Crossover 4/2014, http://systemsintelligence.aalto.fi/being_better_better/ (accessed June 28, 2018).

Hand, D. J. (2005). Discriminant analysis, linear. In *Encyclopedia of Biostatistics*, 2nd edition (eds. P. Armitage and T. Colton). John Wiley & Sons, Chichester, U.K.

Hand, D. J. (2008). *Statistics: A Very Short Introduction*. Oxford University Press, Oxford.

Heitjan, D. F. (1997). Bayesian interim analysis of phase II cancer clinical trials. *Statistics in Medicine*, 16, 1791–1802.

Hendrickson, A. E. and White, P. O. (1964). Promax: A quick method for rotation to oblique simple structure. *British Journal of Mathematical and Statistical Psychology*, 17, 65–70.

Heywood, H. B. (1931). On finite sequences of real numbers. *Proceedings of the Royal Statistical Society, Series A*, 134, 486–501.

Hildreth, L. (2013). Residual analysis for structural equation modeling. D.Phil. Thesis. Graduate Theses and Dissertations, Iowa State University. https://lib.dr.iastate.edu/etd/13400 (accessed August 2, 2018).

Hotelling, H. (1933). Analysis of a complex of statistical variables into principal components. *Journal of Educational Psychology*, 24, 417–441.

Hothorn, T. and Everitt, B. S. (2014). *A Handbook of Statistical Analyses Using R*, 3rd edition. Chapman and Hall/CRC, Boca Raton, Florida.

Houghton, J. D. and Jinkerson, D. L. (2007). Constructive thought strategies and job satisfaction: A preliminary examination. *Journal of Business Psychology*, 22, 45–53.

Howell, D. C. (2012). *Statistical Methods for Psychology*, 8th edition. Wadsworth, Belmont, California.

Howell, D. C. and Huessy, H. R. (1981). Hyperkinetic behavior followed from 7 to 21 years of age. In *Intervention Strategies with Hyperactive Children* (eds. M. Gettleman and M. E. Sharp). Armonk, New York.

Howell, D. C. and Huessy, H. R. (1985). A fifteen year follow-up of a behavioral history of attention deficit disorder (ADD). *Pediatrics*, 76, 185–190.

Hsieh, F. Y. (1987). A simple method of sample size calculation for unequal-sample-size designs that use the logrank or t-test. *Statistics in Medicine*, 6, 577–581.

Hu, L. and Bentler, P. M. (1999). Cutoff criteria for fit indexes in covariance structure analysis: Conventional criteria versus new alternatives. *Structural Equation Modeling: A Multidisciplinary Journal*, 6, 1–55.

Huba, G. J., Wingard, J. A. and Bentler, P. M. (1981). A comparison of two latent variable causal models for adolescent drug use. *Journal of Personality and Social Psychology*, 40, 180–193.

Huck, S. W. and Sandler, H. M (1979). *Rival Hypotheses: Alternative Interpretations of Data Based Conclusions.* Harper & Row, New York.

Husson, F., Lê, S. and Pagès, J. (2017). *Exploratory Multivariate Analysis by Example Using R.* Chapman and Hall/CRC, Boca Raton, Florida.

Hutcheson, G. D., Baxter, J. S., Telfer, K. and Warden, D. (1995). Child witness statement quality: Question type and errors of omission. *Law and Human Behavior*, 19, 631–648.

Jackson, D. L., Gillaspy, Jr., J. A. and Purc-Stephenson, R. (2009). Reporting practices in confirmatory factor analysis: An overview and some recommendations. *Psychological Methods*, 14, 6–23.

Jacobson, G. C. and Dimock, M. A. (1994). Checking out: The effects of bank overdrafts on the 1992 House elections. *American Journal of Political Science*, 38, 601–624.

Jennrich, R. I. (2004). Rotation to simple loadings using component loss functions: The orthogonal case. *Psychometrika*, 69, 257–273.

Jennrich, R. I. (2006). Rotation to simple loadings using component loss functions: The oblique case. *Psychometrika*, 71, 173–191.

Jennrich, R. I. (2007). Rotation methods, algorithms, and standard errors. In *Factor Analysis at 100: Historical Developments and Future Directions* (eds. R. Cudeck and R. C. MacCallum). Lawrence Elbaum, Mahwah, New Jersey.

Jennrich, R. I. and Sampson, P. F. (1966). Rotation for simple loadings. *Psychometrika*, 31, 313–323.

Johnson, V. E. and Albert, J. H. (2013). *Ordinal Data Modeling.* Springer, New York.

Jolliffe, I. T. (1970). Redundant variables in multivariate analysis. D.Phil. Thesis, University of Sussex.

Jolliffe, I. T. (1972). Discarding variables in a principal components analysis. I: Artificial data. *Journal of the Royal Statistical Society, Series C (Applied Statistics)*, 21, 160–173.

Jolliffe, I. T. (1973). Discarding variables in a principal components analysis. II: Real data. *Journal of the Royal Statistical Society, Series C (Applied Statistics)*, 22, 21–31.

Jolliffe, I. T. (1989). Rotation of ill-defined components. *Journal of the Royal Statistical Society, Series C (Applied Statistics)*, 38, 139–148.

Jolliffe, I. T. (2002). *Principal Component Analysis*, 2nd edition. Springer, New York.

Jöreskog, K. G. (1970). A general method for analysis of covariance structures. *Biometrika*, 57, 239–251.

Jöreskog, K. G. (1973). A general method for estimating a linear structural equation system. In *Structural Equation Models in the Social Sciences* (eds. A. S. Goldberger and O. D. Duncan). Seminar Press, New York.

Jöreskog, K. G. (1978). Structural analysis of covariance and correlation matrices. *Psychometrika*, 43, 443–477.

Jöreskog, K. G. (2007). Factor analysis and its extensions. In *Factor Analysis at 100: Historical Developments and Future Directions* (eds. R. Cudeck and R. C. MacCallum). Lawrence Elbaum, Mahwah, New Jersey.

Kaiser, H. F. (1958). The varimax criterion for analytic rotation in factor analysis. *Psychometrika*, 23, 187–200.

Kalbfleisch, J. D. and Prentice, R. L. (2002). *The Statistical Analysis of Failure Time Data*, 2nd edition. John Wiley & Sons, Hoboken, New Jersey.

Kaplan, E. L. and Meier, P. (1958). Nonparametric estimation from incomplete observations. *Journal of the American Statistical Association*, 53, 457–481.

Kaufman, L. and Rousseeuw, P. J. (1990). *Finding Groups in Data: An Introduction to Cluster Analysis.* John Wiley & Sons, New York.

Keele, L. (2008). *Semiparametric Regression for the Social Sciences.* John Wiley & Sons, Chichester, U.K.

Kelsey, J. L. and Hardy, R. J. (1975). Driving of motor vehicles as a risk factor for acute herniated lumbar interverteblar disc. *American Journal of Epidemiology*, 102, 63–73.

Kinsey, A. C., Wardell, B. P. and Martin, C. E. (1948). *Sexual Behavior in the Human Male.* W. B. Saunders, Philadelphia.

Kinsey, A. C., Wardell, B. P., Martin, C. E. and Gebhard, P. H. (1953). *Sexual Behavior in the Human Female.* W. B. Saunders, Philadelphia.

Kleinbaum, D. G., Kupper, L. L., Nizam, A. and Rosenberg, E. S. (2013). *Applied Regression Analysis and Other Multivariable Methods*, 5th edition. Cengage Learning, Boston.

Kline, R. B. (2016). *Principles and Practice of Structural Equation Modeling*, 4th edition. Guilford Press, New York.

Kruskal, J. B. (1964a). Multidimensional scaling by optimizing goodness of fit to a nonmetric hypothesis. *Psychometrika*, 29, 1–27.

Kruskal, J. B. (1964b). Nonmetric multidimensional scaling: A numerical method. *Psychometrika*, 29, 115–129.

Kruskal, J. B. and Wish, M. (1978). *Multidimensional Scaling.* Sage University Paper series on Quantitative Application in the Social Sciences, 07-011. Sage, London.

Krzanowski, W. J. (2000). *Principles of Multivariate Analysis: A User's Perspective*, 2nd edition. Oxford University Press, Oxford.

Krzanowski, W. J. and Marriott, F. H. C. (1994). *Multivariate Analysis. Part 1: Distributions, Ordination and Inference.* Edward Arnold, London.

Laaksonen, S. (2018). *Survey Methodology and Missing Data: Tools and Techniques for Practitioners.* Springer International, Cham, Switzerland.

Labovitz, S. (1970). The assignments of numbers to rank order categories. *American Sociological Review*, 35, 515–524.

Lawley, D. N. and Maxwell, A. E. (1971). *Factor Analysis as a Statistical Method*, 2nd edition. Butterworths, London.

Lebart, L., Morineau, A. and Warwick, K. (1984). *Multivariate Descriptive Statistical Analysis: Correspondence Analysis and Related Techniques for Large Matrices.* John Wiley & Sons, Chichester, U.K.

Lee, Y. J. (1984). Quick and simple approximation of sample sizes for comparing two independent binomial distributions: Different-sample-size case. *Biometrics*, 40, 239–241.

Lehman, D., Wortman, C. and Williams, A. (1987). Long term effects of losing a spouse or a child in a motor vehicle crash. *Journal of Personality and Social Psychology*, 52, 218–231.

Lehtonen, R. and Pahkinen, E. (2004). *Practical Methods for Design and Analysis of Complex Surveys*, 2nd edition. John Wiley & Sons, Chichester, U.K.

Liang, K. and Seger, S. L. (1986). Longitudinal data analysis using generalized linear models. *Biometrika*, 73, 13–22.

Little, R. J. A. (2005). Missing data. In *Encyclopedia of Biostatistics*, 2nd edition (eds. P. Armitage and T. Colton). John Wiley & Sons, Chichester, U.K.

Little, R. J. A. and Rubin, D. B. (2002). *Statistical Analysis with Missing Data*, 2nd edition. John Wiley & Sons, New York.

Longford, N. T. (1993). *Random Coefficient Models.* Oxford University Press, Oxford.

Macdonnell, W. R. (1902). On criminal anthropometry and the identification of criminals. *Biometrika*, 1, 177–227.

Magnello, E. and van Loon, B. (2009). *Introducing Statistics: A Graphic Guide.* Icon Books, London.

Manly, B. F. J. (1986). *Multivariate Statistical Methods: A Primer.* Chapman and Hall, London.

Mardia, K. V., Kent, J. T. and Bibby, J. M. (1979). *Multivariate Analysis.* Academic Press, London.

Marriott, F. H. C. (1974). *The Interpretation of Multiple Observations.* Academic Press, London.

Marriott, F. H. C. (1982). Optimization methods of cluster analysis. *Biometrika*, 69, 417–421.

Matthews, D. E. (2005). Multiple linear regression. In *Encyclopedia of Biostatistics*, 2nd edition (eds. P. Armitage and T. Colton). John Wiley & Sons, Chichester, U.K.

Matthews, J. N., Altman, D. G., Campbell, M. J. and Royston, P. (1990). Analysis of serial measurements in medical research. *British Medical Journal*, 300, 230–235.

Maxwell, S. E. and Delaney, H. D. (2003). *Designing Experiments and Analyzing Data: A Model Comparison Perspective*, 2nd edition. Lawrence Erlbaum, Mahwah, New Jersey.

McCullagh, P. and Nelder, J. A. (1989). *Generalized Linear Models*, 2nd edition. Chapman and Hall, London.

McHugh, R. B. and Le, C. T. (1984). Confidence estimation and the size of a clinical trial. *Contemporary Clinical Trials*, 5, 157–163.

McKay, R. J. and Campbell, N. A. (1982a). Variable selection techniques in discriminant analysis. I: Description. *British Journal of Mathematical and Statistical Psychology*, 35, 1–29.

McKay, R. J. and Campbell, N. A. (1982b). Variable selection techniques in discriminant analysis. II: Allocation. *British Journal of Mathematical and Statistical Psychology*, 35, 30–41.

McNeish, D., An, J. and Hancock, G. R. (2018). The thorny relation between measurement quality and fit index cutoffs in latent variable models. *Journal of Personality Assessment*, 100, 43–52.

Miles, J. and Shevlin, M (2001). *Applying Regression and Correlation.* Sage, London.

Morrison, D. F. (1990). *Multivariate Statistical Methods*, 3rd edition. McGraw-Hill, New York.

Morrison, D. F. (2005). Multivariate analysis of variance. In *Encyclopedia of Biostatistics*, 2nd edition (eds. P. Armitage and T. Colton). John Wiley & Sons, Chichester, U.K.

Mulaik, S. A. (2009). *Linear Causal Modeling with Structural Equations.* Chapman and Hall/CRC, Boca Raton, Florida.

Mulaik, S. A. (2010). *Foundations of Factor Analysis*, 2nd edition. Chapman and Hall/CRC, Boca Raton, Florida.

Murdoch, D. J. and Chow, E. D. (1996). A graphical display of large correlation matrices. *The American Statistician*, 50, 178–180.

Murray, G. and Findlay, J. (1988). Correcting for bias caused by dropouts in hypertension trials. *Statistics in Medicine*, 7, 941–946.

Mustonen, S. (1995). *Tilastolliset monimuuttujamenetelmät* [*Statistical Multivariate Methods*, in Finnish]. Survo Systems, Helsinki. https://www.survo.fi/mustonen/monim.pdf (accessed June 17, 2018).

Mustonen, S. (1996). *Survo ja minä* [*Survo and I*, in Finnish]. Survo Systems, Helsinki. https://www.survo.fi/books/1996/SM_kirja.pdf (accessed June 7, 2018).

Nathanson, J. A. (1971). An application of multivariate analysis in astronomy. *Journal of the Royal Statistical Society, Series C (Applied Statistics)*, 20, 239–249.

Needham, R. M. (1967). Automatic classification in linguistics. *Journal of the Royal Statistical Society, Series D (The Statistician)*, 17, 45–54.

Nelder, J. A. (1977). A reformulation of linear models (with discussion). *Journal of the Royal Statistical Society, Series A*, 140, 48–77.

Nelder, J. A. and Wedderburn, R. W. M. (1972). Generalized linear models. *Journal of the Royal Statistical Society, Series A*, 135, 370–384.

Oakes, M. (1986). *Statistical Inference: A Commentary for the Social and Behavioural Sciences.* John Wiley & Sons, Chichester, U.K.

Oldham, P. D. (1962). A note on the analysis of repeated measurements of the same subjects. *Journal of Chronic Diseases*, 15, 969–977.

Palotie, U., Eronen, A. K., Vehkalahti, K. and Vehkalahti, M. M. (2017). Longevity of 2- and 3-surface restorations in posterior teeth of 25- to 30-year-olds attending Public Dental Service—A 13-year observation. *Journal of Dentistry*, 62, 13–17.

Pearl, J. and Mackenzie, D. (2018). *The Book of Why: The New Science of Cause and Effect.* Basic Books, New York.

Pearson, K. (1901). On lines and planes of closest fit to systems of points in space. *Philosophical Magazine*, 2, 559–572.

Pett, M. A., Lackey, N. R. and Sullivan, J. J. (2003). *Making Sense of Factor Analysis: The Use of Factor Analysis for Instrument Development in Health Care Research.* Sage, London.

Piantadosi, S. (1997). *Clinical Trials: A Methodologic Perspective.* John Wiley & Sons, New York.

Pocock, S. J. (1996). Clinical trials: A statistician's perspective. In *Advances in Biometry* (eds. P. Armitage and H. A. David). John Wiley & Sons, Chichester, U.K.

Proudfoot, J., Ryden, C., Everitt, B. S. et al. (2004). Clinical efficacy of computerized cognitive-behavioural therapy for anxiety and depression in parimary care: Randomized controlled trial. *British Journal of Psychiatry*, 185, 46–54.

Puntanen, S., Styan, G. P. H. and Isotalo, J. (2011). *Matrix Tricks for Linear Statistical Models: Our Personal Top Twenty.* Springer, Heidelberg. http://www.sis.uta.fi/tilasto/matrixtricks/ (accessed August 22, 2018).

Puntanen, S., Styan, G. P. H. and Isotalo, J. (2013). *Formulas Useful for Linear Regression Analysis and Related Matrix Theory: It's Only Formulas But We Like Them.* Springer, Heidelberg.

Rabe-Hesketh, S. and Skrondal, A. (2012). *Multilevel and Longitudinal Modeling Using Stata*, 3rd edition. Volume I: Continuous Responses. Volume II: Categorical Responses, Counts, and Survival. Stata Press, College Station, Texas.

Rawlings, J. O., Pantula, S. G. and Dickey, D. A. (1998). *Applied Regression Analysis: A Research Tool*, 2nd edition. Springer, New York.

Rencher, A. C. and Christensen, W. F. (2012). *Methods of Multivariate Analysis*, 3rd edition. John Wiley & Sons, New York.

Rosenbaum, P. R. (2002). *Observational Studies*, 2nd edition. Springer, New York.

Rosenman, R. H., Brand, R. J., Jenkins, C. D., Friedman, M., Straus, R. and Wurm, M. (1975). Coronary heart disease in the Western Collaborative Group study: Final follow-up experience of 8 1/2 years. *The Journal of the American Medical Association*, 233, 872–877.

Rosling, H., Rosling, O. and Rosling Rönnlund, A. (2018). *Factfulness: Ten Reasons We're Wrong About the World—and Why Things Are Better Than You Think.* Sceptre, London.

Rubin, D. B. (1976). Inference and missing data. *Biometrika*, 63, 581–592.

Rubin, D. B. (2004). *Multiple Imputation for Nonresponse in Surveys.* John Wiley & Sons, New York.

Rubin, D. B. and Schenker, N. (1991). Multiple imputation in health-care databases: An overview and some applications. *Statistics in Medicine*, 10, 585–598.

Saarinen, E. and Hämäläinen, R. P. (2004). Systems intelligence: Connecting engineering thinking with human sensitivity. In *Systems Intelligence—Discovering a Hidden Competence in Human Action and Organizational Life* (eds. R. P. Hämäläinen and E. Saarinen). Helsinki University of Technology, Research Reports A88. http://sal.aalto.fi/publications/pdf-files/systemsintelligence2004.pdf (accessed March 28, 2018).

Sarkar, D. (2008). *Lattice: Multivariate Visualization with R.* Springer, New York.

Schafer, J. L. (1997). *Analysis of Incomplete Multivariate Data.* Chapman and Hall, London.

Schafer, J. L. (1999). Multiple imputation: A primer. *Statistical Methods in Medical Research*, 8, 3–15.

Schafer, J. L. and Graham, J. W. (2002). Missing data: Our view of the state of the art. *Psychological Methods*, 7, 147–177.

Schmid, C. F. (1954). *Handbook of Graphic Presentation.* Ronald Press, New York.

Schmidt, U., Evans, K., Tiller, J. and Treasure, J. (1995). Puberty, sexual milestones and abuse: How are they related in eating disorder patients? *Psychological Medicine*, 25, 413–417.

Schoenfeld, D. A. (1983). Sample-size formula for the proportional-hazards regression model. *Biometrics*, 39, 499–503.

Schuman, H. and Kalton, G. (1985). Survey methods. In *Handbook of Social Psychology*, Vol. I (eds. G. Lindzey and E. Aronson). Random House, New York.

Scott, A. J. and Symons, M. J. (1971). Clustering methods based on likelihood ratio criteria. *Biometrics*, 37, 387–398.

Seeber, G. U. H. (2005). Poisson regression. In *Encyclopedia of Biostatistics*, 2nd edition (eds. P. Armitage and T. Colton). John Wiley & Sons, Chichester, U.K.

Senn, S. J. (1997). *Statistical Issues in Drug Development.* John Wiley & Sons, Chichester, U.K.

Shadish, W. R., Cook, T. D. and Campbell, D. T. (2002). *Experimental and Quasi-Experimental Designs for Generalized Causal Inference.* Houghton Mifflin, Boston.

Shepard, R. N. (1962a). The analysis of proximities: Multidimensional scaling with an unknown distance function. I. *Psychometrika*, 27, 125–140.

Shepard, R. N. (1962b). The analysis of proximities: Multidimensional scaling with an unknown distance function. II. *Psychometrika*, 27, 219–246.

Sibson, R. (1979). Studies in the robustness of multidimensional scaling: Perturbational analysis of classical scaling. *Journal of the Royal Statistical Society, Series B*, 41, 217–229.

Siddhartha, R. D., Fowlkes, E. B. and Hoadley, B. (1989). Risk analysis of the space shuttle: Pre-Challenger prediction of failure. *Journal of the American Statistical Association*, 84, 945–957.

Siegfried, T. (2010). "Odds are, it's wrong: Science fails to face the shortcomings of statistics". *Science News*, 177, 26.

Simon, R. (1991). A decade of progress in statistical methodology for clinical trials. *Statistics in Medicine*, 10, 1789–1817.

Skrondal, A. and Rabe-Hesketh, S. (2004). *Generalized Latent Variable Modeling: Multilevel, Longitudinal, and Structural Equation Models.* Chapman and Hall/CRC, Boca Raton, Florida.

Smith, D. A. and Patterson, E. B. (1984). Applications and a generalization of MIMIC models to criminological research. *Journal of Research in Crime and Delinquency*, 21, 333–352.

Spear, M. E. (1952). *Charting Statistics.* McGraw-Hill, New York.

Spearman, C. (1904). General intelligence, objectively determined and measured. *American Journal of Psychology*, 15, 201–293.

Spicer, C. C., Lawrence, C. J. and Southall, D. P. (1987). Statistical analysis of heart rates and subsequent victims of sudden infant death syndrome. *Statistics in Medicine*, 6, 159–166.

Steinley, D. (2008). Stability analysis in K-means clustering. *British Journal of Mathematical and Statistical Psychology*, 61, 255–273.

Stigler, S. M. (2016). *The Seven Pillars of Statistical Wisdom.* Harvard University Press, Cambridge, Massachusetts.

Tarkkonen, L. and Vehkalahti, K. (2005). Measurement errors in multivariate measurement scales. *Journal of Multivariate Analysis*, 96, 172–189.

Thall, P. F. and Vail, S. C. (1990). Some covariance models for longitudinal count data with overdispersion. *Biometrics*, 46, 657–671.

Therneau, T. M. and Grambsch, P. M. (2000). *Modeling Survival Data: Extending the Cox Model.* Springer, New York.

Thomsen, O. Ø., Wulff, H. R., Martin, A. and Springer, P. A. (1993). What do gastroenterologists in Europe tell cancer patients? *The Lancet*, 341, 473–476.

Thurstone, L. L. (1931). Multiple factor analysis. *Psychological Review*, 39, 406–427.

Thurstone, L. L. (1935). *Vectors of Mind.* The University of Chicago Press, Chicago.

Thurstone, L. L. (1947). *Multiple-Factor Analysis.* The University of Chicago Press, Chicago.

Timm, N. H. (2002). *Applied Multivariate Analysis.* Springer, New York.

Torgerson, W. S. (1952). Multidimensional scaling: I. Theory and method. *Psychometrika*, 17, 401–419.

Torgerson, W. S. (1958). *Theory and Methods of Scaling.* John Wiley & Sons, New York.

Törmänen, J., Hämäläinen, R. P. and Saarinen, E. (2016). Systems Intelligence inventory. *The Learning Organization*, 23, 218–231.

Tourangeau, R., Rips, L. J. and Rasinski, K. (2000). *The Psychology of Survey Response.* Cambridge University Press, New York.

Tufte, E. R. (1983). *The Visual Display of Quantitative Information.* Graphics Press, Cheshire, Connecticut.

Tukey, J. W. (1977). *Exploratory Data Analysis.* Addison Wesley, Reading, Massachusetts.

United Nations (2016). *Demographic Yearbook.* United Nations, New York. https://unstats.un.org/unsd/demographic-social/products/dyb/ (accessed June 3, 2018).

van Buuren, S. (2018). *Flexible Imputation of Missing Data*, 2nd edition. Chapman and Hall/CRC, Boca Raton, Florida.

Vehkalahti, K., Koponen, J. and Kuusi, H. (2018). From Russia with love to *infovis*: Graphic train schedules reflect the history of Finland. *CHANCE*, 31, 46–53.

Vehkalahti, K., Puntanen, S. and Tarkkonen, L. (2007). Effects of measurement errors in predictor selection of linear regression model. *Computational Statistics & Data Analysis*, 52, 1183–1195.

Velleman, P. F. and Wilkinson, L. (1993). Nominal, ordinal, interval, and ratio typologies are misleading. *The American Statistician*, 47, 65–72.

Vetter, B. M. (1980). Working women scientists and engineers. *Science*, 207, 28–34.

Wainer, H. (1997). *Visual Revelations.* Springer, New York.

Wasserstein, R. L. and Lazar, N. A. (2016). The ASA's statement on *p*-values: Context, process, and purpose. *The American Statistician*, 70, 129–133.

Watkins, E. and Williams, R. M. (1998). The efficacy of cognitive-behavioural therapy. In *The Management of Depression* (ed. S. Checkley). Blackwell Science, Oxford.

White, I. R., Royston, P. and Wood, A. M. (2011). Multiple imputation using chained equations: Issues and guidance for practice. *Statistics in Medicine*, 30, 377-399.

Widaman, K. F. (2007). Common factors versus components: Principals and principles, errors and misconceptions. In *Factor Analysis at 100: Historical Developments and Future Directions* (eds. R. Cudeck and R. C. MacCallum). Lawrence Elbaum, Mahwah, New Jersey.

Wilkinson, L. (1992). Graphical displays. *Statistical Methods in Medical Research*, 1, 3–25.

Willerman, L., Schultz, R., Rutledge, J. N. and Bigler, E. D. (1991). In vivo brain size and intelligence. *Intelligence*, 15, 223–228.

Wittes, J. and Wallenstein, S. (1987). The power of the Mantel–Haenszel test. *Journal of the American Statistical Association*, 82, 1104–1109.

Wright, S. (1921). Correlation and causation. *Journal of Agricultural Research*, 20, 557–585.

Wright, S. (1934). The method of path coefficients. *The Annals of Mathematical Statistics*, 5, 161–215.

Yarkoni, T. and Westfall, J. (2017). Choosing prediction over explanation in psychology: Lessons from machine learning. *Perspectives on Psychological Science*, 12, 1100–1122.

Yates, F. (1982). Regression models for repeated measurements. *Biometrics*, 38, 850–853.

Young, G. and Householder, A. S. (1938). Discussion of a set of points in terms of their mutual distances. *Psychometrika*, 3, 19–22.

Yule, W., Berger, M., Butler, S., Newham, V. and Tizard, J. (1969). The WPPSI: An empirical evaluation with a British sample. *British Journal of Educational Psychology*, 39, 1–13.

Index

A

Ability and aspiration model, 325
  observed correlations for, 326
  path diagram for, 326
  results of fitting, 327
Absolute (or overall) fit indices, 324
Acute herniated lumbar intervertebral discs (AHLID), 134
Agglomerative hierarchical clustering, 344–347
  clustering individuals, body measurements, 347–348
  dendrogram, 345, 346
  Euclidean distances, 344–345
  intergroup distance measure, 345, 346
  life expectancy, clustering countries on, 348–352
AHLID, *see* Acute herniated lumbar intervertebral discs
AIC, *see* Akaike's information criterion
Akaike's fit criterion, 128
Akaike's information criterion (AIC), 96–98, 128, 129, 131–133
All subsets regression approach, 94
American Statistical Association (ASA), 18
American Statistician, The, 18
Analysis of variance (ANOVA), 109, 113
  multiple linear regression, 102–109
    balanced design, 105–106
    for experimental designs, 104–105
    fecundity of fruit flies, 102–104
    unbalanced design, 106–109
ANOVA, *see* Analysis of variance
AR-1 autoregressive correlation matrix, 193
ASA, *see* American Statistical Association
Automatic model selection, 95–96
Available-case analysis, 213
Average linkage clustering, 345
Average linkage solution, 347–348

B

Backward elimination
  application of, 96–98
  automatic model selection, 95
Balanced design, 105–106
Bar charts, 26
  for crime percentages, 25
  displaying data as, 29
  life expectancies, 30
  stacked, 50, 51
  Titanic passengers fate, 31–32
Baseline hazard function, 147
Bath tub hazard function, 145
Bayesian approach, 359
Bayesian information criterion (BIC), 359, 361
BDI, *see* Beck Depression Inventory
Beat the Blues (BtB), 219–221
  cognitive behavioral therapy, 181–186
  generalized estimating equations, 196–197

Beck Depression Inventory (BDI), 219–221
 box plots of, 184
 measurement, 219–221
 scatterplot matrix of, 185
Beck Depression Inventory II, 183
Behavioral research, 2, 3
 experiments, 4–5
 multivariate response, 155
 observational studies, 5–6
 quasi-experimental designs, 6
 surveys, 3–4
Between-subject variation, 155
BIC, *see* Bayesian information criterion
Binary response variables, 115–117
Bivariate boxplot, 42–44
 data on electrodes one and two, 47
 time spent looking after car and age, 44
 time spent looking after car and extroversion, 44
Bivariate data
 principal components of, 246–247
 three clusters, 343
Bivariate normal density, 231–232
Body measurement data
 chi-square plot of, 234
 clustering individuals based on, 347–348
 graphical descriptions of, 229–230
 numerical summary statistics, 228
Book of Why, The (Pearl and Mackenzie), 320
Box, George, 13
Boxplots, 33, 36, 70
 of BDI scores, 184
 bivariate, 42–44
 of body measurements, 229
 for BPRS data, 161
 construction of, 34
 count of "and then . . . " statements, 35
 data from risk-taking experiment, 377
 mean summary measures, 163
 for standardized and deletion residuals, 100
BPRS, *see* Brief psychiatric rating scale
Brief psychiatric rating scale (BPRS), 156–159
 analysis of covariance, 164
 boxplots for, 161, 163
 individual response profiles for, 158, 159
 mean response profiles for, 160
BtB, *see* Beat the Blues
Bubbleplot, 40, 42

C

CA, *see* Correspondence analysis
Canonical correlation, 286
Case-control investigation, 6
Categorical data, 24–32
Categorical measurements, 7
Causal relations, 320
Causal relationship, 4
CBT, *see* Cognitive behavioral therapy
Censored observations, 139
CFA, *see* Confirmatory factor analysis
CFI, *see* Comparative fit index
Chaining phenomenon, 347–348
Challenger space shuttle, 57, 58
Characteristic roots, 242
Characteristic vector, 242
Charts, 23–24
Chi-square plot
 multivariate data for normality, 233–236
 one-way MANOVA to data, 378
Chi-square test, 287, 293, 327, 329, 334, 336

CI, *see* Confidence interval
Classical multidimensional scaling, 270–273
  classical composers, 278–280
  distance matrix, 274
  of Finnish road distances, 274–278
  principal components, 273–274
Classical music
  mapping composers of, 278–280
  re-mapping composers of, 281–283
Classical scaling, 269, 270
Classification
  functions, 378–381
  variety of, 341–342
Classification-likelihood clustering approach, 362
Classification maximum likelihood clustering, 357–359
Clinical trials, 15–16
Cluster analysis, 341
  agglomerative hierarchical techniques, 344–352
  classifications, 341–342
  k-means clustering, 352–356
  model-based clustering
    classification maximum likelihood clustering, 357–359
    clustering European countries, 359–362
  scatterplots, 343
Clustering criterion, 354, 358
Cluster solution
  four-cluster solution from complete linkage, 351, 352, 353
  four-group cluster solution, 350, 351
  graphic of, 351
Coefficient of correlation, 56
Cognitive behavioral therapy (CBT), 181–186
Common factors, 295
Communality, 297
Companion Encyclopedia of Psychology, 3
Comparable groups, 4
Comparative fit index (CFI), 323, 324
Complete-case analysis, 212, 217–218
Complete linkage clustering, 345
Complete linkage solution, 347–348, 350
Compound symmetry, 172
Conditional likelihood function, 135
Conditional logistic regression model, 135
Conditional mean imputation, 213
Conditional models, 190
  generalized linear mixed effects models, 194, 203–206
  logistic regression model, 194
Conditioning plots, 48–54
  time against extroversion conditioned on age, 48–49
  time spent after car against extroversion, 47
Confidence interval (CI), 16–18, 124
  for odds ratio, 124–125
  for slope parameter, 65
Confirmatory factor analysis (CFA), 295, 319, 326
  assessing the fit, 322–324
  degrees of freedom, 324
  estimation, 320–321
  examples of
    ability and aspiration, 325–327
    drug usage among students, 327–330
  identification, 321–322
  validation data, 334
Constant regression coefficients, 152
Constructive thought processes, 336
Cook's distance, 99, 101, 102
Coplot, *see* Conditioning plots

Correlation coefficient, 38, 227–228, 246, 300, 309
Correlation matrix
for drug usage data, 262–263
eigenvalues of, 304
Correlation of variables, 248
Correspondence analysis (CA), 269, 284
application of, 286–288
basic details of, 284–286
symmetric map and, 286
work activities and job advantages, 288–291
Correspondence matrix, 285
Corrgram, 332
Covariance matrix, 85, 227, 358, 380
Covariance of variables, 248, 297–298
Covariance structure modeling, *see* Structural equation models (SEM)
Covariance structure, reasons for modeling, 192–193
Cox, David, 140, 146
Cox regression, *see* Cox's proportional hazards model
Cox's proportional hazards model, 140, 146
assumption of proportional hazards, 151
baseline hazard, 148
constant regression coefficients, 152
exponential distribution, 147
heroin addicts treated with methadone, 149–151
partial likelihood, 148
Crime rates
clustering, 355–356
by factor analysis, 312–314
in United States, 255–260
Cross-loadings, 329
Cubic spline, 77–78
Cumulative hazard function, 145–146

D

Data
analysis, role of models in, 11–13
labeling three classes, 379–380
multivariate, 10
replotting, 55
Data collection, 1
Data graphics, 23
Degrees of freedom (DF), 65, 179
Deletion residual, 99–101
Dendrogram, 345
example of, 346
for single, complete, and average linkage, 347–348
Dependent variables, 10
DF, *see* Degrees of freedom
Dimock, M. A., 78, 79
Disattenuated correlation, 326
Discriminant analysis for three groups, 379
Discriminant function analysis, 365
scatterplot of imaging data, 381
thresholds for imaging data, 381
Disparities, 280
Dissimilarity, 267, 268, 272, 278
Distance concept, 267
Distance matrix, 344
Do-it-yourself data, 131–133
Dot plots
for crime data, 26
for drinkers' and abstainers' crime percentages, 27
for 10 percentages, 27
Drosophila melanogaster, 102
Drug usage model, 327
fitting correlated three-factor model, 330
nonzero correlations, 329
path diagram for, 328
Dynamic graphics, 24

E

EFA, *see* Exploratory factor analysis
Eigenvalues, 242–244, 246–250, 271–272, 304
Eigenvector, 242, 244, 246–247, 272, 301
EM algorithm, 359
Embedded Figures Test (EFT), 37, 38, 39
Epilepsy data
  generalized estimating equations, 201–203
  generalized linear mixed effects models, 204–206
Error distribution, 113–114
Error terms, 116
Estimated regression coefficients, 84, 88, 128, 192, 198
Estimated variances, 85
Euclidean distance matrix, 267–268
Euclidean distances, 270–272, 344–345, 347
European countries, clustering, 359–362
Evolutionary tree concept, 345, 347
Exchangeable correlation matrix, 193
Experimental designs, multiple linear regression for, 104–105
Experiments, 4–5
Explanatory variables, 10, 64, 70
  fitted model with gender and extroversion, 88–89
  t-values associated with, 88
  variance inflation factors of, 93
Exploratory data analysis, 75
Exploratory factor analysis (EFA), 295, 319, 334
  estimating factor scores, 311–314
    crime rates by factor analysis, 312–314
  factor analysis model, *see* Factor analysis model
  factor loadings, 304–305
  factor rotation, 305–306
  learning data, 334
  number of factors, 301–302
  principal components analysis *vs.*, 315
  rotation of factors, 304–306
    example of numerical rotation, 311
    graphical rotation, 306–309
    numerical rotation methods, 309–311
Exponential distribution, 147

F

Factfulness, 352
Factor analysis model, 295, 296, 315
  analyzing crime rates by, 312–314
  covariance of variables, 297–298
  estimating parameters in, 299–301
    maximum likelihood factor analysis, 301
    principal factor analysis, 300–301
  example of fitting, 302–304
  multivariate observations, 298
  observed correlation matrix, 298–299
  sample correlation matrix, 298
Factor loadings, 295
  lack of uniqueness, 304–305
Factor pattern coefficients, 309
Factor rotation, 305–306, 310
Factor structure coefficients, 309
Familial andenomatous polyposis (FAP), 117
FAP, *see* Familial andenomatous polyposis
Fecundity of fruit flies, 102–104
Finland, road distances in, 274–278
Fisher, Ronald Aylmer, 4, 5, 102

Fisher's linear discriminant function, 369–374
Fitted logistic regression model, 126–130
Forward selection approach, 95
FSIQ, *see* Full-scale IQ
F-statistics, 85, 220
F-test, 86–87, 103, 218
Full-scale IQ (FSIQ), 366
Functional magnetic resonance imaging (fMRI) investigation, 378

G

Galton, 63
GEE, *see* Generalized estimating equations
General Health Questionnaire (GHQ), 123
  applying logistic regression to, 125–130
  estimated regression coefficient for, 128
  fitted logistic regression model, 126–130
  psychiatric caseness data, 124
Generalized estimating equations (GEE), 193, 195
  to fit marginal models, 196
    Beat the Blues revisited, 196–197
    epilepsy, 201–203
    respiratory illness, 197–201
Generalized linear mixed effects models, 194
  to fit conditional models, 203
    epilepsy, 204–206
    respiratory illness, 203–204
Generalized linear models (GLMs), 113–114, 118, 190, 191
  binary response variables, 115–117
  colonic polyps data, 117
  error distribution, 113–114
  link function, 114
  logistic regression, 115–117
  to longitudinal data, 191–192
  Poisson error distributions, 119–120
  response variables, 117–119
  variance function, 114, 119
GHQ, *see* General Health Questionnaire
GLMs, *see* Generalized linear models
Graphical deception, 55–59
  breast cancer death rate, 55
  graphical distortion, 56–57
  lie factor, 56
Graphical descriptions, body measurement data, 229–230
Graphical displays, 23, 55
  of longitudinal data, 157–159
  prime objective of, 24
Graphical distortion, suggestions for avoiding, 56–57
Graphical methods, advantages of, 23–24
Graphical rotation, 306–309
Graphic design, 24
Graphs, 23–24
Grouped multivariate data, 365
  more than two groups, 374–381
    classification functions, 378–381
    multivariate analysis of variance, 374–378
  two-group multivariate data, 366–374
    Fisher's linear discriminant function, 369–374
    Hotelling's T^2 test, 366–368

H

Hat matrix, 99
Hazard function, 140
  bath tub, 145
  cumulative, 145–146
  definition, 144

Heights data
  pulse rates and, 63
  simple linear regression model and, 66–67
Heywood case, 301
Hierarchical classifications, 344, 345
Histogram, 32–33
Homogeneous groups, 341
Hotelling's T^2 test, 365, 366–368
Hritz, Ronald, 374

I

Identity matrix, 193
Independent variables, 10
Inertia, 285
Infographics, 24
Information visualization, *see* Infographics
Infovis, *see* Infographics
Intelligence, 295
Intercluster distance or dissimilarity, 345, 346
Interval/quasi-interval data, 32–37
Interval scale, 8
Intraclass correlation, 172

J

Jacobson, G. C., 78, 79
Jacobs, Stanley, 374

K

Kaplan–Meier estimator, 141
  for survival function, 141–142, 143
Keele, L., 80
k-factor analysis model, 300–301, 304–305
Kinesiology, simple linear regression, 67–69, 70
Kinsey, Alfred Charles, 3
k-means clustering technique, 352–355
  algorithms steps, 354
  clustering crime rates, 355–356
  minimization of WGSS, 354–355
  within-group sum of squares, 353–354

L

Lagrange multiplier technique, 243
Latent variables, 295
Least-squares criterion, 77
Least-squares estimation process, 84
Lie factor, 56
Life expectancies
  alternative display of, 30
  bar chart showing, 30
  clustering countries basis of, 348–352
Likelihood ratio statistic, 323
Likelihood ratio test, 180
Linear mixed effects models, 169
  cognitive behavioral therapy, 181–186
  compound symmetry, 172
  intraclass correlation, 172
  maximum likelihood, 173
  random intercept and slope model, 170, 172–173
  random intercept model, 170, 171
  rat data, 174
    independence model to, 174–176
    linear mixed models to, 176–181
  for repeated measures data, 170
  variance component model, 171–172
Linear regression fit, 38
Linear spline function, 77
Link function, 114, 116
Listwise deletion, 212
Local independence assumption, 170
Locally weighted regression, 73–75
  lowess fit, 75–76
  scatterplot smoothers, 75–81
  spline smoothers, 76–78

Log-eigenvalue diagram, 250–251
Logistic discrimination approach, 373
Logistic regression, 114, 115–117, 123
general health questionnaire data and, 125–130
matched case–control study, 134–135
model, 116, 194
logit link function, 198
simulation of random effects, 195
odds and odds ratios, 123–125
parsimonious logistic regression model, 130–134
Logit function, 116
Log-rank test, 142
Longitudinal data, 155, 169
graphical displays of, 157–159
sources of correlation, 170
summary measure analysis, 159–160
applying, 162–164
choosing, 160–161
dealing with missing values, 164–166
incorporating pre-treatment outcome values, 164
possible, 162
Lowess fit, 75

M

Magnitude criterion, 272
Mahalanobis distances, 232
Manifest variables, 295
MANOVA, *see* Multivariate analysis of variance
MAR, *see* Missing at random
Marginal models, 190
covariance structure, 192–193
generalized estimating equations, 193, 196–203
generalized linear models, 191–192
for longitudinal data, 190–191
sandwich estimator, 192
Matched case–control study, 134–135
Maximum likelihood (ML) estimator, 358
linear mixed effects models and, 173
Maximum likelihood factor analysis, 299, 301, 302
correlation matrix of, 307
two-factor solution, 303, 308
MCAR, *see* Missing completely at random
MDS, *see* Multidimensional scaling
Mean imputation, 213, 218–219
Means, in imaging data, 380
Measurement
error, for verbal ability, 12
interval scale, 8
model, 333
nominal or categorical, 7
ordinal scale, 8
quality, 334
ratio scales, 9
response and explanatory variables, 10
types, 7
Misfit indices, 324
Missing at random (MAR), 211, 216
Missing completely at random (MCAR), 210–211, 212
Missing data
available-case analysis, 213
complete-case analysis, 212
by design, 209–210
listwise deletion, 212
mechanisms, 210–212
Missing not at random (MNAR), 211–212
Missing values, 10–11, 209
approaches to dealing, 212–213
Beat the Blues, 219–221
imputing, 213–214
missing data, *see* Missing data

multiple imputation, *see* Multiple imputation
multiply imputed data, 215–216
summary measure approach, 164–166
ML, *see* Maximum likelihood
MNAR, *see* Missing not at random
Model
in analysis of data, 11–13
appropriate, 12–13
selection, 359
statistical, 13
Model-based clustering, 356–362
classification maximum likelihood clustering, 357–359
clustering European countries, 359–362
Monotonic regression, 281
Monotonic transformations, 280
Multicollinearity, 92–93
Multidimensional scaling (MDS), 269
classical, 270–273
connection to principal components, 273–274
mapping composers of classical music, 278–280
nonmetric, 280–281
re-mapping composers of classical music, 281–283
road distances in Finland, 274–278
Multipanel dot plot, 50, 52
Multiple imputation, 214–215
application of, 216–219
complete-case analysis, 217–218
of data set, 219
mean imputation, 218–219
combined estimates and standard errors in, 215–216
Multiple linear regression, 83
analysis of variance, 102–109
balanced design, 105–106
experimental designs, 104–105
fecundity of fruit flies, 102–104
unbalanced design, 106–109
error terms, 84
example of, 85–90
goals of, 83
parsimonious model, 90–94
automatic model selection, 95–96
backward elimination, 96–98
regression diagnostics, 98–102
for response variable, 83–84
results from fitting, 87
Multiple R-squared, 66
Multiply imputed data, 215–217
Multivariate analysis, 225, 230
Multivariate analysis of variance (MANOVA), 365, 374–378
one-way, 375–376
principal test statistics for, 376
risk-taking investigation, 376
tests applied in, 377
Multivariate data, 225
assessing, for normality, 233–236
multivariate normal density function, 230–233
numerical summary statistics for, 227–228
summary statistics for, 226
Multivariate hypothesis, 368
Multivariate normal density function, 230–233
Mustonen, Olli, 278, 280
Mustonen, Seppo, 278
Mutually exclusive, 341

N

Nominal measurements, 7
Nonlinear model, 69
Nonmetric multidimensional scaling, 280–281

Nonmetric scaling, 269
Non-normal responses, 189
conditional models, 194–195
generalized linear mixed effects models to fit, 203–206
marginal models, 190–193
generalized estimating equations to fit, 196–203
modeling, 190
Nonparametric smoothers, 75
Normal distribution, 36
Normality, assessing multivariate data for, 233–236
Normed residuals, 324
Nuisance parameters, 169
Numerical rotation methods, 309–311
example of, 311
oblimin rotation, 310
oblique rotation, 309
orthogonal rotation, 309, 310
promax rotation, 310
quartimax rotation, 310
varimax rotation, 310
Numerical summary statistics
body measurements data, 228
for multivariate data, 227–228

O

Oblimin rotation, 310
Oblique rotation, 309
Observational studies, 5–6
Occam's razor, 13
Odds ratios, 116, 117, 123
confidence interval for, 124–125
estimation, 125
One-factor model, 308
One-way MANOVA, 375–376
Ordinal scale measurements, 8
On the Origin of Species by Natural Selection (Darwin), 345
O-rings, 57, 58
Orthogonal rotation, 309, 310
Outcome variables, *see* Dependent variables
Overdispersion, 119
Overinterpretation, 257

P

Parsimonious logistic regression model, 130–134
Parsimonious model
multiple linear regression, 90–94
automatic model selection, 95–96
backward elimination, 96–98
Partial likelihood, 148
Partialling out or controlling for other variables, 84
Path analysis, 335
Path diagram, 320
ability and aspiration model, 326
drug usage model, 328
hypothesized structural equation model, 336
hypothesized Systems Intelligence factor model, 333
PCA, *see* Principal components analysis
Percentages of degrees, 28, 29
Performance IQ (PIQ), 366
Perot, H. Ross, 78, 80
Pie charts
for crime rates, 25
for drinkers and abstainers, 24, 25
for 10 percentages, 26
scientific use of, 26
Pinker, Steven, 341
PIQ, *see* Performance IQ
Poisson distribution, 114, 118
Poisson regression, 118
definition, 114
to polyps data, 119, 120

Population-average models, *see* Marginal models
Population covariance of two variables, 227
Postdoctoral position, 50, 53
Predicted value, estimated variance of, 65
Principal components, 240–241, *see also* Principal components analysis
  of bivariate data
    with correlation coefficient, 246–247
  connection to, 273–274
  covariance or correlation matrix, 244–246
  extracting, 242–243
  finding sample, 241–244
  first, 240–243, 252–254
  observed covariance matrix, 248–249
  rescaling, 248
  scores, calculating, 251–252
  second, 241–242, 254
Principal components analysis (PCA), 239
  application of, 240, 255–260
    crime rates in United States, 255–260
    drug usage by American college students, 260–264
    head size of brothers, 252–255
  basic goal of, 239–240
  choosing number of components, 249–251
  correlation coefficient, 246–247
  covariance or correlation matrix, 244–245
  exploratory factor analysis *vs.*, 315
  principal components, 240–241
  rescaled coefficients, 248
  sample principal components, 241–244
  selecting subset of variables, 264–265
Principal coordinates, 273, 285
Principal factor analysis, 299, 300–301
Principal inertia, 285–286, 287
Probability plots, 36–37, 70, 72
Profiles, 285
Promax rotation, 310
Proportion of degrees, 28
Proudfoot, Judy, 183
Proximity matrices, 269, 272
Pulse rates, 63
  heights data and, 63
  simple linear regression model and, 66–67
p-values, 16–18
  in hypothesis testing, 18
  statistical null hypothesis, 18
  test of knowledge about, 17

Q

Quadratic discriminant function, 374
Quartimax rotation, 310
Quasi-experimental designs, 6
Quasi-likelihood approach, 120

R

Random effects, 170
Random intercept and slope model, 170, 172–173, 179–180
Randomization, 4
  primary benefit, 5
  in scientific experiments, 5
Rank order, 280
Rat data
  body weights of rats, 174
  independence model and, 174–176
  interaction model, 180, 181
  linear mixed models and, 176–181
  long form of, 175

Rat data (*Continued*)
plot of individual rat growth, 177
random intercept and random slope model, 179, 180
scatterplot matrix of, 178
weight against time for, 176
Ratio scale, 9
"Raw" regression coefficients, 87
Reduced covariance matrix, 300
Regression analysis, 63
Regression coefficients, 64, 87
Regression diagnostics, 69–73, 98–102
fitting linear regression model, 69–70
plotting residual values, 70
residual plots
idealized, 71
for oxygen uptake and expired ventilation data, 73, 74
for pulse rates and heights data, 72
Regression equation, 65
Regression mean square (RGMS), 65
Regression model, functional form for, 74
Regression sum of squares, 105–107
Reification, 259
Relative (or incremental) fit indices, 324
REML, *see* Restricted maximum likelihood
Repeated measures data, 155
Researcher
in experiment, 4
missing values, 10–11
Residual covariances, visual inspection of, 323
Residual mean square (RMS), 65, 85
Residual plots, 80
idealized, 71
for oxygen uptake and expired ventilation data, 73, 74
for pulse rates and heights data, 72
Respiratory illness data, 197–201
exchangeable correlation structure, 199–200
generalized estimating equations, 197–201
generalized linear mixed effects models, 203–204
logistic regression model, 198–199
Response and explanatory variables, 10
Response variables, 10, 70, 117–119, 155
overdispersion and quasi-likelihood, 119–120
Poisson regression, 118–119
variation in, 84
Restricted maximum likelihood (REML), 173
RGMS, *see* Regression mean square
RMS, *see* Residual mean square
RMSEA, *see* Root mean square error of approximation
Road distances in Finland, 274–278
Root mean square error of approximation (RMSEA), 323, 324
Row and column masses, 285

S

Sagan, Carl, 23
Sample size
calculation of, 14–15
determining, 14–16
risk, 16
Sandwich estimator, 192
Scatterplot, 37–40, 73
BDI scores, 185
bivariate boxplot, 42–44
bubbleplot, 40–42
challenger vote and Perot vote, 79, 81

completion time against EFT, 38–39
constructing, 40
course evaluation data, 92
crime rate data, 257
factor scores, 314
fitted linear regression for pulse and heights data, 67
imaging data with three classes labeled, 380
matrix, 45–47
of body measurements, 230
definition, 45
measurements of skin resistance, 46
of violent crime rates, 54
measures in rat growth data, 178
number of polyps and age, 118
of pulse rate against height, 76, 78
three-dimensional, 49
three discriminant functions, 381
time spent looking after car and age, 40, 41
time spent looking after car data, 86
use of, 37–38
variables birth weight from SID data, 372
women applicants percentage, 43
Scatterplot smoothers, 75–76, 78
Scientific research, 2
Scree diagram, 250–251
Scree plot, 256, 258, 261, 264, 318
SE, *see* Standard errors
SEM, *see* Structural equation models
Sexual Behavior in the Human Female, 3
Sexual Behavior in the Human Male, 3
Shepard diagram, 282, 283
SI, *see* Systems Intelligence
Side-by-side boxplots, 35, 36
Significance tests, 16–18
Similarity matrix, 344
Similarity measure, 269
Simple linear regression, 64
expired ventilation against oxygen, 68
kinesiology, 67–69, 70
model, 64
pulse rates and heights data, 66–67
regression coefficients, 64–65
technical details of, 64–66
Simple structure, 306
Simultaneous equation models, 335
Single imputation, 213–214
Single linkage clustering, 345
Single linkage solution, 347, 348
Singular value decomposition (SVD), 284
Slope parameter, confidence interval for, 65
Slope parameter estimate, estimated variance of, 65
Social class, 295
Specific or unique variance, 297
Specific variates, 296–297
Spline smoothers, 76–78
SRMR, *see* Standardized root mean squared residual
Stacked bar chart, 50, 51
Standard errors (SE), 326
Standardized residual, 99
Standardized root mean squared residual (SRMR), 323, 324
Statistical Abstract of the USA, 255
Statistical graphics, 23, 59
Statisticians, 1–2, 15
Statistics
definition, 1
misuse of, 1–2
models, 11–13
techniques, regression analysis, 63

Stein, Gertrude, 2
Stepwise regression method, 95
Stratified proportional hazards model, 148
Stress criterion, 281
Structural equation models (SEM), 295–296, 319
  assessing the fit, 322–324
  estimation, 320–321
  example of, 335–337
  identification, 321–322
  path analysis, 335
Structural relationships, 244
Student's t-test, 365
Subject-specific- models, *see* Conditional models
Sudden infant death (SID) syndrome, 369–372
  data, 370
  Fisher's Linear Discriminant Function on, 373
  scatterplot of variables birth weight, 372, 374
Summary measure analysis, of longitudinal data, 159–160
  applying, 162–164
  choosing, 160–161
  incorporating pre-treatment outcome values, 164
  missing values using, 164–166
  possible, 162
Summary statistics, for multivariate data, 226
Surveys, 3–4
Survival analysis, 139
Survival data
  age at first sexual intercourse for women, 140
  from behavioral study, 139–140
  Cox's proportional hazards model, 146–152
  hazard function, 144–146
  survival function, 140–144
  techniques for analysis, 139
Survival function, 140
  age at first sexual intercourse for women, 142–144
  definition, 141
  estimated, 141, 146
  Kaplan–Meier estimator for, 141–142, 143
  log-rank test, 142
  nonparametric, 141
SVD, *see* Singular value decomposition
Symmetric map, 286
Systems Intelligence (SI)
  data, corrgram for, 332
  definition, 331
  factors of, 331
  inventory, factorial validity of, 333–334
  path diagram for hypothesized, 333

T

TAU, *see* Treatment as Usual
Test statistic T^2, 367
Three-dimensional scatterplot, 49
Time-to-event data, 139
TLI, *see* Tucker–Lewis index
Trace criterion, 272
Tracking phenomenon, 157–158
Treatment as Usual (TAU), 182, 183
Trellis graphics, 51
  definition, 49
  examples of, 49–50
t-statistics, 94
Tucker–Lewis index (TLI), 323, 324
Tufte, Edward R., 29, 55–56
Tukey, John, 24
Two-factor model, 303
Type III sums of squares, 109
Type I sums of squares, 109

U

Unbalanced design, regression models, 106–109

Univariate and multivariate tests for equality of means of two variables, 368–369
Unstructured correlation matrix, 193

V

Variables, response and explanatory, 10
Variance component model, 171–172
Variance-covariance matrix, 227
Variance function, 114
Variance inflation factors, 93
Varimax rotation, 310
Verbal IQ (VIQ), 366
Viennese Classics, 279
VIQ, *see* Verbal IQ
Visual relationships, 23

W

Wechsler Adult Intelligence Scale-Revised test, 366
Wechsler Intelligence Scale for Children (WISC), 37
WGSS, *see* Within-group sum of squares
WISC, *see* Wechsler Intelligence Scale for Children
Within-group sum of squares (WGSS), 353–354
Within-subject variation, 155